IEE ELECTROMAGNETIC WAVES SERIES 34

Series Editors: Professor P. J. B. Clarricoats
Professor Y. Rahmat-Samii
Professor J. R. Wait

ELECTRODYNAMIC THEORY of SUPERCONDUCTORS

Other volumes in this series:

Volume 1 **Geometrical theory of diffraction for electromagnetic waves** G. L. James
Volume 2 **Electromagnetic waves and curved structures** L. Lewin, D. C. Chang and E. F. Kuester
Volume 3 **Microwave homodyne systems** R. J. King
Volume 4 **Radio direction-finding** P. J. D. Gething
Volume 5 **ELF communications antennas** M. L. Burrows
Volume 6 **Waveguide tapers, transitions and couplers** F. Sporleder and H. G. Unger
Volume 7 **Reflector antenna analysis and design** P. J. Wood
Volume 8 **Effects of the troposphere on radio communications** M. P. M. Hall
Volume 9 **Schumann resonances in the earth-ionosphere cavity** P. V. Bliokh, A. P. Nikolaenko and Y. F. Flippov
Volume 10 **Aperture antennas and diffraction theory** E. V. Jull
Volume 11 **Adaptive array principles** J. E. Hudson
Volume 12 **Microstrip antenna theory and design** J. R. James, P. S. Hall and C. Wood
Volume 13 **Energy in electromagnetism** H. G. Booker
Volume 14 **Leaky feeders and subsurface radio communications** P. Delogne
Volume 15 **The handbook of antenna design, Volume 1** A. W. Rudge, K. Milne, A. D. Olver, P. Knight (Editors)
Volume 16 **The handbook of antenna design, Volume 2** A. W. Rudge, K. Milne, A. D. Olver, P. Knight (Editors)
Volume 17 **Surveillance radar performance prediction** P. Rohan
Volume 18 **Corrugated horns for microwave antennas** P. J. B. Clarricoats and A. D. Olver
Volume 19 **Microwave antenna theory and design** S. Silver (Editor)
Volume 20 **Advances in radar techniques** J. Clarke (Editor)
Volume 21 **Waveguide handbook** N. Marcuvitz
Volume 22 **Target adaptive matched illumination radar** D. T. Gjessing
Volume 23 **Ferrites at microwave frequencies** A. J. Baden Fuller
Volume 24 **Propagation of short radio waves** D. E. Kerr (Editor)
Volume 25 **Principles of microwave circuits** C. G. Montgomery, R. H. Dicke, E. M. Purcell (Editors)
Volume 26 **Spherical near-field antenna measurements** J. E. Hansen (Editor)
Volume 27 **Electromagnetic radiation from cylindrical structures** J. R. Wait
Volume 28 **Handbook of microstrip antennas** J. R. James and P. S. Hall (Editors)
Volume 29 **Satellite-to-ground radiowave propagation** J. E. Allnutt
Volume 30 **Radiowave propagation** M. P. M. Hall and L. W. Barclay (Editors)
Volume 31 **Ionospheric radio** K. Davies
Volume 32 **Electromagnetic waveguides: theory and application** S. F. Mahmoud
Volume 33 **Radio direction finding and superresolution** P. J. D. Gething

ELECTRODYNAMIC THEORY of SUPERCONDUCTORS

by Shu-Ang Zhou

Peter Peregrinus Ltd. on behalf of the Institution of Electrical Engineers

Published by: Peter Peregrinus Ltd., London, United Kingdom

Peter Peregrinus Ltd.,
Michael Faraday House,
Six Hills Way, Stevenage,
Herts. SG1 2AY, United Kingdom

British Library Cataloguing in Publication Data

A CIP catalogue record for this book
is available from the British Library

ISBN 0 86341 257 2

Printed in England by Short Run Press Ltd., Exeter

Contents

Preface

Currently, rapid development of high-technology and applied physical sciences is requiring interdisciplinary research in many areas. In the areas of classical electromagnetics and mechanics, the situation has developed for many years that researchers and engineers in mechanical engineering seldom consider electromagnetic phenomena, while physicists and electrical engineers often ignore the mechanical effects. However, such traditionally separated subjects of electromagnetics and mechanics are now being fused because of the needs of modern science and technology, requiring novel utilization of the effects of electromagnetic-mechanical interaction for electronic instruments and optimal balance between basic mechanical structures and the overall electromagnetic behavior of engineering devices.

The emerging interdisciplinary subject of electrodynamics of deformable solids, which deals with the deformation, motion, heat and electromagnetic phenomena in material solids, has recently been studied extensively. However, little work has so far been carried out systematically on electrodynamics of superconductors, although the interest of world industry in engineering applications of superconductivity has been greatly stimulated by recent discovery of high temperature oxide superconductors.

This research monograph is, therefore, intended to be an introductory text for applied physicists, engineers and post-graduate students who are likely to be involved in the subject of electrodynamics of superconductors. The summary of knowledge in the text is based on a literature review of several hundreds relevant references and on several years of research results of the author and his collaborators and is thought to be the first of its kind.

The text will start from basic concepts and principles, and give a rather detailed derivation of important theoretical results that is easy to understand and may be used for study by interested readers having a reasonably good mathematical background. Greater emphasis will be put on general methodology, which is of particular utility due to the rapid broadening of engineering fields and their interdisciplinary nature in recent years. SI units are used throughout the text, which is thought to be particularly convenient for applied physicists, electrical and mechanical engineers studying large-scale phenomena, especially in engineering applications.

The text is divided into three chapters. The first chapter introduces essential concepts and principles involved in classical electromagnetics and continuum mechanics, and shows step-by-step how the effects of electromagnetic-mechanical interactions can be taken into account consistently in the continuum modelling of electrodynamic phenomena in deformable solids. Emphasis will be put on the unity of electric, magnetic and mechanical phenomena. Readers may find that some of derivations of the subject are in a new form which may be hopefully interesting and stimulating. Standard examples of solving problems concerning classical materials such as normal conductors, dielectrics, piezoelectric and ferromagnetic solids are, however, not provided in this chapter since there already exist several excellent research monographs that deal with these materials.

The second chapter introduces the basic concepts of superconductivity and electrodynamic theory of superconductors. Since the text is intended for readers such as electrical and mechanical engineers who might not be familiar with the phenomena of superconductivity, each topic will be treated and discussed in sufficient detail. Emphasis will be put on the presentation of phenomenological theories, such as the London theory, the Ginzburg-Landau theory and electrodynamic theory of Josephson junctions. A brief introduction on the micro-mechanism of superconductivity and the BCS theory will, however, also be provided. Examples illustrating the basic concepts, physics and mathematical models of superconductors are selected and given in the text.

Besides electromagnetic properties of superconductors, mechanical behavior of the superconductor has been of interest and concern to researchers for many years because of large mechanical forces expected in future applications of superconducting devices, such as magnets for plasma confinement and energy storage, superconducting generators, electrical transmission lines, magnetic levitating trains, electromagnetic-propulsion ships, or still unimagined new devices. Also, dynamic effects in superconductors may introduce interesting acoustic phenomena in superconductors, which are of importance for potential technological applications. The last chapter of this book is, therefore, devoted to the introduction of some theoretical models developed recently for the study of electromagnetic-mechanical interaction in superconductors. In this chapter, a continuum theory of elastic superconductors is, first, introduced to describe the effects of elastic deformation on superconducting properties as well as the elastic behavior of superconductors in superconducting state. We then introduce a macroscopic theory for magnetoelastic superconductors, in which magnetoelastic properties of ferromagnetic superconductors are studied and the phenomenon of the coexistence of ferromagnetism and superconductivity is discussed. Finally, a nonequilibrium theory of thermoelastic superconductors is proposed for the study of the electromagnetic and mechanical behavior of superconductors in nonequilibrium states. Because of the enormous volume of literature on the subjects of electromagnetics, continuum mechanics and superconductivity, it is difficult to cite all the relevant articles, so the author hopes to have the readers' understanding if they discover that some of their own work has gone unmentioned.

It is a great pleasure of mine to thank my Chinese teachers at Fudan University in Shanghai and my Swedish teachers at the Royal Institute of Technology in Stockholm, who introduced me to this scientific field. Also, I must thank Professor K. Miya for his initiation, encouragement, fruitful discussions and selfless support of my researches in the area of applied electromagnetics in materials during my one-year post-doctoral research at his research Laboratory at the University of Tokyo. The author is also most thankful to Prof. A.C. Eringen at Princeton University, Prof. G.A. Maugin at University of Paris VI, Prof. D.R. Axelrad at McGill University, Dr. J.P. Nowacki at Polish Academy of Science, and Prof. A.A.F. van de Ven at Eindhoven University of Technology for their constant encouragement, useful suggestions and helpful discussions of my research work. The very useful comments and suggestions of Prof. A.L. Cullen and other referees of a preliminary version of the manuscript are specially acknowledged. My thanks are also extended to the publisher, especially to Mr. J.D. StAubyn at Peter Peregrinus Ltd. for his efforts and continuous interest in my research monograph.

Finally, I must give my hearty thanks to my parents and my wife, Yatong XU, who have given me so much of their time, energy, understanding and encouragement during my many years of day and night study.

Stockholm, 1991. Shu-Ang ZHOU

To my parents
and
my teachers

Chapter 1

Introduction to electrodynamics of solids

Electrodynamics of solids is a subject covering electromagnetic phenomena as well as their interaction with mechanical phenomena in solid materials which may be deformed or in motion under electromagnetic and/or mechanical loadings. The purpose of this chapter is to give a brief introduction to this subject, its basic concepts and theoretical principles for the study of phenomena of electromagneto-thermoelastic interaction in material solids. The basic principles presented here are not only valid for conventional electromagnetic solids but also for superconductors. Particular emphasis will be put on electro- and magneto-quasistatic problems which are of practical interest for many engineering applications of electromagnetic-mechanical devices involving velocities small compared to the velocity of light. Starting from elementary knowledge of classical electrodynamics and continuum mechanics, the subject will be presented systematically and concisely in a way that both electrical and mechanical engineers who have just begun this subject may benefit and can get a quick and global view about this seemingly complicated subject of electromagnetic-mechanical interaction in material solids. Illustratively, several simplified material models are formulated for some materials as thermoelastic conductors, thermopiezoelectric insulators, soft ferromagnetic elastic insulators, and soft ferromagnetic thermoelastic conductors, which are useful for solving a variety of engineering problems of practical interest. Material modelling of superconductors will, however, be given in later chapters.

1.1 Charges and currents

1.1.1 Charges and charge density

Charge is the source and the object of action of an electromagnetic field. In nature, charge is found in two forms called positive charge and negative charge. The numerical value of a charge can only be an integral multiple of the elementary charge which has the numerical value $|e| = 1.60219\times10^{-19}$ C (Coulomb). Examples of charged particles are, for instance, electrons and protons. The electron is the material carrier of an elementary negative charge (having a mass of 9.1×10^{-31} kg), which is usually assumed as a structureless point particle. The proton is the material carrier of an elementary positive charge (having a mass of 1.7×10^{-27} kg), which is not considered as a point particle. The entire proton charge is practically concentrated in a sphere of radius of about 10^{-15} m. However, since an elementary charge is very small and most of macroscopic phenomena in electromagnetics involve a huge number of electric charges, such a discrete nature is not manifested in any way in the domain of macroscopic phenomena. Hence, we may assume the charge to be continuously distributed in space and disregard its discreteness. In the continuum model, we introduce a volume density of charges by

$$\rho_e = \lim_{\Delta V \to 0} \frac{1}{\Delta V} \sum_\alpha e_\alpha \qquad (\mathrm{C/m^3}) \tag{1.1.1}$$

where e_α are the elementary charges within the volume ΔV. Here, we speak of the volume ΔV as an infinitely small volume in the physical sense, which means that it is very small and hence its position in space is characterized with a sufficiently high accuracy by the coordinate of a point lying inside this volume. In addition, to be valid the limit $\Delta V \to 0$ must represent a volume large enough to contain a large number of charges e_α, yet small enough to appear infinitesimal when compared with the characteristic dimensions of the system considered. For example, a very small cube with each side of 1 micron has a volume of 10^{-18} $\mathrm{m^3}$, which can still contain about 100 billion atoms. Thus, it is expected that the smoothed-out function ρ_e defined by eqn.(1.1.1) may yield accurate macroscopic results for nearly all practical purposes.

1.1.2 **Current density**

If charges contained in a volume ΔV are moving with velocities which may differ in magnitude and in direction, the motion of a charge will result in a transport of the charge in the direction of its velocity. Consequently, various movements of charges contained in the volume ΔV result in a certain average transport of the charge contained in this volume. The intensity of this transport of charges may be characterized by introducing a vector quantity **J** called the current density defined by

$$\mathbf{J} = \lim_{\Delta V \to 0} \frac{1}{\Delta V} \sum_{\alpha} e_{\alpha} \mathbf{v}_{\alpha} \qquad (\mathrm{A/m^2}) \tag{1.1.2}$$

where $\mathbf{v}_{\alpha}$ is the velocity of the charge e_{α}.

The unit of current (Ampere) and the unit of charge (Coulomb) can be found to be related by 1 Coulomb = 1 Ampere-second. The direction of current density of positive charges coincides with the direction of their average velocity, while for negative charges, the current density has a direction opposite to that of the average velocity. We may, therefore, write

$$\mathbf{J} = \lim_{\Delta V \to 0} \frac{1}{\Delta V} \left\{ \sum_{\alpha} e^{+}_{\alpha} \mathbf{v}^{+}_{\alpha} + \sum_{\beta} e^{-}_{\beta} \mathbf{v}^{-}_{\beta} \right\} \tag{1.1.3}$$

where we have separated the contributions from the positive charges e^{+}_{α} with velocity $\mathbf{v}^{+}_{\alpha}$ and from the negative charges e^{-}_{β} with velocity $\mathbf{v}^{-}_{\beta}$. If we define a charge-average velocity $\mathbf{v}^{+}$ and $\mathbf{v}^{-}$ by

$$\mathbf{v}^{+} = \frac{\sum_{\alpha} e^{+}_{\alpha} \mathbf{v}^{+}_{\alpha}}{\sum_{\alpha} e^{+}_{\alpha}} \quad \text{and} \quad \mathbf{v}^{-} = \frac{\sum_{\beta} e^{-}_{\beta} \mathbf{v}^{-}_{\beta}}{\sum_{\beta} e^{-}_{\beta}} \tag{1.1.4}$$

the total current density may be expressed as

$$\mathbf{J} = \rho^{+}\mathbf{v}^{+} + \rho^{-}\mathbf{v}^{-} \tag{1.1.5}$$

where ρ^{+} and ρ^{-} are respectively the density of positive charges and the density of negative charges. The net charge density is given by $\rho_e = \rho^{+} +$

ρ^-. In many cases, the net charge may be zero. However, a current can still be present due to the difference between charge-average velocities.

The electric current describing the flow of, for instance, positive charges with an average velocity **v** across a planar surface S with its unit normal vector **n** parallel to **v** can be found by

$$I = \int_S \mathbf{J}\cdot\mathbf{n}dS = \frac{\rho v \Delta t S}{\Delta t} = \frac{\Delta q}{\Delta t} \quad \text{(A)} \tag{1.1.6}$$

where Δq is the total charge flowing across the surface S during the time interval Δt. Eqn.(1.1.6) shows that the electric current defines the rate of flow of charge.

1.1.3 **Conservation law of charge**

Charges are conservative, i.e., charges can neither be created nor be destroyed. Such a postulate has been confirmed by experimental evidence. The mathematical statement of this postulate on the conservation of charge is expressed by the equation of continuity that may be obtained by the following consideration. Consider a volume V enclosed by a stationary surface S. The conservation of charge means that the net outflow of charge current through the surface S must equal the rate at which the total charge in the volume V decreases, i.e.

$$\int_S \mathbf{J}\cdot\mathbf{n}dS = -\frac{\partial}{\partial t}\int_V \rho_e dV \tag{1.1.7}$$

where **n** is the unit outward normal vector of the surface S.

For a smooth continuous distribution of charges, since charges are conserved for any volume element, including an arbitrarily small one located anywhere, one gets the local equation of conservation of charge as

$$\nabla \cdot \mathbf{J} + \frac{\partial \rho_e}{\partial t} = 0 \tag{1.1.8}$$

which is the equation of continuity. This conservation law of charge must be satisfied at all times and under any circumstances.

At a discontinuous surface where surface charges accumulate, one has the following interface condition

$$\mathbf{n} \cdot [\![\mathbf{J}]\!] = -\frac{\partial \alpha_f}{\partial t} \tag{1.1.9}$$

where α_f is the surface charge density. Here, $[\![A]\!] = A^+ - A^-$ denotes the jump of the quantity A across the interface S with $\mathbf{n}$ being the unit normal vector always drawn from S^- to S^+.

For steady currents, charge density does not vary with time, and therefore, eqn.(1.1.8) is reduced to

$$\nabla \cdot \mathbf{J} = 0 \tag{1.1.10}$$

which implies that the stream lines of steady currents are closed loops, unlike those of electrostatic field that originate and end on charges. The integral form of eqn.(1.1.10) is

$$\int_S \mathbf{J} \cdot \mathbf{n} dS = 0 \tag{1.1.11}$$

a particular form of which reads

$$\sum_i I_i = 0. \tag{1.1.12}$$

This equation expresses a well-known law in electric circuit theory, that is the Kirchhoff current law which states that the algebraic sum of all the currents flowing out of a junction in an electric circuit is zero.

1.1.4 **Coulomb's law**

Charges in space interact with each other. The interaction between charges in free space (vacuum) was first studied quantitatively by Coulomb in 1785. He found an experimental law for describing the force between two point charges in free space, which is now called the Coulomb law. Coulomb's law was originally formulated as an action at a distance with the magnitude

$$F = \frac{1}{4\pi\varepsilon_o}\frac{q_1 q_2}{r^2} \qquad \text{(N)} \tag{1.1.13}$$

where $\varepsilon_o = 8.854\times10^{-12}$ Farad/meter (1 F = 1 $C^2/N{\cdot}m$) is called the permittivity of free space (vacuum) and r the distance between the two point charges. The direction of the force is along the line between the two charges. For both charges of the same sign, the force is repulsive, and for charges of opposite sign, the force is attractive. In addition, the Coulomb forces acting on the two charges are equal and opposite even though the individual charges may differ greatly in magnitude. Coulomb's law is known experimentally to be valid at a distance down to the order of 10^{-17} m (Jackson (1975) and Matveev (1986)).

A comparison of the Coulomb force F_e and the gravitational force F_g between two charged particles, for instance, an electron and a proton, may be made by

$$\frac{F_e}{F_g} = \frac{1}{4\pi\varepsilon_o}\frac{e^2}{Gm_e m_p} = 2.2\times10^{39} \tag{1.1.14}$$

where $G = 6.7\times10^{-11}$ $m^3/kg{\cdot}sec^2$ is the gravitational constant. m_e and m_p are respectively the mass of the electron and the mass of the proton. It is shown that the electrical force of attraction between an electron and a proton is 2.2×10^{39} times greater than the gravitational force of attraction. Since the gravitational constant G is extremely small, the gravitational interaction can become considerable only for very large masses. For this reason gravitational forces are in general negligible in the mechanics of atoms and molecules.

1.2 Electric and magnetic fields

1.2.1 Electric field

Coulomb's law was interpreted as a long-range interaction, i.e. it was assumed that one body acts on another as if without intermediaries. Such a long-range interaction differs from a short-range interaction which is defined according to bodies interacting only due to a continuous "transfer of forces" in the space between them.

To describe the interaction between material bodies, one may either formulate the action at a distance between the interacting bodies or separate the interaction process into the production of a field by one system and the action of the field on another system. These two descriptions are physically indistinguishable in the static case. However, if the bodies are in motion, it is both physically and mathematically advantageous to ascribe physical reality to the field itself. In what follows, we shall formulate electromagnetic and mechanical interactions as a field theory.

In classical electrodynamics, electric and magnetic fields are fundamental fields of forces that originate from charges. The electric field in free space is defined in terms of the force produced on a test charge q by the equation

$$\mathbf{E} = \lim_{q \to 0} \frac{\mathbf{F}}{q} \qquad \text{(N/C)} \tag{1.2.1}$$

where **F** is the force on the test charge q. The limit q->0 is introduced in order that the test charge will not influence the behavior of the sources of the field, which will then be independent of the presence of the test body. Practically, such a definition is entirely suitable only for macroscopic phenomena.

According to Coulomb's law, the electric field **E** at a position **r** due to a charge q_0 at the origin of the radius vector may be found immediately by

$$\mathbf{E} = \frac{q_0}{4\pi\varepsilon_0} \frac{\mathbf{r}}{r^3} \tag{1.2.2a}$$

or

$$\mathbf{E} = -\frac{q_0}{4\pi\varepsilon_0} \nabla\left(\frac{1}{r}\right) . \tag{1.2.2b}$$

The force acting on a test charge q due to the interaction between the charge q_0 and the charge q can now be expressed by **F** = q**E**.

Essentially, Coulomb's law describes quantitatively the interaction between two charges. Naturally, we shall ask the question how to describe interactions among large numbers of charges. To answer this question, a linear superposition principle of fields in free space had been proved experimentally with sufficient accuracy in the classical domain of sizes and attainable field strengths. It states that the total electric field produced by N charges q_α located respectively at $\mathbf{x}_\alpha$ (α = 1, 2,..., N) in free space is simply the sum of the electric fields caused by individual charges, i.e.

$$\mathbf{E}(\mathbf{x}) = \sum_{\alpha=1}^{N} \frac{q_\alpha}{4\pi\varepsilon_o} \frac{(\mathbf{x}-\mathbf{x}_\alpha)}{|\mathbf{x}-\mathbf{x}_\alpha|^3} \,. \tag{1.2.3}$$

For a continuous distribution of charges with charge density ρ_e in a volume V, the electric field can be found as

$$\mathbf{E}(\mathbf{x}) = \int_V \frac{\rho_e(\mathbf{x}')}{4\pi\varepsilon_o} \frac{(\mathbf{x}-\mathbf{x}')}{|\mathbf{x}-\mathbf{x}'|^3} dV' \,. \tag{1.2.4}$$

This expression for $\mathbf{E}$ is valid for the position point $\mathbf{x}$ outside the charged region as well as inside. If the charge density ρ_e and its first derivatives are continuous, the gradient of the electric field exists and is continuous at interior points of the charged region. Detail discussions about the mathematics involved are referred to the work of Tiersten (1990).

1.2.2 **Electric potential**

By taking the curl of eqn.(1.2.3) or (1.2.4), one may find $\nabla\times\mathbf{E}=0$, which implies that the electrostatic field generated by charges is irrotational. Thus, we may introduce a scalar electric potential ϕ defined by

$$\mathbf{E} = -\nabla\phi \tag{1.2.5}$$

and can easily find

$$\phi(\mathbf{x}) = \sum_{\alpha=1}^{N} \frac{q_\alpha}{4\pi\varepsilon_o} \frac{1}{|\mathbf{x}-\mathbf{x}_\alpha|} \tag{1.2.6}$$

for the discrete distribution of N point charges q_α located respectively at $\mathbf{x}_\alpha$ (α = 1, 2,..., N) in free space, and

$$\phi(\mathbf{x}) = \int_V \frac{\rho_e(\mathbf{x}')}{4\pi\varepsilon_o} \frac{1}{|\mathbf{x}-\mathbf{x}'|} dV' \qquad (1.2.7)$$

for the continuous distribution of charges with charge density ρ_e in the volume V. It can be shown that the expression (1.2.7) for the electric potential ϕ is valid for the position point $\mathbf{x}$ outside the charged region as well as inside. If the charge density ρ_e and its first derivatives are continuous, the second spatial derivatives of ϕ exist and are continuous at interior points of the charged region.

The scalar electric potential ϕ has a physical interpretation when we consider the work done on a test charge q in transporting it from infinity to a position point $\mathbf{x}$ in a region of localized electric field described by ϕ (which vanishes at infinity). Since the force acting on the charge q at any point is $\mathbf{F} = q\mathbf{E}$, the work done in moving the charge from infinity to the position $\mathbf{x}$ is

$$W = -\int_\infty^{\mathbf{x}} \mathbf{F}\cdot d\mathbf{l} = \int_\infty^{\mathbf{x}} q\nabla\phi\cdot d\mathbf{l} = q\phi(\mathbf{x}) \qquad (1.2.8)$$

which shows that $q\phi(\mathbf{x})$ can be interpreted as the potential energy of the test charge in the electrostatic field. The unit of the electric potential ϕ is volt (V).

1.2.3 **Gauss' theorem**

With the aid of the superposition principle, we may introduce the following Gauss' theorem as an integral form of Coulomb's law

$$\int_S \varepsilon_o \mathbf{E}\cdot\mathbf{n} dS = \int_V \rho_e dV \qquad (1.2.9)$$

where V is a space volume bounded by a closed surface S.

Eqn.(1.2.9) can be proved simply by writing

$$\int_S \varepsilon_o \mathbf{E}\cdot\mathbf{n} dS = \int_V \varepsilon_o \nabla\cdot\mathbf{E} dV = \int_V \varepsilon_o \nabla\cdot \int \frac{\rho_e(\mathbf{x}')}{4\pi\varepsilon_o} \frac{(\mathbf{x}-\mathbf{x}')}{|\mathbf{x}-\mathbf{x}'|^3} dV' dV$$

$$= \int\int_V \frac{\rho_e(\mathbf{x}')}{4\pi} \nabla^2 \{\frac{-1}{|\mathbf{x}-\mathbf{x}'|}\} dV'dV = \int\int_V \rho_e(\mathbf{x}')\delta(\mathbf{x}-\mathbf{x}')dV'dV = \int_V \rho_e dV \tag{1.2.10}$$

where $\delta(\mathbf{x}-\mathbf{x}')$ is the Dirac delta function.

For a continuous distribution of volume charge density ρ_e, we have the following local differential form of Gauss´ theorem which reads

$$\nabla \cdot (\varepsilon_o \mathbf{E}) = \rho_e \,. \tag{1.2.11}$$

This important result is one of Maxwell's equations in free space.

In the case of a discrete set of point charges q_α located at $\mathbf{x}_\alpha$ (α=1,2,...) in the free space, we may write

$$\rho_e = \sum_\alpha q_\alpha \delta(\mathbf{x}-\mathbf{x}_\alpha) \,, \tag{1.2.12}$$

and can find from eqn.(1.2.9)

$$\int_S \varepsilon_o \mathbf{E}\cdot\mathbf{n} dS = \sum_\alpha q_\alpha \tag{1.2.13}$$

where the sum is over only those charges inside the closed surface S.

It is shown that Gauss´ theorem (1.2.9) (or (1.2.13)) expresses the physical fact that the total outward flux of the electric field over any closed surface in free space is equal to the total charge enclosed in the surface divided by ε_o.

1.2.4 **Magnetic field and Lorentz force**

So far, we have shown that interaction between fixed point charges is defined completely by Coulomb's law. However, Coulomb's law is incapable of describing the interaction between moving charges. Such a conclusion is based on relativistic properties of space and time and the relativistic equation of motion rather than on the specific features of Coulomb interaction. The interaction between moving charges is due not only to a Coulomb force but also to another kind of force called the magnetic force.

When a small test charge q is moving in a magnetic field (to be defined), experiments show that it experiences a force that cannot be expressed in terms of **E**, but can be expressed by defining a new vector field quantity, the magnetic (induction) field **B**, such that

$$\mathbf{F}_m = q\mathbf{v} \times \mathbf{B} \qquad (1.2.14)$$

where **v** is the velocity vector of the test charge. The magnetic induction field **B** has the unit of tesla (1 tesla = 1 weber/meter^2 = 1 volt-second/meter^2) in SI unit. In general, when both electric and magnetic fields are present, the total electromagnetic force on a charge q moving with velocity **v** in the fields can be expressed by

$$\mathbf{F} = q\mathbf{E} + q\mathbf{v} \times \mathbf{B} \ . \qquad (1.2.15)$$

This expression is known as the Lorentz force equation. Its validity has been well established by experiments. In classical electrodynamics, the Lorentz force equation may be considered as a fundamental postulate of our electromagnetic model. However, in a special theory of relativity, the Lorentz force expression may be obtained from the requirement of the invariance of the (special) relativistic equation of motion.

The Lorentz force for a charge moving in free space can be measured directly in a laboratory. A generalization of it to a charge moving in a material medium is not, however, subject to direct experimental confirmation and, therefore, it is considered as an assumption for the moment. Quantitatively, the magnetic interaction shown in eqn. (1.2.15) can be compared with electric interaction only at sufficiently high velocities of charged particles. If, however, Coulomb interaction is absent for some reason, magnetic interaction can manifest itself at very low velocities, for instance, when an electric current flows in a conductor, which has an immensely high density of free electrons and where the electric field of moving charges is neutralized by the electric field of the opposite charges of the conductor, i.e. it is screened. Thus, the magnetic force in good conductors may become dominant since the product of charge density and velocity can be large even when the velocity is very small.

1.3 The laws of electrodynamics

1.3.1 Ampere´s circuital law in free space

In this section, we shall present the well-known laws of electrodynamics, based on which Maxwell's equations for free space are found. First of all, we introduce the integral form of Ampere´s circuital law in free space by

$$\int_L \left(\frac{\mathbf{B}}{\mu_o}\right)\cdot d\mathbf{l} = \int_S \mathbf{J}\cdot d\mathbf{S} + \frac{\partial}{\partial t}\int_S \varepsilon_o \mathbf{E}\cdot d\mathbf{S} \qquad \text{(A)} \qquad (1.3.1)$$

where $\mu_o = 4\pi \times 10^{-7}$ H/m (Henry per meter) is the permeability of free space. Here, L is any closed line that bounds a two-sided surface S in free space. The positive direction of the typical element $d\mathbf{S}$ may be taken to either side of S, but the positive integration sense about L must agree with the right-hand rule relative to $d\mathbf{S}$.

By applying Stokes´ theorem to the line integral on the left-hand side of eqn.(1.3.1), we may derive the local differential form of Ampere´s circuital law in free space as

$$\nabla \times \left(\frac{\mathbf{B}}{\mu_o}\right) = \mathbf{J} + \frac{\partial(\varepsilon_o \mathbf{E})}{\partial t}. \qquad (1.3.2)$$

This equation states that the curl of $\mathbf{B}/\mu_o$ at any point in a region is the sum of the electric current density $\mathbf{J}$ and the displacement current density $\partial(\varepsilon_o\mathbf{E})/\partial t$ at that point. The displacement current term is historically the contribution of Maxwell who provided that missing link to unify the theories of electricity and magnetism and predicted the propagation of electromagnetic wave in free space in the absence of charges and currents.

In steady states of electromagnetic systems, the displacement current vanishes and the Ampere circuital law (1.3.1) is reduced to

$$\int_L \left(\frac{\mathbf{B}}{\mu_o}\right)\cdot d\mathbf{l} = \int_S \mathbf{J}\cdot d\mathbf{S} \qquad (1.3.3)$$

which states that the circulation of the magnetic field divided by μ_o around any closed path is equal to the free current flowing through the surface bounded by the path. The integral form of the Ampere circuital law is sometimes useful to determine the magnetic induction field **B** caused by a steady current I when there is a closed path around the current such that the magnitude of **B** is constant over the path.

1.3.2 Non-existence of magnetic charges

In contrast to Gauss' theorem for electric field, there exists a hypothesis of non-existence of magnetic charges (also called magnetic monopoles), the mathematical expression of which is given by

$$\int_S \mathbf{B}\cdot d\mathbf{S} = 0 \, . \tag{1.3.4}$$

This hypothesis is also often called Gauss' law for magnetic field since, so far, there is no experimental evidence for the existence of magnetic charges or monopoles (Jackson (1975) and Johnk (1988)).

The differential form of Gauss' law for magnetic field is written

$$\nabla\cdot\mathbf{B} = 0 \, . \tag{1.3.5}$$

This equation shows that we may introduce a vector field **A** defined by

$$\mathbf{B} = \nabla\times\mathbf{A} \tag{1.3.6}$$

so that eqn.(1.3.2) becomes

$$\nabla\times(\nabla\times\mathbf{A}) = \mu_o\mathbf{J} + \frac{1}{c^2}\frac{\partial\mathbf{E}}{\partial t} \tag{1.3.7}$$

where $c = \sqrt{\mu_o\varepsilon_o}$ is the speed of light in free space, which has the numerical value of 2.998×10^8 meter/second. The vector field **A** so defined is called the magnetic vector potential. Its SI unit is Weber per meter (Wb/m). The magnetic vector potential **A** can be related to the magnetic flux Φ through a given surface S that is bounded by contour L

$$\Phi = \int_S \mathbf{B}\cdot d\mathbf{S} = \int_L \mathbf{A}\cdot d\mathbf{l} \qquad \text{(Wb)} \tag{1.3.8}$$

with the aid of Stokes´theorem.

1.3.3 Biot-Savart law

For steady currents, the displacement current term disappears in eqn.(1.3.7). Thus, one may get

$$\nabla^2 \mathbf{A} = -\mu_o \mathbf{J} \tag{1.3.9}$$

by using the Coulomb gauge $\nabla\cdot\mathbf{A} = 0$. This is a vector Poisson´s equation. Thus, for a steady current distribution confined in a finite volume V, the magnetic vector potential **A** can be found as

$$\mathbf{A}(\mathbf{x}) = \frac{\mu_o}{4\pi}\int_V \frac{\mathbf{J}(\mathbf{x}')}{|\mathbf{x}-\mathbf{x}'|}dV' \qquad \text{(Wb/m).} \tag{1.3.10}$$

This expression for the magnetic vector potential **A** is valid for the position point **x** outside the current region as well as inside. If the current density **J** and its first derivatives are continuous, the second derivatives of the magnetic vector potential **A** exist and are continuous at interior points of the current region.

By taking the curl of eqn.(1.3.10), we derive the following result which is usually called Biot-Savart law for a steady current distribution:

$$\mathbf{B}(\mathbf{x}) = \frac{\mu_o}{4\pi}\int_V \frac{\mathbf{J}\times\mathbf{r}}{r^3}\, dV' \tag{1.3.11}$$

with $\mathbf{r} = \mathbf{x} - \mathbf{x}'$ and $r = |\mathbf{x} - \mathbf{x}'|$.

It is seen that the Biot-Savart law satisfies automatically the condition of the non-existence of magnetic charges by simply making the divergence of **B** for eqn.(1.3.11), the result of which gives $\nabla\cdot\mathbf{B} = 0$.

For a closed current circuit L' with current I', eqn.(1.3.11) may be written as

$$\mathbf{B}(\mathbf{x}) = \frac{\mu_0}{4\pi} \int_{L'} \frac{I'd\mathbf{l}' \times \mathbf{r}}{r^3} . \tag{1.3.12}$$

In addition, for the steady currents, one has $\nabla \cdot \mathbf{J} = 0$ which implies that any current flowing into an arbitary volume must flow out.

1.3.4 **Magnetic force on current circuit**

Using the Lorentz force expression (1.2.15), the magnetic force acting on a line current element $d\mathbf{l}$ with the current I located at $\mathbf{x}$ can be calculated by

$$dF = Id\mathbf{l} \times \mathbf{B}(\mathbf{x}) . \tag{1.3.13}$$

Thus, the total magnetic force acting on the current circuit L with the steady current I in the magnetic field generated by another current circuit L' with the steady current I' can be found as

$$\mathbf{F}_{(L)} = \int_L Id\mathbf{l} \times \mathbf{B}(\mathbf{x}) = \frac{\mu_0}{4\pi} \int_L Id\mathbf{l} \times \int_{L'} \frac{I'd\mathbf{l}' \times \mathbf{r}}{r^3} . \tag{1.3.14}$$

One may also calculate the total magnetic force acting on the current circuit L' due to the magnetic field generated by the current circuit L. It is found that the force $\mathbf{F}_{(L')}$ has the same magnitude but opposite direction of the force $\mathbf{F}_{(L)}$, i.e., $\mathbf{F}_{(L')} = -\mathbf{F}_{(L)}$, which means that the interaction forces between two steady current circuits satisfy Newton's third law. Obviously, in order for the current circuit L (or L') to be in equilibrium, additional mechanical force has to be acted on it so that the net force acting on the current circuit L (or L') will be zero.

1.3.5 **Faraday's law of induction**

The Faraday law of induction comes from the experimental results on the generation of electromotive forces (e.m.f.) by varying magnetic fields. It states that the electromotive force induced in a closed circuit at rest relative to the observer is equal to the negative rate of increase of the magnetic flux linking the circuit. Mathematically, the integral form of the Faraday law of induction can be expressed by

$$\varepsilon^{ind} = -\frac{\partial}{\partial t}\int_S \mathbf{B} \cdot d\mathbf{S} = \int_L \mathbf{E} \cdot d\mathbf{l} \tag{1.3.15}$$

where ε^{ind} is the induced electromotive force. Since ε^{ind} is measured in volts, it is also called the induced voltage. The negative sign appearing in eqn.(1.3.15) is due to the fact that the direction of the induced e.m.f. is determined by the Lenz law which states that induced current is in such a direction as to oppose the magnetic flux variation causing it.

By applying Stokes' theorem to the line integral on the right-hand side of eqn.(1.3.15), one may derive the local differential form of Faraday's law

$$\nabla \times \mathbf{E} = -\frac{\partial \mathbf{B}}{\partial t} \tag{1.3.16}$$

which applies to every point in space, whether it is in free space or in a material medium. Eqn.(1.3.16) shows that the electric field in a region of time-varying magnetic induction field is non-conservative and cannot be expressed simply as the gradient of a scalar potential. However, for static fields, the term $\partial\mathbf{B}/\partial t = 0$ in eqn.(1.3.16), and thus Faraday's law states that the line integral of a static electric field $\mathbf{E}$ about any closed path is always zero. This means that all static electric fields are conservative.

The Faraday law of induction expressed mathematically by eqn.(1.3.15) (or (1.3.16)) presents a new physical phenomenon: a varying electric field is created not only by electric charges but also by a varying magnetic field, and is also supposed to be valid even in the presence of material media.

1.3.6 Maxwell's equations in free space

According to Faraday's and Ampere's laws, electric and magnetic effects are not isolated phenomena; the variation in one must produce a change in the other. Thus, we shall often speak of electromagnetic field, with the electric and the magnetic fields representing different aspects of the same phenomenon. The complete mathematical description of the electromagnetic phenomenon in free space was first proposed by Maxwell based on the experimental laws for isolated electric charges and electric current circuits in free space. It is now known as Maxwell's equations in free space, which can be summarized as follows:

$$\nabla \times \mathbf{B} = \mu_o \mathbf{J} + \frac{1}{c^2}\frac{\partial \mathbf{E}}{\partial t} \qquad \text{(Ampere's circuital law)} \qquad (1.3.17)$$

$$\nabla \cdot \mathbf{B} = 0 \qquad \text{(Nonexistence of magnetic charges)} \qquad (1.3.18)$$

$$\nabla \times \mathbf{E} = -\frac{\partial \mathbf{B}}{\partial t} \qquad \text{(Faraday's law of induction)} \qquad (1.3.19)$$

$$\nabla \cdot (\varepsilon_o \mathbf{E}) = \rho_e \qquad \text{(Gauss' theorem for electric fields)} \qquad (1.3.20)$$

together with

$$\nabla \cdot \mathbf{J} + \frac{\partial \rho_e}{\partial t} = 0 \qquad \text{(Conservation law of charges).} \qquad (1.3.21)$$

It is seen that in free space all electromagnetic phenomena with electric charges and electric currents as sources may be described by only two fields, the electric field **E** and the magnetic induction field **B**.

1.4 Electric and magnetic multipoles in free space

In this section, we shall introduce the concepts of electric and magnetic multipoles in free space and their interactions with externally applied electric and magnetic fields. The multipole concept is of interest since it may be used in modelling of electromagnetic materials as we shall show in later sections.

1.4.1 Electric multipoles in free space

We shall start with the introduction of the concept of electric multipoles for localized charge distribution. We consider an array of M point electric charges located in a small volume centered at **x**' as shown in **Figure** 1.4.1, which ideally assumes that all charges are treated as geometrical points in space. In the laboratory, this approximation can be achieved by making any distances of separation that are involved very large compared to the dimensions of the charged particles. The electric potential ϕ (defined by $\mathbf{E} = -\nabla\phi$) for such an array of electric charges can be found as

$$\phi(\mathbf{x}) = \sum_{\alpha=1}^{M} q^{\alpha} G^{e}(\mathbf{x}, \mathbf{x}'+\mathbf{d}^{\alpha}) \tag{1.4.1}$$

where $\mathbf{x}'+\mathbf{d}^{\alpha}$ denotes the position vector of the α-th point charge q^{α}, and G^e is the electric Green's function in free space, defined by the following equation

$$\nabla^2 G^e(\mathbf{x}, \mathbf{x}') + \frac{1}{\varepsilon_0}\delta(\mathbf{x} - \mathbf{x}') = 0 \tag{1.4.2}$$

with $\delta(\mathbf{x} - \mathbf{x}')$ being a Dirac delta function.
The fundamental solution of eqn.(1.4.2) may be found as

$$G^e(\mathbf{x}, \mathbf{x}') = \frac{1}{4\pi\varepsilon_0|\mathbf{x} - \mathbf{x}'|} \tag{1.4.3}$$

which can be explained physically as the electric potential at the position $\mathbf{x}$ produced by a unit point charge located at $\mathbf{x}'$ in free space.
By expending the electric Green's function in a Taylor series about $(\mathbf{x},\mathbf{x}')$, eqn.(1.4.1) can be expressed as

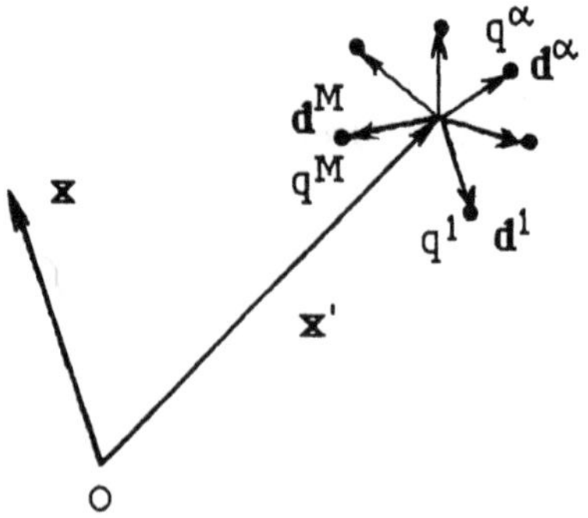

Figure 1.4.1 An array of point electric charges.

$$\phi(\mathbf{x}) = \sum_{k=0}^{\infty} \frac{(-1)^k}{k!} p^e_{s_1 \ldots s_k} G^e_{,s_1 \ldots s_k}(\mathbf{x} - \mathbf{x}') \tag{1.4.4}$$

where one has introduced the point electric multipole moment of order k (k=0,1,2,...) defined by

$$p^e_{s_1 \ldots s_k} = \sum_{\alpha=1}^{M} q^\alpha d^\alpha_{s_1} \ldots d^\alpha_{s_k} \tag{1.4.5}$$

which is located at the position $\mathbf{x}'$.

For k=0, one has the net charge of the array of point electric charges

$$p^e = \sum_{\alpha=1}^{M} q^\alpha \tag{1.4.6}$$

which may also be called the electric monopole moment of the charge array. If the point $\mathbf{x}$ is far away from $\mathbf{x}'$ and if the net charge of the array of charges is not zero, the electric monopole term in eqn.(1.4.4) will be the dominant term in the electrical potential, which means that the whole electric charge distribution will act as if it were a point electric charge at $\mathbf{x}'$.

For k=1, one arrives at the point electric dipole moment

$$\mathbf{p}^e = \sum_{\alpha=1}^{M} q^\alpha \mathbf{d}^\alpha \tag{1.4.7}$$

which involves only the properties of the charge distribution and does not involve the location of the field point $\mathbf{x}$. If the field point $\mathbf{x}$ is very far away and if the electric monopole moment p^e vanishes, the dipole term will then be the leading term in the expansion of ϕ and the point electric dipole moment $\mathbf{p}^e$ will be the dominant feature of the charge array.

By introducing the electric center of gravity of the positive and negative charges, one can rewrite eqn.(1.4.7) in the following form

$$\mathbf{p}^e = \sum_{\text{positive}} q^\alpha \mathbf{d}^\alpha + \sum_{\text{negative}} q^\alpha \mathbf{d}^\alpha = Q^+\mathbf{x}^+ + Q^-\mathbf{x}^- \tag{1.4.8}$$

where $\mathbf{x}^+$ and $\mathbf{x}^-$ denotes the position vectors of the electric center of gravity of the positive charges and the negative charges respectively. Q^+ and Q^- are the total positive charge and the total negative charge respectively.

In the case of a zero net charge, eqn.(1.4.8) is reduced to

$$\mathbf{p}^e = Q\mathbf{d} \tag{1.4.9}$$

where Q denotes the total positive charge and $\mathbf{d} = \mathbf{x}^+ - \mathbf{x}^-$ represents the vectorial distance between the centers of gravity of the positive and negative charges.

In a simple case, the electric charge system consists of only two point charges +q and -q at a distance $\mathbf{d}$. Such a system is usually called a (physical) electric dipole, the moment of which is equal to q$\mathbf{d}$, where the vector $\mathbf{d}$ is pointing from the negative to the positive charge.

A mathematical abstraction derived from the above defined physical dipole is the ideal or point dipole, which is defined as that obtained by replacing the distance $\mathbf{d}$ and the charge Q respectively by $\mathbf{d}/n$ and nQ, the limit approached as the number n tends to infinity is the ideal dipole.

It is worth while to mention that apart from these permanent or intrinsic dipole moments, a temporary or induced dipole moment may arise in a molecule, having for instance originally a zero electric dipole moment, when it is brought into an external electric field, since in the presence of this field, there are forces acting on those charges in the molecule which distort the molecular charge distribution from its originally symmetric form. In addition, molecules, especially when distorted by the field, may have higher-order electric multipoles, such as quadrupole moments, etc. (Stogryn (1966) and Böttcher (1973)).

For higher multipoles, it is noticed that though the electric monopole moment p^e is a unique property of the charge array, the electric dipole moment, the electric quadrupole moment, etc. are in general not unique properties of the charge array and depend on the choice of origin. However, if the monopole moment p^e vanishes, the dipole moment $\mathbf{p}^e$ is independent of the origin $\mathbf{x}'$. In general, only the lowest non-vanishing moment is unique unless $\mathbf{x}'$ is specified (Böttcher (1973)).

It is shown by eqn.(1.4.4) that the electric potential at a field point $\mathbf{x}$ outside of a small volume containing the array of point electric charges could be described by a set of point electric multipole moments of the charge system located at $\mathbf{x}'$.

1.4.2 **Interaction of electric multipoles with external field**

For a given array of M point electric charges located at $\mathbf{x}+\mathbf{d}^\alpha$ ($\alpha=1,2,\ldots,M$), the interaction energy of the array of electric charges with an external electric field $\mathbf{E}^o$ (its corresponding scalar potential is denoted by ϕ^o) reads

$$U^{eo} = \sum_{\alpha=1}^{M} q^\alpha \phi^o(\mathbf{x} + \mathbf{d}^\alpha). \tag{1.4.10}$$

If the scalar potential ϕ^o varies slowly in the region of the electric charge array, one can make a Taylor expansion of $\phi^o(\mathbf{x} + \mathbf{d}^\alpha)$ around the point $\mathbf{x}$ and write

$$U^{eo} = p^e \phi^o(\mathbf{x}) + \sum_{k=1}^{\infty} \frac{1}{k!} p^e_{s_1 \ldots s_k} \phi^o{}_{,s_1 \ldots s_k}(\mathbf{x}) \tag{1.4.11}$$

where p^e denotes the total charge of the point charge array, and $p^e_{s_1 \ldots s_k}$ the point electric multipole moments defined in eqn.(1.4.5) but now located at $\mathbf{x}$.

The interaction energy of a point electric multipole moment of order k located at $\mathbf{x}$ in the external field can thus be introduced by

$$U^{eo(k)} = \frac{1}{k!} p^e_{s_1 \ldots s_k} \phi^o{}_{,s_1 \ldots s_k}(\mathbf{x}) \tag{1.4.12}$$

which, for a point electric dipole moment, reads

$$U^{eo(1)} = - \mathbf{p}^e \cdot \mathbf{E}^o . \tag{1.4.13}$$

The total translational force acting on the array of electric charges in the external electric field can then be found by

$$F^e_n = p^e E^o_n(x) + \sum_{k=1}^{\infty} \frac{1}{k!} p^e_{s_1 \ldots s_k} E^o_{n,s_1 \ldots s_k}(x) \, . \tag{1.4.14}$$

It is shown that if the net charge of the array of electric charges is not zero ($p^e \neq 0$), the leading term of eqn.(1.4.14) will be the force on the point electric monopole moment. Obviously, there is no torque acting on the electric monopole moment. If, however, the net charge vanishes ($p^e=0$), the electric force on the electric dipole moment $\mathbf{p}^e$ will be the dominant term in eqn.(1.4.14) and can be expressed by

$$\mathbf{F}^e = (\mathbf{p}^e \cdot \nabla)\mathbf{E}^o \, . \tag{1.4.15}$$

This expression shows that the electric dipole force is proportional to the gradient of the external electric field and, therefore, that it vanishes in a uniform external electric field. There is, however, a torque acting on the electric dipole moment in a uniform field as well as in a non-uniform field. This torque may be derived by the negative derivative of the energy (see (1.4.13)) with respect to the angle between the external field and the dipole moment vector, i.e.,

$$\tau^e = \mathbf{p}^e \times \mathbf{E}^o \, . \tag{1.4.16}$$

It is seen that the torque tends to orient the electric dipole moment in the direction of the external electric field.

1.4.3 **Magnetic multipoles in free space**

We shall now introduce the concept of point magnetic multipoles. It is known that although the existence of magnetic charges (monopoles) is compatible with the requirements of quantum mechanics (see Dirac (1931), Amaldi (1968), etc.), their reliable experimental discovery seems not yet to have been achieved. Therefore, we still admit here that the divergence of the magnetic field **B** is zero and that the sources of magnetic field **B** are only electric currents, until the magnetic monopole is discovered.

In analogy with electrostatics, we may consider the similar thing for an arbitrary localized current distribution. Suppose that there is a steady current distribution **J** within a small volume V in space. Denoting a certain point in the volume by a position vector **x**', which can be usually taken as the center of gravity of the volume "matter", shown in **Figure** 1.4.2, the magnetic vector potential **A** (defined by **B** = ∇ × **A** with the Coulomb gauge ∇·**A** = 0) at the field point **x** can be found for the steady current distribution as

$$\mathbf{A}(\mathbf{x}) = \int_V \mathbf{J}(\mathbf{x}'+\mathbf{d})G^m(\mathbf{x}, \mathbf{x}'+\mathbf{d})dV(\mathbf{d}). \tag{1.4.17}$$

In this equation, G^m denotes the magnetic Green's function in free space and is defined by the following equation

$$\nabla^2 G^m(\mathbf{x}, \mathbf{x}') + \mu_0 \delta(\mathbf{x} - \mathbf{x}') = 0 \tag{1.4.18}$$

which has the fundamental solution in the infinite space given by

$$G^m(\mathbf{x}, \mathbf{x}') = \frac{\mu_0}{4\pi|\mathbf{x} - \mathbf{x}'|} \; . \tag{1.4.19}$$

If the current I flows in a closed loop L whose line element is **dl**, eqn.(1.4.17) can be written as

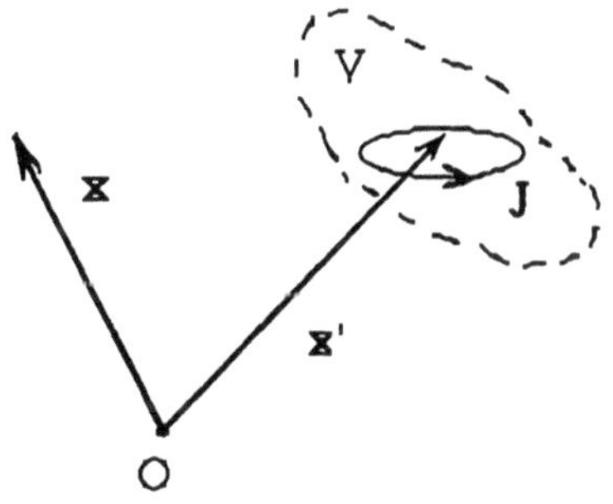

Figure 1.4.2 A volume distribution of currents in space.

$$\mathbf{A}(\mathbf{x}) = \frac{\mu_o}{4\pi} \int_L \frac{I d\mathbf{l}''}{|\mathbf{x} - \mathbf{x}''|} \tag{1.4.20}$$

where $\mathbf{x}''$ is the position point taken on the current loop L.

By expanding the magnetic Green's function $G^m(\mathbf{x}, \mathbf{x}'+\mathbf{d})$ in a Taylor series about $(\mathbf{x}, \mathbf{x}')$ for eqn.(1.4.17), the magnetic vector potential $\mathbf{A}$ can then be expressed in the following form

$$A_i(\mathbf{x}) = \sum_{k=0}^{\infty} \frac{(-1)^k}{k!} p^m{}_{is_1 \ldots s_k} G^m{}_{,s_1 \ldots s_k}(\mathbf{x} - \mathbf{x}') \tag{1.4.21}$$

where one has formally introduced the point magnetic multipole moment of order k (k=0,1,2,...) located at $\mathbf{x}'$, defined by

$$p^m{}_{is_1 \ldots s_k} = \int_V J_i \, d_{s_1} \ldots d_{s_k} \, dV(\mathbf{d}) \, . \tag{1.4.22}$$

For k=0, one has the point magnetic monopole term

$$\mathbf{p}^m = \int_V \mathbf{J} \, dV \tag{1.4.23}$$

which is zero since the localized current in the volume follows closed paths. This means that the leading term in the expansion of the magnetic potential (1.4.21) is always the magnetic dipole term under the constraint of nonexistence of magnetic charges (monopoles).

For k=1, by noting the zero divergence of the electrical current density vector $\mathbf{J}$, the point magnetic dipole term may be written as

$$A_i(\mathbf{x}) = - p^m{}_{is} G^m{}_{,s}(\mathbf{x} - \mathbf{x}') = - \varepsilon_{iks} \, m_k \, G^m{}_{,s}(\mathbf{x} - \mathbf{x}') \tag{1.4.24}$$

or

$$\mathbf{A}(\mathbf{x}) = \mathbf{m} \times \nabla' G^m(\mathbf{x} - \mathbf{x}') \tag{1.4.25}$$

where ∇' denotes the gradient operator acting on $\mathbf{x}'$, and $\mathbf{m}$ is the point magnetic dipole moment for the localized current distribution by

$$\mathbf{m} = \frac{1}{2}\int_V \mathbf{d} \times \mathbf{J}\, dV(\mathbf{d}) \, . \tag{1.4.26}$$

If the current is confined in an arbitrary current loop on a plane, and the current I flows in a closed circuit L whose element is **dl**, eqn.(1.4.26) becomes

$$\mathbf{m} = \frac{1}{2}\int_L \mathbf{d} \times \mathbf{dl} \tag{1.4.27}$$

the magnitude of which is $|\mathbf{m}| = I \times$ (Area of the loop) regardless of the shape of the circuit. In analogy to electrostatics, it can also be shown that the magnetic dipole moment is a unique property of the current distribution and is independent of the choice of origin of coordinate systems since the magnetic monopole does not exist.

1.4.4 **Interaction of magnetic multipoles with external field**

We now consider the magnetic interaction energy for a localized volume distribution of electric current which is subject to the influence of an external magnetic field $\mathbf{B}^o$ (= $\nabla \times \mathbf{A}^o$). It can be found that the magnetic interaction energy reads

$$U^{mo} = \int_V \mathbf{J} \cdot \mathbf{A}^o(\mathbf{x} + \mathbf{d})dV(\mathbf{d}) \tag{1.4.28}$$

where **x** is a certain position point in the volume V.

If a steady current I flows in a closed loop L whose line element is **dl** and the magnetic vector potential $\mathbf{A}^o$ is produced by another current circuit L_0 with its flowing steady current I_0, eqn.(1.4.28) becomes

$$U^{mo} = \frac{\mu_0}{4\pi}\int_{L_0}\int_L \frac{I_0 I\ \mathbf{dl}\cdot\mathbf{dl}'}{|\mathbf{x} - \mathbf{x}'|} \tag{1.4.29}$$

which may also be identified as the magnetostatic energy of the system of the two current circuits.

In eqn.(1.4.28), if the volume V is supposed to be small and the magnetic vector potential $\mathbf{A}^o$ does not vary much over the current distribution, one can make a Taylor expansion of $\mathbf{A}^o(\mathbf{x} + \mathbf{d})$ around the point $\mathbf{x}$ and write

$$U^{mo} = \sum_{k=1}^{\infty} \frac{1}{k!} p^{m}{}_{is_1...s_k} A^{o}{}_{i,s_1...s_k}(\mathbf{x}) \tag{1.4.30}$$

where $p^{m}{}_{is_1...s_k}$ are the point magnetic multipole moments of order k (=1,2,...) defined in eqn.(1.4.22) and located at $\mathbf{x}$. The leading term of this expansion is the magnetic interaction energy of the point magnetic dipole moment $\mathbf{m}$ in the external magnetic field $\mathbf{B}^o$, which is

$$U^{mo(1)} = \mathbf{m}\cdot(\nabla \times \mathbf{A}^o) = \mathbf{m}\cdot\mathbf{B}^o \, . \tag{1.4.31}$$

The total translational force acting on the electric current volume may then be found as

$$F^{m}_{n} = \sum_{k=1}^{\infty} \frac{1}{k!} p^{m}{}_{is_1...s_k} A^{o}{}_{i,ns_1...s_k}(\mathbf{x}) \tag{1.4.32}$$

which, in the dipole approximation, reads

$$\mathbf{F}^m = \nabla(\mathbf{m}\cdot\mathbf{B}^o) = (\mathbf{m}\cdot\nabla)\mathbf{B}^o \tag{1.4.33}$$

since $\mathbf{m}$ is a constant vector and the curl of the external magnetic field $\mathbf{B}^o$ at $\mathbf{x}$ is zero. It can be seen that the magnetic force on a magnetic dipole vanishes for a uniform external magnetic field.

The torque on the magnetic dipole moment $\mathbf{m}$ in the external magnetic field reads

$$\tau^m = \mathbf{m} \times \mathbf{B}^o \tag{1.4.34}$$

which exists even in a uniform external magnetic field.

One may notice that eqns.(1.4.33) and (1.4.34) are completely analogous to the corresponding eqns.(1.4.15) and (1.4.16) in electrostatics. However, the analogy breaks down in the energy eqns.(1.4.31) and (1.4.13), where a sign difference appears. This is due to the fact that, to maintain the constant current distribution in applying the external magnetic field, certain amount of energy has to be supplied to the current system (Jackson (1975)). It is common to define a quantity, called the potential energy of a magnetic dipole in an external magnetic field, by

$$U'^{m} = -\mathbf{m}\cdot\mathbf{B}^{o}. \tag{1.4.35}$$

In such a case, one may write the expression for the force in complete analogy to that used in mechanics, namely, as the negative gradient of a potential energy. In other words, one may write

$$\mathbf{F}^{m} = -\nabla U'^{m} \tag{1.4.36}$$

which is completely in agreement with eqn.(1.4.33).

1.5 Maxwell's equations for materials at rest

Electromagnetic phenomena in the presence of material bodies are more complicated to describe than those in free space since materials in nature are composed of atoms or molecules, each made up of positively and negatively charged particles having various configurations in free space, varying states of motion and their statistical character. The earliest work dealing with the electromagnetic phenomena in matter were given by Lorentz (1916) who used a volume average procedure to derive a set of macroscopic electromagnetic equations for the matter at rest. It was found that only two field variables were not sufficient to describe all electromagnetic pheno-mena in matter. Two more field variables, the polarization **P** and the magnetization **M**, were then proposed to describe the behaviour of materials. Subsequently, there have been many attempts to provide derivations of increasing rigor, such as the works of Van Vleck (1932), Rosenfeld (1951), Mazur and Nijboer (1953), De Groot and Vlieger

(1965), Russakoff (1970), Robinson (1973), etc. In this section, we shall give the derivation of macroscopic Maxwell's equations for materials at rest with the use of the concept of electric and magnetic multipoles. The material body considered here is assumed to be rigid. The effects of electromagnetic-mechanical interaction in deformable solids will, however, be treated later in this chapter.

1.5.1 Dipole model of electromagnetic solids

We begin with the introduction of the macroscopic Maxwell's equations for materials at rest with the use of a dipole model by assuming that, on average, the dominant features of the material that are of interest to us are simply those associated with the electric and magnetic dipole moments.

We consider a material made up of a large number of atoms or molecules with bounded charges and free charges. In most materials, the electric monopole term (the net charge) from the bounded charges of the atom or molecule is zero in the absence of applied fields since the materials have equal amounts of positive and negative charges. In some simple materials, the multipole moments of each atom or molecule are also all zero in the absence of applied fields. However, permanent multipole moments may exist in some materials even in the absence of applied fields. In the presence of an applied electric field, the materials are, in general, electrically polarized. Such an electric polarization in materials is mainly due microscopically to electronic polarization, ionic (or atomic) polarization and orientational polarization, which can be explained as follows.

The electronic polarization is due to the small displacement of negative electron cloud relative to the positive charge (the nucleus) of the atom in an external electric field since the field tends to shift the positive and negative charges in opposite directions. Such a polarization process does not change the true mass center of the atom because the nucleus is much heavier than electrons and, therefore, the predominant displacement is that of the electronic charges. Thus, highest frequency phenomena, such as optical phenomena, are almost exclusively due to the electronic polarization since this process involves only the motion of light electrons.

The ionic polarization is caused in ionic crystals (such as NaCl) in which an external electric field displaces the positive (Na^+) ions relative to the negative (Cl^-) ions. This polarization process in general results in a certain deformation of the crystal lattice, and is essentially a low frequency phenomenon because it involves the motion of ions with masses generally more than 10^4 greater than the mass of the electron. In addition the ionic polarization may be produced by external forces which deform the crystal sublattices of positive ions and of negative ions differently so that the positive ions are also displaced relatively to the negative ions, and present, therefore, the piezoelectric effect of the ionic crystals.

The orientational polarization is due to the reorientation of electric dipole (permanent or induced) moments caused by the torques acting on the dipole moments in the presence of an external electric field, which make the dipole moments tend to orient themselves along the direction of the applied electric field since the stable state of the system is at its minimum energy. The orientational polarization is counteracted by the thermal movement of the atoms or molecules. Therefore, it is strongly temperature-dependent.

To describe macroscopically the effect of electrical polarization in the material, we shall assume that the macroscopic electric behavior of a material body with a volume V bounded by a closed surface S may be modeled by a distribution of the macroscopic volume free charge density ρ_e, the macroscopic surface free charge density α_f and the macroscopic electric polarization (dipole) density $\mathbf{P}$ (measured in C/m^2) in the material body. In this model, we do not consider the surface density of polarization since it plays no significant part in the mathematical model of electromagnetism and it is difficult to devise experiments to involve it in an essential way (King and Prasad (1986)).

From a macroscopic point of view, the macroscopic electric potential field in the static case can be expressed, by noting eqn.(1.4.4), in the dipole model as

$$\phi(\mathbf{x}) = \int_V \rho_e(\mathbf{x}')G^e(\mathbf{x}-\mathbf{x}')dV' + \int_S \alpha_f(\mathbf{x}')G^e(\mathbf{x}-\mathbf{x}')dS' - \int_V P_j(\mathbf{x}')G^e{}_{,j}(\mathbf{x}-\mathbf{x}')dV' \tag{1.5.1}$$

where the position vector $\mathbf{x}$ may be either inside the material body or outside the material body. It is noticed that, for the field point $\mathbf{x}$ being outside the material body, the electric field is well defined (by $\mathbf{E} = -\nabla\phi$) and can be checked by measuring the force acting on a test charge there. The electric field inside the material body is, however, not subject to direct measurement. The justification of the macroscopic model can, thus, be only based on its agreements with experiments.

1.5.2 **Gauss' theorem in material medium**

To examine how Gauss' theorem (1.2.9) is to be modified in the presence of material medium, let us consider a larger closed surface S_0 enclosing the material body V (see **Figure** 1.5.1) and, then, make a surface integral of the electric field on S_0. One has

$$\int_{S_0} \mathbf{E}(\mathbf{x})\cdot\mathbf{n}dS = -\int_{S_0}\int_{V} \rho_e(\mathbf{x}')G^e_{,i}(\mathbf{x}-\mathbf{x}')n_i(\mathbf{x})dV'dS - \int_{S_0}\int_{S} \alpha_f(\mathbf{x}')G^e_{,i}(\mathbf{x}-\mathbf{x}')n_i(\mathbf{x})dS'dS$$

$$+ \int_{S_0}\int_{V} P_j(\mathbf{x}')G^e_{,ij}(\mathbf{x}-\mathbf{x}')n_i(\mathbf{x})dV'dS. \qquad (1.5.2)$$

Eqn.(1.5.2) can be further written as

$$\int_{S_0} \mathbf{E}(\mathbf{x})\cdot\mathbf{n}dS = -\int_{V_0}\int_{V} \rho_e(\mathbf{x}')\nabla^2 G^e(\mathbf{x}-\mathbf{x}')dV'dV - \int_{V_0}\int_{S} \alpha_f(\mathbf{x}')\nabla^2 G^e(\mathbf{x}-\mathbf{x}')dS'dV$$

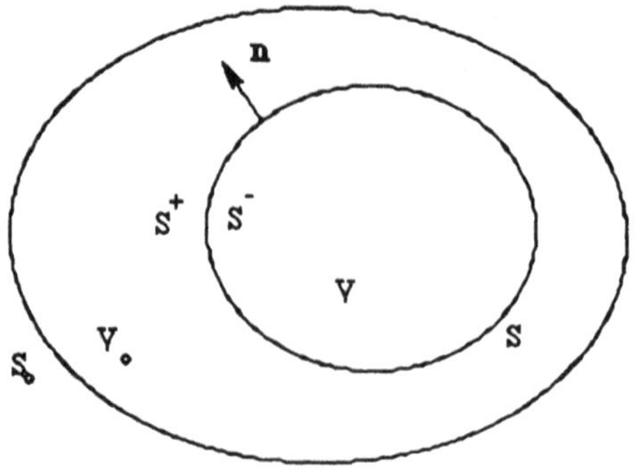

Figure 1.5.1 A material body with volume V is enclosed by a surface S_0.

$$- \int_{V_o} \int_{S} \mathbf{P}^{-}(\mathbf{x}')\cdot\mathbf{n}(\mathbf{x}')\nabla^2 G^e(\mathbf{x} - \mathbf{x}')dS'dV + \int_{V_o} \int_{V} \nabla'\cdot\mathbf{P}(\mathbf{x}')\nabla^2 G^e(\mathbf{x} - \mathbf{x}')dV'dV \tag{1.5.3}$$

by applying Gauss' divergence theorem for the surface integrals on S_o on the right-hand side of eqn.(1.5.2). Here the superscript "-" of $\mathbf{P}^-$ means its value being taken on the material side (S^-) of the surface S. By noting eqn.(1.4.2), we can then arrive at

$$\int_{S_o} \varepsilon_0 \mathbf{E}(\mathbf{x})\cdot\mathbf{n}(\mathbf{x})dS = \int_{V} \rho_e(\mathbf{x})dV + \int_{S} \alpha_f(\mathbf{x})dS + \int_{S} \mathbf{P}^{-}(\mathbf{x})\cdot\mathbf{n}(\mathbf{x})dS - \int_{V} \nabla\cdot\mathbf{P}(\mathbf{x})dV. \tag{1.5.4}$$

We now write the surface integral on the left-hand side of eqn.(1.5.4) in the following form

$$\int_{S_o} \varepsilon_0 \mathbf{E}(\mathbf{x})\cdot\mathbf{n}(\mathbf{x})dS = \int_{V_o - V - S} \nabla\cdot(\varepsilon_0 \mathbf{E})dV + \int_{V-S} \nabla\cdot(\varepsilon_0 \mathbf{E})dV + \int_{S} \varepsilon_0(\mathbf{E}^{+} - \mathbf{E}^{-})\cdot\mathbf{n}dS \tag{1.5.5}$$

with the use of Gauss' divergence theorem for the regions V_o-V-S and V-S. By identifying eqn.(1.5.4) with (1.5.5), we can then find that, in the presence of the material medium, the local differential form of Gauss' theorem (1.2.11) is modified to be

$$\nabla\cdot\mathbf{D} = \rho_e \qquad \text{in V} \tag{1.5.6}$$

with the introduction of an electric displacement vector **D**, defined by

$$\mathbf{D} = \varepsilon_0 \mathbf{E} + \mathbf{P} \tag{1.5.7}$$

where **P** is the electric polarization density vector, which vanishes in free space. The boundary condition at the interface between the material body and free space (or different media) may also be found as

$$\mathbf{n}\cdot[\,\mathbf{D}\,] = \alpha_f \qquad \text{on S.} \tag{1.5.8}$$

1.5.3 Ampere's circuital law in material medium

We shall now consider the problem how Ampere's circuital law in free space is modified by the presence of a material body. We study the case when the material under the exertion of applied fields is in steady state. In the material, bounded charges are moving in closed paths, localized around the centers of atoms or molecules, and the free charges are moving continuously, forming the macroscopic current flow in the material. The localized microscopic current loops due to moving bounded charges result in the presence of magnetic multipole moments in the material as we have seen in section 1.4.3, and, thus, lead to the magnetization of the material. In general, the magnetization of the material is due microscopically to the orbital current magnetization, the spin magnetization and the orientational magnetization.

The orbital current magnetization appears in all atoms placed in an external magnetic field due to the interaction effect of the field on the electrons in orbits in the atom, explained qualitatively by Lenz's law of electricity. According to this law, atom's electronic motion, considered as a current loop, will be changed in such a sense that a magnetic moment will be induced in a direction opposite to that of the applied field. Such an induced magnetic moment is, therefore, called the diamagnetic moment. Obviously diamagnetism is a property of all matter.

The spin magnetization is due to the intrinsic angular momentums of electron and nucleus, i.e., the electron spin and the nucleus spin. The nucleus spin is very small compared with electron spin and is normally neglected in the consideration of usual macroscopic magnetic properties of bulk materials.

The orientational magnetization comes from the preferred direction of orientation of magnetic moments due to torques acting on them caused by external magnetic fields or by anisotropic fields inherent to crystalline materials due to exchange interaction among atomic moments.

To describe macroscopically the effect of magnetization of the material, we shall assume that the macroscopic magnetic behavior of the material body with a volume V bounded by a closed surface S may be modeled by a distribution of the macroscopic volume current density $\mathbf{J}$, the macroscopic surface current density $\mathbf{K}_f$ and the macroscopic magnetization (magnetic dipole density) $\mathbf{M}$ (measured in A/m) in the material body.

The magnetic vector potential in the presence of the material body can thus be found, by noting eqn.(1.4.25), in the dipole model as

$$\mathbf{A}(\mathbf{x}) = \int_V \mathbf{J}(\mathbf{x}')G^m(\mathbf{x}-\mathbf{x}')dV' + \int_S \mathbf{K}_f(\mathbf{x}')G^m(\mathbf{x}-\mathbf{x}')dS' + \int_V \mathbf{M}(\mathbf{x}')\times\nabla' G^m(\mathbf{x}-\mathbf{x}')dV'. \tag{1.5.9}$$

This expression for **A** is valid for the field point **x** outside the material volume V as well as inside. Similar to an electric field, it is noticed that for the field point **x** being outside the material body, the magnetic induction field **B** is well defined (by $\mathbf{B} = \nabla \times \mathbf{A}$) and can be checked by measuring the force acting on a test current element (loop) there. The magnetic field inside the material body is, however, not subject direct measurement. The justification of the macroscopic model can, thus, be only based on its agreement with experiment. Some discussions on the proper definition of the magnetic field **B** as well as the electric field **E** in respectively a magnetized region and a polarized region are referred to in the work of Tiersten (1990).

Consider now an arbitrary closed line loop L, which may be partly (L_0) outside and partly (L_1) inside the material volume V as shown in **Figure** 1.5.2. We make a line integral of the magnetic induction field **B** (= $\nabla\times\mathbf{A}$) from eqn.(1.5.9) on L, and, after some manipulations, we can find

$$\int_L \mathbf{B}\cdot d\mathbf{l} = \int_{S_{L1}} \mu_0(\mathbf{J} + \nabla \times \mathbf{M}).\mathbf{n}_s dS + \int_{L_s} \mu_0(\mathbf{K}_f + \mathbf{M}^- \times \mathbf{n})\cdot\mathbf{n}_s dl_s \tag{1.5.10}$$

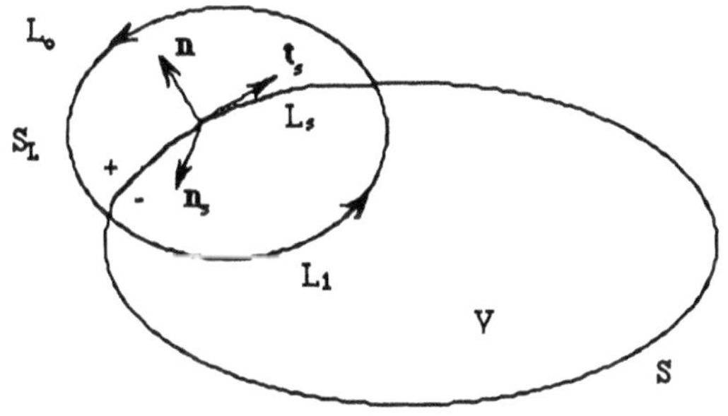

Figure 1.5.2 Configuration of a closed line loop and a material body.

with the aid of eqn.(1.4.18). Here, L_s denotes the line from the intersect $S_L \cap S$ with S_L being the open surface on the closed line loop L. S_{L1} is the part of the open surface S_L and is on the closed line loop $L_1 \cup L_s$. $\mathbf{n}_s$ is the unit normal vector on the open surface S_L, and **n** is the unit normal vector of the closed material surface S. We note that the unit direction vector $\mathbf{t}_s$ of the line element $d\mathbf{l}_s$ has its relation $\mathbf{t}_s = \mathbf{n} \times \mathbf{n}_s$ on the smoothly continuous line L_s (see **Figure** 1.5.2).

We write the line integral on the left-hand side of eqn.(1.5.10) in the following form

$$\int_L \mathbf{B} \cdot d\mathbf{l} = \int_{S_{Lo}} (\nabla \times \mathbf{B}) \cdot \mathbf{n}_s dS + \int_{S_{L1}} (\nabla \times \mathbf{B}) \cdot \mathbf{n}_s dS + \int_{L_s} [\mathbf{n} \times (\mathbf{B}^+ - \mathbf{B}^-)] \cdot \mathbf{n}_s dl_s \tag{1.5.11}$$

with the use of Stokes' theorem for the closed lines $L_o + L_s^+$ and $L_1 + L_s^-$.

By identifying eqn.(1.5.10) with (1.5.11), we find that, in the presence of the material medium, Ampere´s circuital law becomes

$$\nabla \times \mathbf{H} = \mathbf{J} \qquad \text{in V} \tag{1.5.12}$$

with the introduction of a magnetic intensity field **H** defined by

$$\mathbf{H} = \frac{1}{\mu_o}\mathbf{B} - \mathbf{M} \tag{1.5.13}$$

where **M** is zero in free space.

The boundary condition at the interface between the material body and the free space is found to be

$$\mathbf{n} \times [\,\mathbf{H}\,] = \mathbf{K}_f \qquad \text{on S}. \tag{1.5.14}$$

which may also be used as the interface condition at an interface between two different media.

In general, for dynamic cases, the local differential form of Ampere´s circuital law (1.3.2) is found to be modified as

$$\nabla \times \mathbf{H} = \mathbf{J} + \frac{\partial \mathbf{D}}{\partial t} \tag{1.5.15}$$

compatible with the equation of conservation law of charges (1.1.8) and Gauss' theorem for electric field (1.5.6). The second term on the right-hand side of eqn.(1.5.15) is called the displacement current by Maxwell. This term is essential for the existence of electromagnetic waves propagating in material media.

1.5.4 **Maxwell's equations for materials at rest**

To summarize, we have obtained the following set of macroscopic Maxwell's equations for materials at rest:

$$\nabla \times \mathbf{H} = \mathbf{J} + \frac{\partial \mathbf{D}}{\partial t} \tag{1.5.16}$$

$$\nabla \cdot \mathbf{B} = 0 \tag{1.5.17}$$

$$\nabla \times \mathbf{E} = -\frac{\partial \mathbf{B}}{\partial t} \tag{1.5.18}$$

$$\nabla \cdot \mathbf{D} = \rho_e \tag{1.5.19}$$

with the equation of conservation of charges

$$\nabla \cdot \mathbf{J} + \frac{\partial \rho_e}{\partial t} = 0 . \tag{1.5.20}$$

The interface conditions at an interface between two different media are

$$\mathbf{n} \cdot [\,\mathbf{D}\,] = \alpha_f \tag{1.5.21}$$

$$\mathbf{n} \times [\,\mathbf{E}\,] = 0 \tag{1.5.22}$$

$$\mathbf{n} \cdot [\,\mathbf{B}\,] = 0 \tag{1.5.23}$$

$$\mathbf{n} \times [\,\mathbf{H}\,] = \mathbf{K}_f \tag{1.5.24}$$

$$\mathbf{n} \cdot [\,\mathbf{J}\,] = -\frac{\partial \alpha_f}{\partial t} \tag{1.5.25}$$

where $[\![A]\!] = A^+ - A^-$ denotes the jump of the quantity A across the interface S with **n** being the unit normal vector always drawn from S^- to S^+. Here α_f is the surface density of free charge, and $\mathbf{K}_f$ is the surface density of free current.

In particular, at an interface between two perfect dielectrics, the surface density of charge α_f vanishes and, consequently, the normal component of **D** is continuous across the interface. At an interface between two conductors of finite conductivity, the surface density of free current $\mathbf{K}_f$ cannot exist. Thus, the tangential component of **H** is continuous at an interface separating two materials with finite conductivity. Some more detail discussions about interface conditions are referred to in the work of Johnk (1988).

So far, we have said nothing about the macroscopic polarization **P** and the macroscopic magnetization **M** except for their definitions respectively as the electric dipole moment per unit volume and the magnetic dipole moment per unit volume. Since microscopic electric and magnetic dipoles in materials are influenced by applied fields, we can expect that there must exist some macroscopic relations among the macroscopic polarization **P**, the macroscopic magnetization **M** and the applied fields. In the case of neglecting the effects of deformation as well as temperature field of the material, **P** and **M** are simply functions of the field quantities **E** and **H** (or **B**). Such functional relations are also called the constitutive relations. Here, we shall simply present some well-known linear constitutive relations used in classical electromagnetetics without going into the details of constitutive theories that will be discussed later in this chapter.

For linear isotropic rigid materials, **P** and **M** are found experimentally to be directly proportional to **E** and **H** respectively, i.e.,

$$\mathbf{P} = \varepsilon_o \chi_e \mathbf{E} \qquad (\text{or} \quad \mathbf{D} = \varepsilon \mathbf{E}) \tag{1.5.26}$$

and

$$\mathbf{M} = \chi_m \mathbf{H} \qquad (\text{or} \quad \mathbf{B} = \mu \mathbf{H}) \tag{1.5.27}$$

where constants χ_e and χ_m are called the electric susceptibility and the magnetic susceptibility respectively. $\varepsilon=\varepsilon_o(1+\chi_e)$ and $\mu=\mu_o(1+\chi_m)$ are called the dielectric permittivity and the magnetic permeability of the material respectively.

In addition, when conduction currents are present in the material medium, one has to include Ohm's law (1827) expressed by

$$\mathbf{J} = \sigma \mathbf{E} \tag{1.5.28}$$

where **J** is the macroscopic current density, and σ the electric conductivity of the material. According to the values of the conductivity σ, a simple classification of materials may be made. The dielectrics are the materials with low electric conductivity less than 10^{-5} S/m, semiconductors are materials having electric conductivity between 10^{-5} and 10^3 S/m, and conductors are characterized by an electric conductivity higher than 10^3 S/m. Mainly, conductors are metals. The best conductors among them are copper and silver which have electric conductivity of the order of 10^7 S/m.

Also, some simple classification of magnetic materials may be made according to the values of the magnetic susceptibility, χ_m. The diamagnetic material has a small negative temperature-independent susceptibility χ_m with a magnitude of the order of 10^{-5}, and since it is negative, the induced magnetic moment in the diamagnetic material is directed oppositely to the applied magnetic field. This kind of magnetism is a direct consequence of Lenz's law applied to the motions of the elementary charges (generally electrons) of the system. All materials have diamagnetic contributions to their susceptibilities. The paramagnetic material made of atoms or molecules having permanent magnetic moments has a positive temperature-dependent susceptibility χ_m with a magnitude of the order of 10^{-4} for aluminum and copper, and the order of 10^{-3} for $FeSO_4$, $NiSO_4$ and $CrCl_3$ at room temperature and varying with temperature approximately as $1/T$. This kind of magnetic behavior is explained as a consequence of two opposing effects: one, the tendency of the applied field to orient the magnetic moments in the direction of the field, and the other, the tendency of thermal agitation to preserve a random orientation of the magnetic moments. In most of practical calculations for engineering applications, we may neglect such weak induced magnetization in diamagnetic and paramagnetic materials. The soft ferromagnetic material has a positive temperature-dependent susceptibility χ_m with a magnitude of the order of 10^4 or higher for iron, nickel, cobalt and their alloys. In general, ferromagnetic materials may present permanent macroscopic magnetization, and have nonlinear constitutive relations and hysteresis behaviors (Chikazumi (1964)).

1.6 Electromagnetic energy and Poynting's theorem

Electromagnetic energy can be transported through free space or within material media by means of electromagnetic waves or electric current flows. The determination of electromagnetic power losses during the energy transport in materials is of particular interest for the study of problems concerning Joule heating in normal conductors, AC losses in superconductors, and transmission losses in waveguides, such as optical fibers and superconducting cables. In this section, some problems concerning electromagnetic energies and their transport in material media will be discussed and formulated.

1.6.1 Electric field energy for charges

To study the electric field energy for charges, we first consider a test charge q in an electrostatic field **E**. Since the force acting on the test charge q in the electric field is **F** = q**E**, the work done to displace the charge by **δl** is then

$$\delta W = - q\mathbf{E} \cdot \delta \mathbf{l} \qquad (1.6.1)$$

where the minus sign appears because we are calculating the work done on the charge against the action of the field.

If the charge is displaced from point 1 to point 2 along a trajectory L (see **Figure** 1.6.1), the work done per unit charge thus reads

$$W' = - \int_{(1)}^{(2)} \mathbf{E} \cdot d\mathbf{l} \ . \qquad (1.6.2)$$

Since the static electric field is irrotational, an electrostatic potential ϕ may be introduced such that $\mathbf{E} = - \nabla\phi$. Thus, eqn.(1.6.2) becomes

$$W' = \phi(2) - \phi(1) \qquad (1.6.3)$$

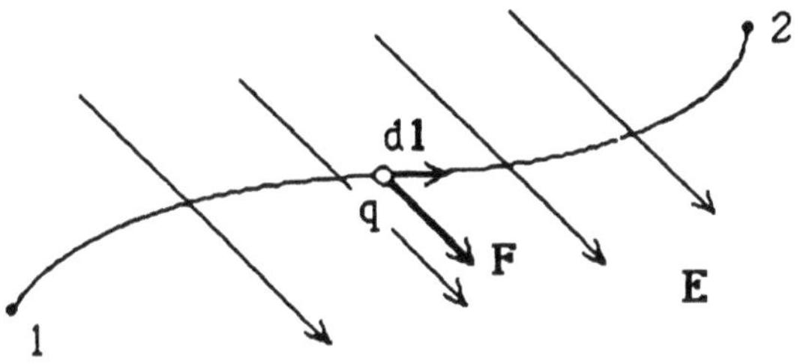

Figure 1.6.1

which shows that ϕ can be interpreted as the potential energy per unit charge in the electrostatic field.

By defining $\phi(\infty) = 0$, we may also introduce an electric field potential for a unit point charge by

$$\phi(\mathbf{x}) = \int_{\mathbf{x}}^{\infty} \mathbf{E} \cdot d\mathbf{l} \, . \tag{1.6.4}$$

In particular, the electric potential at a distance r from a point charge q' is given by

$$\phi(r) = \frac{1}{4\pi\varepsilon_0} \frac{q'}{r} \, . \tag{1.6.5}$$

Thus, we can see that the interaction energy between two point charges q_1 and q_2 may be found by

$$W = \frac{1}{4\pi\varepsilon_0} \frac{q_1 q_2}{r} \tag{1.6.6}$$

which is equal to the work done by displacing the point charge q_2, in the field produced by the point charge q_1, from infinity to the position having a distance r from the point charge q_1.

Due to the symmetry of the problem, eqn.(1.6.6) can be written as

$$W = \frac{1}{2} [\phi(r, q_1) q_2 + \phi(r, q_2) q_1] . \tag{1.6.7}$$

In general, the interaction energy associated with a collection of point charges $q^{(n)}$ (n=1, 2,..., N) may be expressed by

$$W = \frac{1}{2}\sum_{n=1}^{N} q^{(n)}\phi^{(n)} \tag{1.6.8}$$

where $\phi^{(n)}$ is the electric potential at the position of the nth charge due to all the other charges.

For a continuous distribution of charges in a space volume V, the work required to assemble the continuous charge distribution from infinitesimally small charge elements at infinity is given by

$$W^e = \frac{1}{2}\int_V \phi\rho_e dV. \tag{1.6.9}$$

It is noticed that formula (1.6.9) is not simply a generalization of eqn.(1.6.8) for the continuous case. Since eqn.(1.6.9) takes into account the self-energy, which is infinity for a point charge, it is not allowable to use eqn.(1.6.9) for point charges (Böttcher (1973)).

With the use of the equation $\nabla\cdot\mathbf{D} = \rho_e$ and $\mathbf{E} = -\nabla\phi$, eqn.(1.6.9) may be expressed as

$$W^e = \frac{1}{2}\int_V \mathbf{E}\cdot\mathbf{D}dV + \frac{1}{2}\int_S \phi\mathbf{D}\cdot d\mathbf{S} \tag{1.6.10}$$

where S is a closed surface enveloping the volume V.

If one assumes that all charges are located in a finite region of space, at large distance r from the charges, the integral is of the order of 1/r and tends to zero as the surface of integration approaches infinity. Hence, for the entire space, one has

$$W^e = \frac{1}{2}\int_\infty \mathbf{E}\cdot\mathbf{D}\, dV. \tag{1.6.11}$$

1.6.2 **Electrostatic energy for material media**

In the presence of dielectric media, eqn.(1.6.9) cannot be simply used due to the polarization effect in the dielectric media. The reason is that with dielectric media work is done not only to bring real charge into position, but also to produce a certain state of polarization in the medium. If one considers a small change in energy δW^e due to some sort of change $\delta\rho_e$ in the charge density ρ_e existing in all space, one can find (Jackson (1975)) that the work done to accomplish this change is

$$\delta W^e = \int_\infty \phi \delta \rho_e \, dV \tag{1.6.12}$$

where ϕ is the potential due to the charge density ρ_e already present. Since $\nabla \cdot \mathbf{D} = \rho_e$, the change $\delta\rho_e$ can be related to a change in the displacement vector $\mathbf{D}$ such that eqn.(1.6.12) becomes

$$\delta W^e = \int_\infty \mathbf{E} \cdot \delta \mathbf{D} \, dV \tag{1.6.13}$$

extended over all space. Formally, this equation can be written as

$$\delta W^e = \delta \int_\infty [\int_0^{\mathbf{D}} \mathbf{E} \cdot d\mathbf{D}] dV \tag{1.6.14}$$

which gives the electrostatic energy for dielectrics

$$W^e = \int_\infty [\int_0^{\mathbf{D}} \mathbf{E} \cdot d\mathbf{D}] dV \, . \tag{1.6.15}$$

For linear dielectric media, eqn.(1.6.15) becomes

$$W^e = \frac{1}{2} \int_\infty \mathbf{E} \cdot \mathbf{D} \, dV . \tag{1.6.16}$$

It is shown that eqn.(1.6.11) (or eqn.(1.6.9)) is valid macroscopically only for linear dielectric media (including vacuum).

In deriving the energy expression, it is assumed that the medium is held at rest and hence no work is done in motion against forces. This implies that the virtual process of assembling the charges in the dielectrics is a process with particular constraints. The resultant energy expression is nevertheless general, since no nonconservative forces are involved.

We shall now consider the energy of electrostatic field of conductors. It is known that a fundamental property of conductors in electrostatic cases is that the electric field inside a conductor must be zero. Hence, any charges in a conductor must be located on its surface. Thus, the problem of the electrostatics of conductors amounts to determining the electric field in the free space outside the conductors and the distribution of charges on their surfaces. The total energy W^c of the electrostatic field of charged conductors is then

$$W^c = \frac{\varepsilon_0}{2} \int_\infty E^2 \, dV \tag{1.6.17}$$

where the integral is taken over all space outside the conductors.

With the use of the result that the total charge on a conductor is

$$Q = \varepsilon_0 \int_S \mathbf{E} \cdot d\mathbf{S} = -\varepsilon_0 \int_S \frac{\partial \phi}{\partial n} dS \tag{1.6.18}$$

where S is the surface of the conductor, we can get

$$W^c = \frac{1}{2} \sum Q_i \phi_i \tag{1.6.19}$$

where ϕ_i denotes the constant value of the electric potential on the surface of the ith conductor.

1.6.3 Magnetic field energy for currents and material media

We now consider the magnetostatic energy for a set of elementary closed current loops L_i, with each loop carrying a steady current I_i (i=1, 2,..., N). They are confined to a finite region of space. By noting eqns.(1.4.20) and (1.4.29), we may find that the total magnetic field energy of the N current circuits can be expressed by

$$W^m = \frac{1}{2} \sum_{i=1}^{N} \sum_{j=1}^{N} L_{ij} I_i I_j \tag{1.6.20}$$

where the self-energy of each particular circuit (i.e., for i=j) is taken into account. The factor 1/2 occurs because in the double summation each circuit is counted twice, and since $L_{ij} = L_{ji}$. The coefficient L_{ij} ($i \neq j$) is known as the coefficient of mutual inductance between the circuit i and circuit j, defined by

$$L_{ij} = \frac{\mu_0}{4\pi} \int_{L_i} \int_{L_j} \frac{d\mathbf{l} \cdot d\mathbf{l}'}{|\mathbf{x} - \mathbf{x}'|} \qquad (\mathrm{H}) \tag{1.6.21}$$

which is evidently symmetric. For i=j, the quantity L_{ii} (here i is not summed) is called the coefficient of self-inductance. Eqn.(1.6.21) is the Neumann formula for mutual inductance. It is seen that the coefficient of mutal (or self-) inductance is a property of the geometric shape and the physical arrangement of the coupled circuits, and it is independent of currents carried by the circuits.

By introducing the flux Φ_i threading the ith circuit due to all circuits

$$\Phi_i = \int_{S_i} \mathbf{B} \cdot d\mathbf{S} = \int_{L_i} \mathbf{A} \cdot d\mathbf{l} \,, \tag{1.6.22}$$

eqn.(1.6.20) may also be written as

$$W^m = \frac{1}{2} \sum_{i=1}^{N} I_i \Phi_i \,. \tag{1.6.23}$$

Comparison of eqn.(1.6.23) and eqn.(1.6.20) shows

$$\Phi_i = \sum_{j=1}^{N} L_{ij} I_j \,. \tag{1.6.24}$$

Eqn.(1.6.23) can be generalized to determine the magnetic field energy of a continuous distribution of current with a space volume V, which reads

$$W^m = \frac{1}{2} \int_V \mathbf{J} \cdot \mathbf{A} \, dV. \tag{1.6.25}$$

This expression is valid for current distribution in linear conductive media (including free space). A more general form of the magnetic field energy for continuous current distribution in nonlinear magnetic media is

$$W^m = \int_V [\int_0^A \mathbf{J} \cdot d\mathbf{A}]\, dV. \qquad (1.6.26)$$

It is often desirable to express the magnetic energy in terms of field quantities **B** and **H** instead of current density **J** and vector potential **A**. By using the relations $\mathbf{B} = \nabla \times \mathbf{A}$ and $\mathbf{J} = \nabla \times \mathbf{H}$, one may find

$$W^m = \frac{1}{2}\int_\infty \mathbf{H} \cdot \mathbf{B}\, dV \qquad (1.6.27)$$

for linear magnetic media, and

$$W^m = \int_\infty [\int_0^B \mathbf{H} \cdot d\mathbf{B}]\, dV \qquad (1.6.28)$$

for nonlinear magnetic media. Here, the integration is taken over all space. It is seen that eqns.(1.6.27) and (1.6.28) are analogous to those of electrostatic energy in respectively eqns.(1.6.16) and (1.6.15).

1.6.4 **Poynting's theorem for electrodynamic systems**

To formulate electromagnetic energy transport problems, we consider an arbitrary volume V enclosed by a surface ∂V at rest in free space. The volume V can be either an empty space volume or a material volume. Thus, an energy integral can be obtained from Maxwell's equations (1.5.16)-(1.5.19)

$$\int_V \frac{\partial w}{\partial t} dV + \int_V \mathbf{J} \cdot \mathbf{E}\, dV = - \int_{\partial V} \mathbf{S} \cdot \mathbf{n}\, dS \qquad (1.6.29)$$

where **S** is called the Poynting vector defined by

$$\mathbf{S} = \mathbf{E} \times \mathbf{H} \qquad (\mathrm{W/m^2}) \tag{1.6.30}$$

which is usually interpreted as the electromagnetic field energy flux (flow) per unit time through a unit area of a given surface, oriented normally to the Poynting vector. $\partial w/\partial t$ denotes the rate of change of electromagnetic field energy density defined by

$$\frac{\partial w}{\partial t} = \mathbf{E}\cdot\frac{\partial \mathbf{D}}{\partial t} + \mathbf{H}\cdot\frac{\partial \mathbf{B}}{\partial t} . \tag{1.6.31}$$

Eqn.(1.6.29) is known as the integral form of Poynting's theorem. It is a mathematical identity compatible with Maxwell's equations. In general, eqn.(1.6.29) does not represent the equation of conservation of energy since there might be other types of energy present, such as thermal and mechanical energies.

By using eqns.(1.5.7) and (1.5.13), eqn.(1.6.31) may also be written as

$$\frac{\partial w}{\partial t} = \frac{\partial}{\partial t}\{\frac{1}{2}(\varepsilon_o E^2 + \mu_o H^2)\} + \mathbf{E}\cdot\frac{\partial \mathbf{P}}{\partial t} + \mu_o \mathbf{H}\cdot\frac{\partial \mathbf{M}}{\partial t} . \tag{1.6.32}$$

For linear electromagnetic solids with the constitutive relations given by (1.5.26) and (1.5.27), we may find

$$w = \frac{1}{2}(\varepsilon E^2 + \mu H^2) \tag{1.6.33}$$

which is the stored electromagnetic field energy density.

Thus, in the special case of linearity, Poynting's equation (1.6.29) states that the net inward power-flux

$$P(t) = -\int_{\partial V} \mathbf{S}\cdot\mathbf{n}\, d S \tag{1.6.34}$$

supplied by the electromagnetic field over a closed surface ∂V, equals the sum of the time rate of increase of electromagnetic field energy inside V, plus the total Ohmic loss (that may be converted into mechanical or heat energy) in V, assuming V contains no generators. If

V contains a power generator, an additional term $\int \mathbf{J}_g \cdot \mathbf{E} dV$ with $\mathbf{J}_g$ being the source current density, independent of the field, should be added on the left-hand side of eqn.(1.6.29).

1.6.5 Poynting's theorem for quasi-static systems

For electromagnetic quasistatic systems, Poynting's theorem may be expressed as

$$\int_V \mathbf{H} \cdot \frac{\partial \mathbf{B}}{\partial t} dV + \int_V \mathbf{J} \cdot \mathbf{E} dV = - \int_{\partial V} \mathbf{S} \cdot \mathbf{n} dS \tag{1.6.35}$$

for magneto-quasistatic systems, and

$$\int_V \mathbf{E} \cdot \frac{\partial \mathbf{D}}{\partial t} dV + \int_V \mathbf{J} \cdot \mathbf{E} dV = - \int_{\partial V} \mathbf{S} \cdot \mathbf{n} dS \tag{1.6.36}$$

for electro-quasistatic systems.

If AC fields of periodic variation with a time period t_p are considered, the power dissipated per cycle in a conducting magnetizable medium can be obtained in the magneto-quasistatic approximation by

$$\int_0^{t_p} P(t) dt = - \int_0^{t_p} \int_{\partial V} (\mathbf{E} \times \mathbf{H}) \cdot \mathbf{n} dS dt = \int_0^{t_p} \int_V [\mathbf{J} \cdot \mathbf{E} + \mathbf{H} \cdot \frac{\partial \mathbf{M}}{\partial t}] dV dt . \tag{1.6.37}$$

When the magnetization of the material is negligible, we may find

$$\int_0^{t_p} P(t) dt = \int_0^{t_p} \int_V \mathbf{J} \cdot \mathbf{E} dV dt . \tag{1.6.38}$$

In such a case, we may take $\mathbf{J} \cdot \mathbf{E}$ to be the instantaneous power density dissipated in the material medium.

If an electromagnetic system carrying only direct currents, we may find

$$P = -\int_{V} \mathbf{S}\cdot\mathbf{n}dS = \int_{V} \mathbf{J}\cdot\mathbf{E}dV \qquad (1.6.39)$$

from eqn.(1.6.29), assuming V contains no generators. It is shown that in a DC system the net power-flux entering a closed surface ∂V constructed about the current-carrying conductor is a measure of the Ohmic loss in the conductor.

1.6.6 **Poynting's theorem for time-harmonic systems**

For steady time-harmonic electromagnetic systems, it is often convenient to introduce complex notation in the analysis of the electromagnetic fields. In the complex notation, the physically real electric and magnetic intensity fields are defined by

$$\mathbf{E}(\mathbf{x}, t) = \mathrm{Re}\{\mathbf{E}^c(\mathbf{x})e^{i\omega t}\} \quad \text{and} \quad \mathbf{H}(\mathbf{x}, t) = \mathrm{Re}\{\mathbf{H}^c(\mathbf{x})e^{i\omega t}\} \qquad (1.6.40)$$

where the frequency ω is the same for all field components at all positions in space. $\mathbf{E}^c(\mathbf{x})$ and $\mathbf{H}^c(\mathbf{x})$ are complex functions of only space position point $\mathbf{x}$, and they are determined from the following complex form of Maxwell's equations for linear materials

$$\nabla \times \mathbf{H}^c = \mathbf{J}^c + i\omega\mathbf{D}^c \qquad (1.6.41)$$

$$\nabla \cdot \mathbf{B}^c = 0 \qquad (1.6.42)$$

$$\nabla \times \mathbf{E}^c = -i\omega\mathbf{B}^c \qquad (1.6.43)$$

$$\nabla \cdot \mathbf{D}^c = \rho_e^c . \qquad (1.6.44)$$

From this set of equations, a complex form of Poynting's theorem can be introduced by

$$\int_{V} \frac{i\omega}{2}(\mathbf{B}^c\cdot\mathbf{H}^{c*} - \mathbf{E}^c\cdot\mathbf{D}^{c*})dV + \int_{V} \frac{1}{2}\mathbf{J}^{c*}\cdot\mathbf{E}^c dV = -\int_{\partial V} \frac{1}{2}(\mathbf{E}^c \times \mathbf{H}^{c*})\cdot\mathbf{n}dS \qquad (1.6.45)$$

where the quantity, for instance, $\mathbf{H}^{c*}$ denotes the complex conjugate of $\mathbf{H}^c$. The physical meanings of the real and imaginary parts of eqn.(1.6.45) can be explained from the following discussion.

By writing

$$\mathbf{E}^c(\mathbf{x}) = \mathbf{E}_R(\mathbf{x}) + i\mathbf{E}_I(\mathbf{x}) \qquad \text{and} \qquad \mathbf{H}^c(\mathbf{x}) = \mathbf{H}_R(\mathbf{x}) + i\mathbf{H}_I(\mathbf{x}) \tag{1.6.46}$$

we have

$$\mathbf{E}(\mathbf{x}, t) = \mathrm{Re}\{\mathbf{E}^c(\mathbf{x})e^{i\omega t}\} = \mathbf{E}_R(\mathbf{x})\cos\omega t - \mathbf{E}_I(\mathbf{x})\sin\omega t \tag{1.6.47}$$

$$\mathbf{H}(\mathbf{x}, t) = \mathrm{Re}\{\mathbf{H}^c(\mathbf{x})e^{i\omega t}\} = \mathbf{H}_R(\mathbf{x})\cos\omega t - \mathbf{H}_I(\mathbf{x})\sin\omega t\,. \tag{1.6.48}$$

Substituting them into eqn.(1.6.30), we find

$$\mathbf{S} = \mathbf{E} \times \mathbf{H} = (\mathbf{E}_R \times \mathbf{H}_R)\cos^2\omega t + (\mathbf{E}_I \times \mathbf{H}_I)\sin^2\omega t - [(\mathbf{E}_R \times \mathbf{H}_I)+(\mathbf{E}_I \times \mathbf{H}_R)]\sin\omega t\cos\omega t. \tag{1.6.49}$$

By taking the time-average of the Poynting vector

$$\bar{\mathbf{S}} = \frac{1}{t_p}\int_0^{t_p} \mathbf{S}(\mathbf{x},t)dt \tag{1.6.50}$$

with $t_p = 2\pi/\omega$, we can obtain from eqn.(1.6.49) that

$$\bar{\mathbf{S}} = \frac{1}{2}(\mathbf{E}_R \times \mathbf{H}_R + \mathbf{E}_I \times \mathbf{H}_I) = \frac{1}{2}\mathrm{Re}\{\mathbf{E}^c \times \mathbf{H}^{c*}\} \tag{1.6.51}$$

which shows that the real part of the term on the r.h.s. of eqn.(1.6.45) represents the time-average power flux entering the closed surface ∂V. Thus, for isotropic linear electromagnetic materials with the constitutive relations (1.5.26)-(1.5.28) having pure real ε, μ and σ, we can obtain by taking the real part of eqn.(1.6.45)

$$-\int_{\partial V} \bar{\mathbf{S}}\cdot\mathbf{n}dS = \int_V \overline{\sigma E^2}dV \tag{1.6.52}$$

where $\overline{\sigma E^2} = \frac{1}{2}\sigma\mathbf{E}^c\cdot\mathbf{E}^{c*}$ denotes the time-average of the Joule heat per unit volume. Eqn.(1.6.52) shows that the time-average power flux entering a closed surface ∂V equals the time-average power dissipated as heat inside the volume V bounded by ∂V, provided that there are no power sources in V.

Similarly we can get by taking the imaginary parts of eqn.(1.6.45)

$$-\int_{\partial V} \frac{1}{2}\mathrm{Im}\{\mathbf{E}^c \times \mathbf{H}^{c*}\}\cdot\mathbf{n}dS = \int_V 2\omega(\overline{w^m} - \overline{w^e})dV \tag{1.6.53}$$

in which $\overline{w^m}$ and $\overline{w^e}$ denote respectively the time-average of the stored energy densities of the magnetic field and of the electric field in V, i.e.

$$\overline{w^m} = \frac{1}{2}\mu\overline{H^2} = \frac{1}{4}\mu\mathbf{H}^c\cdot\mathbf{H}^{c*} \tag{1.6.54}$$

and

$$\overline{w^e} = \frac{1}{2}\varepsilon\overline{E^2} = \frac{1}{4}\varepsilon\mathbf{E}^c\cdot\mathbf{E}^{c*}. \tag{1.6.55}$$

Eqn.(1.6.53) states that the imaginary part of the complex power flux entering the closed surface of the volume V is a measure of 2ω times the difference of the time-average energies stored in the magnetic and electric fields, provided that there is no power source in the volume V. In the general case where there is a power generator with the source current density $\mathbf{J}_g$ in the volume V, we have the following complex Poynting theorem:

$$-\int_V \frac{1}{2}\mathbf{J}_g^{c*}\cdot\mathbf{E}^c dV = \int_V \frac{i\omega}{2}(\mathbf{B}^c\cdot\mathbf{H}^{c*} - \mathbf{E}^c\cdot\mathbf{D}^{c*})dV$$

$$+ \int_V \frac{1}{2}\mathbf{J}^{c*}\cdot\mathbf{E}^c dV + \int_{\partial V} \frac{1}{2}(\mathbf{E}^c \times \mathbf{H}^{c*})\cdot\mathbf{n}dS \tag{1.6.56}$$

where the term on the l.h.s. represents the net complex power supplied to the volume V.

1.7 Maxwell's equations for moving media

1.7.1 Basic principles of relativity and Lorentz transformation

Classical electromagnetic theory for systems at rest can be used to deal with many engineering electromagnetic phenomena which involve no relative motion, but it may fail to describe electromagnetic phenomena involving relative motion. The studies on the electrodynamics for moving bodies had in fact led to the emergence of Einstein's theory of relativity in the beginning of the twentieth century. Here, we shall give a short summary of key points on the special theory of relativity relevant to the development of electrodynamics for moving media since there already exist many books and articles on this subject of relativity (see, for instance, Rosser (1964), Møller (1972) and Bladel (1984)).

The fundamental postulates of Einstein's special theory of relativity are:

1) *The postulate of relativity:* All inertial systems of coordinates are equally suitable for the description of all physical phenomena.
2) *The postulate of the constancy of the speed of light:* The speed of light c in free space is the same for all observers and is independent of the motion of the source.

Here, we notice that Einstein had generalized the Galilean relativity principle, which states that the laws of mechanics formulated are absolutely identical in all inertial systems, to all physical phenomena, especially to electromagnetism. This means that the laws of electrodynamics as well as the laws of mechanics are the same in all inertial frames. The second postulate, however, indicates the fact that the velocity of propagation of interaction between two bodies is finite, which is contradictary to the assumption of instantaneous propagation of interaction used in classical Newtonian mechanics. This indicates that equations of classical Newtonian mechanics have to be modified to be consistent with the second postulate of the special theory of relativity, which is firmly based on accurate experimental observations gathered in the process of the development of physical science.

In what follows, we shall show how the electrodynamics for rigid stationary media is generalized to study electromagnetic phenomena for moving rigid bodies with the aid of the special theory of relativity.

We then introduce some approximations that are of sufficient accuracy and are of convenience for engineering applications.

To describe physical phenomena in nature, it is necessary to introduce a system of reference where a system of coordinates is used to indicate the position of, for instance, a particle in space, and clocks fixed in this system serve to indicate the time. According to the relativity principle of Einstein's special theory of relativity, all the laws of nature are identical in all inertial systems of reference. In other words, equations expressing the laws of nature are invariant with respect to transformations of coordinates and time from one inertial system to another, i.e, they should have the same form in all inertial systems. Electrodynamics as one of the fields of physics has also to satisfy the relativity principle, To formulate a relativistic electrodynamics which satisfies Einstein's special relativity principles, an elegant and convenient way is the four-dimensional formulation with the use of a transformation in a 4-dimensional space, which was proposed by Minkowski.

The transformation of coordinates in 4-space, which is quantitatively in accordance with the second postulate, is called the Lorentz transformation. The most general Lorentz transformation can be interpreted as a rigid rotation of axes in the 4-space with coordinates $(x_1, x_2, x_3, x_4) = (x, y, z, ict)$. The rigidity expresses the basic condition for the Lorentz transformation between two coordinate systems S and S'

$$s^2 = x_i x_i = x'_i x'_i \qquad (i = 1, 2, 3, 4). \tag{1.7.1}$$

In general, the transformation equations relating the two sets of coordinates may be written as

$$x'_i = a_{ij} x_j \tag{1.7.2}$$

in which the coefficients a_{ij} satisfy the relation $a_{ij}a_{kj} = \delta_{ik}$ with δ_{ik} being the Kronecker delta in 4-space. In particular, for the two coordinate systems S and S' in relative motion with velocity v along their common xx' direction, the coefficients a_{ij} of the Lorentz transformation may be written as

$$(a_{ij}) = \begin{bmatrix} \gamma & 0 & 0 & i\beta\gamma \\ 0 & 1 & 0 & 0 \\ 0 & 0 & 1 & 0 \\ -i\beta\gamma & 0 & 0 & \gamma \end{bmatrix} \qquad (1.7.3)$$

where $\beta = v/c$ and $\gamma = (1-\beta^2)^{-1/2}$.

1.7.2 Covariance of Maxwell's equations

According to the special theory of relativity, the fundamental laws of electrodynamics, Maxwell's equations, must hold in any inertial frame of reference. In fact, the electromagnetism described by Maxwell's equations for free space is already covariant with respect to the Lorentz transformation. The invariance of form (or covariance) of Maxwell's equations may be shown by introducing the following four-tensor expressions:

$$(Z_{ij}) = \begin{bmatrix} 0 & H_z & -H_y & -icD_x \\ -H_z & 0 & H_x & -icD_y \\ H_y & -H_x & 0 & -icD_z \\ icD_x & icD_y & icD_z & 0 \end{bmatrix} \qquad (1.7.4)$$

and

$$(F_{ij}) = \begin{bmatrix} 0 & E_z & -E_y & icB_x \\ -E_z & 0 & E_x & icB_y \\ E_y & -E_x & 0 & icB_z \\ -icB_x & -icB_y & -icB_z & 0 \end{bmatrix}. \qquad (1.7.5)$$

The pair of Maxwell's equations, $\nabla \times \mathbf{H} = \mathbf{J} + \partial\mathbf{D}/\partial t$ and $\nabla\cdot\mathbf{D} = \rho_e$, may then be expressed by

$$\frac{\partial Z_{ij}}{\partial x_j} = G_i \qquad (i = 1, 2, 3, 4) \qquad (1.7.6)$$

with the 4-current vector $(G_j) = (J_x, J_y, J_z, ic\rho_e)$, and the another pair of Maxwell's equations, $\nabla \times \mathbf{E} = -\partial \mathbf{B}/\partial t$ and $\nabla \cdot \mathbf{B} = 0$, may be expressed by

$$\frac{\partial F_{ij}}{\partial x_j} = 0 \qquad (i = 1, 2, 3, 4). \tag{1.7.7}$$

The equation of conservation of charge, $\nabla \cdot \mathbf{J} = -\partial \rho_e/\partial t$, reads

$$\frac{\partial G_i}{\partial x_i} = 0 \qquad (i = 1, 2, 3, 4). \tag{1.7.8}$$

We thus arrive at a set of covariant Maxwell's equations (1.7.6), (1.7.7) and (1.7.8). It can been seen from eqn.(1.7.8) that relativistically charge density and current are simply different aspects of the same physical quantity.

Now, with the use of the conventional transformation rules for the four-vectors and the four-tensors in the 4-space, i.e.

$$G'_i = a_{ik} G_k \quad \text{and} \quad F'_{ij} = a_{ik} a_{jn} F_{kn} \qquad (i, j, k, n = 1, 2, 3, 4), \tag{1.7.9}$$

we can find that the invariance of Maxwell's equations under the Lorentz transformation results in the following relations between the field quantities ($\mathbf{E}$, $\mathbf{B}$, ...) defined in an inertial laboratory frame of reference $S\{(\mathbf{x},t)\}$ (which may be supposed to be at rest) and the field quantities ($\mathbf{E}'$, $\mathbf{B}'$, ...) defined in a primed inertial frame of reference $S'\{(\mathbf{x}',t')\}$ which is moving with a uniform velocity $\mathbf{v}$ relative to the laboratory frame S :

$$\mathbf{E}'_{\parallel} = \mathbf{E}_{\parallel}, \qquad \mathbf{E}'_{\perp} = \gamma(\mathbf{E} + \mathbf{v} \times \mathbf{B})_{\perp} \tag{1.7.10}$$

$$\mathbf{B}'_{\parallel} = \mathbf{B}_{\parallel}, \qquad \mathbf{B}'_{\perp} = \gamma(\mathbf{B} - \frac{1}{c^2}\mathbf{v} \times \mathbf{E})_{\perp} \tag{1.7.11}$$

$$\mathbf{D}'_{\parallel} = \mathbf{D}_{\parallel}, \qquad \mathbf{D}'_{\perp} = \gamma(\mathbf{D} + \frac{1}{c^2}\mathbf{v} \times \mathbf{H})_{\perp} \tag{1.7.12}$$

$$\mathbf{H}'_{\parallel} = \mathbf{H}_{\parallel}, \qquad \mathbf{H}'_{\perp} = \gamma(\mathbf{H} - \mathbf{v} \times \mathbf{D})_{\perp} \tag{1.7.13}$$

$$\mathbf{J}' = \mathbf{J} - \gamma \frac{\mathbf{v}}{v^2}[\rho_e v^2 - (\mathbf{v} \cdot \mathbf{J})(1 - 1/\gamma)] \tag{1.7.14}$$

$$\rho'_e = \gamma[\rho_e - \frac{1}{c^2}(\mathbf{v} \cdot \mathbf{J})] \tag{1.7.15}$$

where the subscripts $\parallel$ and $\perp$ are used respectively to represent directions parallel and perpendicular to the direction in which S' is moving relative to S with the uniform velocity $\mathbf{v}$.

1.7.3 Minkowski's electrodynamic model

Within the framework of Minkowski's phenomenological electrodynamics for moving material bodies, the electromagnetic properties of the materials and their constitutive equations are assumed to be known in the frame S' in which the material body is at rest. For a large class of linear (rigid) electromagnetic materials, their constitutive equations may be written as

$$\mathbf{B}' = \mu\mathbf{H}', \qquad \mathbf{D}' = \varepsilon\mathbf{E}', \qquad \mathbf{J}' = \sigma\mathbf{E}' \tag{1.7.16}$$

where μ, ε and σ are respectively the constants of the magnetic permeability, the electric permittivity and the electric conductivity. The constitutive equations valid for the material body moving with uniform velocity $\mathbf{v}$ relative to the laboratory frame S can thus be found

$$\mathbf{B} = \mu\mathbf{H} - \alpha\, \mathbf{v} \times \mathbf{E} \tag{1.7.17}$$

$$\mathbf{D} = \varepsilon\mathbf{E} + \alpha\, \mathbf{v} \times \mathbf{H} \tag{1.7.18}$$

$$\gamma(\mathbf{J} - \rho_e\mathbf{v})_\parallel = \sigma(\mathbf{E} + \mathbf{v} \times \mathbf{B})_\parallel \tag{1.7.19a}$$

$$(\mathbf{J} - \rho_e\mathbf{v})_\perp = \sigma\gamma(\mathbf{E} + \mathbf{v} \times \mathbf{B})_\perp \tag{1.7.19b}$$

with $\alpha = \varepsilon\mu - \varepsilon_o\mu_o$.

In the first-order approximation theory, where the terms of the order v^2/c^2 and higher are neglected, eqn.(1.7.19) is reduced to

$$\mathbf{J} = \rho_e\mathbf{v} + \sigma(\mathbf{E} + \mathbf{v} \times \mathbf{B}) \tag{1.7.20}$$

and eqns.(1.7.10)-(1.7.15) may be reduced to

$$\mathbf{E}' = \mathbf{E} + \mathbf{v} \times \mathbf{B}\,, \qquad \mathbf{B}' = \mathbf{B} - \frac{1}{c^2}\mathbf{v} \times \mathbf{E} \tag{1.7.21}$$

$$\mathbf{D}' = \mathbf{D} + \frac{1}{c^2}\mathbf{v} \times \mathbf{H}\,, \qquad \mathbf{H}' = \mathbf{H} - \mathbf{v} \times \mathbf{D} \tag{1.7.22}$$

$$\mathbf{J}' = \mathbf{J} - \rho_e \mathbf{v}\,, \qquad \rho'_e = \rho_e\,. \tag{1.7.23}$$

It is worth mentioning that, for quasi-static problems, which will be discussed later, the concept of "Cerenkov electrodynamics", where a so-called "Cerenkov transformation" is postulated from the Lorentz transformation by replacing the light velocity c in vacuum by the phase velocity $\bar{c} = c/n$ (n is the index of refraction) of the light in media, was introduced to reconcile the theory of electro- and magneto-quasi-statics on the one hand, with relativity and high-frequency phenomena on the other hand. In this theory, the constitutive equations (1.7.16) are invariant under the "Cerenkov transformation" (Ollendorff (1974, 1977) and Schieber (1986)).

1.8 Electromagneto-quasistatic approximations

1.8.1 Galilean approximation

It has been shown that Maxwell's equations of classical electrodynamics are invariant under Lorentz transformation. However, the mechanical balance laws of classical Newtonian mechanics are not invariant under Lorentz transformation, but are invariant under Galilean transformation, defined by

$$\mathbf{x}' = \mathbf{x} - \mathbf{v}t\,, \qquad t' = t \tag{1.8.1}$$

where $\mathbf{v}$ is the uniform velocity of the inertial frame $S'\{(\mathbf{x}', t')\}$ with respect to the inertial frame $S\{(\mathbf{x}, t)\}$. The second relation in eqn.(1.8.1) states that the parameter describing the time is the same in all inertial systems, which means that the time is an absolute quantity in the Newtonian description of mechanics.

Thus, a consistent formulation of electromagneto-mechanics should be to modify the mechanical equations relativistically, which obviously will result in a set of relativistic equations too complicated for engineering applications. However, if we realize that, for most problems in engineering applications, classical mechanics is sufficiently accurate and relativistic effects are negligible in most cases. Where the velocity involved in the kinematical processes considered is much smaller than the velocity of light, we may well introduce a so-called Galilean approximation in which the mechanical equations and Maxwell's equations are both treated classically. In particular, we modify Maxwell's equations to their quasi-static forms which may then be invariant under Galilean transformation. In this way, we may consistently treat the classical mechanics and electrodynamics within the scope of Galilean relativity with sufficient accuracy for studying "low-frequency" electromagneto-mechanical phenomena. Actually, the "low-frequency" approximation can often be used to describe time-varying phenomena with frequencies up to a few giga hertz.

1.8.2 **Maxwell's equations at magneto-quasistatic approximation**

At the magneto-quasistatic approximation, the following set of Maxwell's equations may be introduced

$$\nabla \times \mathbf{H} = \mathbf{J} \tag{1.8.2}$$

$$\nabla \cdot \mathbf{B} = 0 \tag{1.8.3}$$

$$\nabla \cdot \mathbf{J} = 0 \tag{1.8.4}$$

$$\nabla \times \mathbf{E} + \frac{\partial \mathbf{B}}{\partial t} = 0 \tag{1.8.5}$$

with $\mathbf{B} = \mu_o(\mathbf{H} + \mathbf{M})$. Here, the displacement current $\mathbf{D}$ and free charge density ρ_e have been omitted in the magneto-quasistatic approximation. We notice that, even with time-varying sources, the magnetic intensity field $\mathbf{H}$ and the magnetic induction field $\mathbf{B}$ are determined as if the system were magnetostatic. Then, the electric field $\mathbf{E}$ is found from the resulting flux density by using eqn.(1.8.5). This is the origin of the term magneto-quasistatic approximation. Such an approximation is of sufficient accuracy for describing "low-frequency" phenomena in good conductors.

Now, noting the following relations for the differential operators in the two coordinate systems S' and S

$$\nabla' = \nabla , \qquad \frac{\partial}{\partial t'} = \frac{\partial}{\partial t} + \mathbf{v} . \nabla , \tag{1.8.6}$$

a consistent set of field transformation relations, which satisfy the invariance of the magneto-quasi-static equations under the Galilean transformation, may be found as

$$\mathbf{H}' = \mathbf{H} \tag{1.8.7}$$

$$\mathbf{B}' = \mathbf{B} \tag{1.8.8}$$

$$\mathbf{J}' = \mathbf{J} \tag{1.8.9}$$

$$\mathbf{E}' = \mathbf{E} + \mathbf{v} \times \mathbf{B} \tag{1.8.10}$$

from which the transformation relation for the magnetization $\mathbf{M}' = \mathbf{M}$ may also be found.

It may be of interest to also give the integral form of Faraday's law of induction for the magneto-quasistatic system, which reads

$$\int_L \mathbf{E}' \cdot d\mathbf{l} = - \frac{d}{dt} \int_S \mathbf{B} \cdot d\mathbf{S} \tag{1.8.11}$$

where S is a material surface enclosed by a contour L, which is moving in space, and $\mathbf{E}'$ the electric field measured in the moving medium. Introducing a convective time derivative defined by

$$\frac{D_c \mathbf{B}}{Dt} = \frac{\partial \mathbf{B}}{\partial t} + \mathbf{v}(\nabla \cdot \mathbf{B}) + \nabla \times (\mathbf{B} \times \mathbf{v}) \tag{1.8.12}$$

we may find

$$\int_L \mathbf{E}'\cdot d\mathbf{l} = -\int_S \frac{D_c}{Dt}(\mathbf{B})\cdot d\mathbf{S} \tag{1.8.13}$$

from which we can derive the following form of differential equation

$$\nabla \times (\mathbf{E}' - \mathbf{v} \times \mathbf{B}) = -\frac{\partial \mathbf{B}}{\partial t} . \tag{1.8.14}$$

By noting eqn.(1.8.10), the argument of the curl in eqn.(1.8.14), $\mathbf{E}'-\mathbf{v}\times\mathbf{B}$ actually represents the electric field $\mathbf{E}$ measured by a stationary observer in the laboratory frame of reference S. This implies also the fact that the differential formulation of Faraday's law of induction is independent of the motion of the medium inside field.

The interface conditions at an interface between two different material media moving in space can be found, for the magneto-quasistatic system, as

$$\mathbf{n} \times [\, \mathbf{E} + \mathbf{v} \times \mathbf{B} \,] = 0 \tag{1.8.15}$$

$$\mathbf{n} \cdot [\, \mathbf{B} \,] = 0 \tag{1.8.16}$$

$$\mathbf{n} \times [\, \mathbf{H} \,] = \mathbf{K}_f \tag{1.8.17}$$

$$\mathbf{n} \cdot [\, \mathbf{J} \,] + \nabla_T \cdot \mathbf{K}_f = 0 \tag{1.8.18}$$

where $\nabla_T\cdot\mathbf{K}_f$ denotes a tangential (two-dimensional) surface divergence of $\mathbf{K}_f$. For example, if the interface coincides with the xy plane with $\mathbf{K}_f = (K_{fx}, K_{fy}, 0)$, one then has

$$\nabla_T \cdot \mathbf{K}_f = \frac{\partial K_{fx}}{\partial x} + \frac{\partial K_{fy}}{\partial y} . \tag{1.8.19}$$

It is worth mentioning that $\mathbf{K}_f$ vanishes at an interface between two conductors of finite conductivities. In such a case, eqn.(1.8.18) becomes $\mathbf{n}\cdot[\, \mathbf{J} \,] = 0$, i.e. the normal component of current density is continuous across the interface between two conductors of finite conductivities.

1.8.3 Maxwell's equations at electro-quasistatic approximation

In the case of electro-quasistatics, Maxwell's equations are

$$\nabla \times \mathbf{E} = 0 \tag{1.8.20}$$

$$\nabla \cdot \mathbf{D} = \rho_e \tag{1.8.21}$$

$$\nabla \cdot \mathbf{J} = -\frac{\partial \rho_e}{\partial t} \tag{1.8.22}$$

$$\nabla \times \mathbf{H} = \mathbf{J} + \frac{\partial \mathbf{D}}{\partial t} \tag{1.8.23}$$

with $\mathbf{D} = \varepsilon_0\mathbf{E} + \mathbf{P}$. Here, the magnetic induction term has been dropped from Faraday's law in the electro-quasistatic approximation. We also notice that even with time-varying sources, the electric field **E** and the electric displacement **D** are determined as if the system were static. Then, the current density **J** is determined from the equation of conservation of charges. The magnetic field **H** (if it is of interest) may be found from eqn.(1.8.23). Such an approximation is of sufficient accuracy for describing "low-frequency" phenomena in dielectrics.

Similarly, a consistent set of field transformation relations for the electro-quasistatic system, which satisfy the invariance of the electro-quasistatic equations under Galilean transformation, may be found as

$$\mathbf{E}' = \mathbf{E} \tag{1.8.24}$$

$$\mathbf{D}' = \mathbf{D} \tag{1.8.25}$$

$$\rho'_e = \rho_e \tag{1.8.26}$$

$$\mathbf{J}' = \mathbf{J} - \rho_e\mathbf{v} \tag{1.8.27}$$

$$\mathbf{H}' = \mathbf{H} - \mathbf{v} \times \mathbf{D} \tag{1.8.28}$$

from which the transformation relation for the polarization $\mathbf{P}' = \mathbf{P}$ may also be found.

For the electro-quasistatic system, we may write the integral form of Ampere's circuital law as

$$\int_L \mathbf{H}'\cdot d\mathbf{l} = \int_S \mathbf{J}'\cdot d\mathbf{S} + \frac{d}{dt}\int_S \mathbf{D}\cdot d\mathbf{S} \tag{1.8.29}$$

and the integral form of the conservation law of charge reads

$$-\int_{\partial V} \mathbf{J}'\cdot d\mathbf{S} = \frac{d}{dt}\int_V \rho_e dV \tag{1.8.30}$$

where V is a material volume moving in space. $\mathbf{H}'$ and $\mathbf{J}'$ are respectively the magnetic field and the free current density measured in the moving medium. Eqns.(1.8.29) and (1.8.30) may be shown similarly by carrying out the time derivatives of surface and volume integrals and by noting eqns.(1.8.27) and (1.8.28). It should, however, be noticed that the given integral forms of Faraday's law of induction, Ampere's circuital law, and the conservation law of charge in eqns.(1.8.11), (1.8.29) and (1.8.30) are valid only for quasistatic systems, in which the velocity of media are small compared with the velocity of light.

The interface conditions at an interface between different material media in the electro-quasistatic approximation are

$$\mathbf{n}\times[\,\mathbf{E}\,] = 0 \tag{1.8.31}$$

$$\mathbf{n}\cdot[\,\mathbf{D}\,] = \alpha_f \tag{1.8.32}$$

$$\mathbf{n}\cdot[\,\mathbf{J} - \rho_e\mathbf{v}\,] = -\nabla_T\cdot\mathbf{K}_f - \frac{\partial\alpha_f}{\partial t} \tag{1.8.33}$$

$$\mathbf{n}\times[\,\mathbf{H} - \mathbf{v}\times\mathbf{D}\,] = \mathbf{K}_f\,. \tag{1.8.34}$$

For interfaces between two perfect dielectrics, the free surface charge α_f and the free surface current $\mathbf{K}_f$ vanish if no excess charge is supplied to the interface by external agent (such as rubbing it with cat's fur). In such a case, the normal component of the dielectric displacement vector $\mathbf{D}$ is continuous across the interface separating two perfect dielectrics.

1.9 Mass and motion of continuous media

1.9.1 Mass and mass density

Besides charges, materials have mass which characterizes the property of inertia. In classical mechanics, mass is assumed to be conserved, i.e. the mass of a material body is the same at all time. In continuum mechanics it is further assumed that the mass is an absolutely continuous function of volume. In other words, it is assumed that a positive quantity ρ, the mass density, can be defined at every point in the body as the limit

$$\rho(\mathbf{P}) = \lim_{\Delta V \to 0} \frac{\Delta m}{\Delta V} \tag{1.9.1}$$

where Δm is the total mass contained in the small volume ΔV which is shrinking down upon the position point $\mathbf{P}$. It is seen that the definition of the mass density is similar to the definition of charge density presented in section 1.1.1. Such a concept of the material continuum as a mathematical idealization of the real material is applicable to problems in which fine structures, such as atomic or molecular structures of the matter, can be ignored. If the mass density is known at every point in a material body, the total mass of the body can be obtained by means of an integral taken over the volume of the body, i.e.

$$m = \int_V \rho dV. \tag{1.9.2}$$

The inertial properties of a point mass are fully defined by the value of its mass. The inertial properties of a body of finite dimensions are, however, defined by the law for the distribution of the density throughout the volume of the body. In the case of a rigid body, the inertial characteristics can be completely defined by the total mass of the body, the position of the center of mass (center of gravity), and the moment of inertia tensor at the center of mass.

1.9.2 Motion of continuum media

In continuum mechanics, the material body is described mathematically by specifying the positions of the material points of the idealized continuous material medium at time t in a region V of space with the use of a chosen frame of reference. Here, for simplicity, we shall choose the same Cartesian coordinate system to describe the reference configuration of a continuum body at time $t = t_0$ with conventional right-handed set of orthogonal axes O-XYZ, and the current configuration of the body at the current time t with the same set of orthogonal axes but denoted by O-xyz. A general analysis of deformation based on different coordinate systems is referred to in the classical work of continuum mechanics (Eringen (1967), etc.).

Let us now suppose that, at time $t = t_0$ (for convenience, one may set $t_0=0$), the material body occupies a region of space V_0 bounded by the surface ∂V_0, and the material points are denoted by the position vector $\mathbf{X}(X,Y,Z)$. After motion and deformation have taken place, at time $t > 0$, the material body occupies a region V bounded by ∂V in space, and the material particle (point) labelled $\mathbf{X}$ is moved to a position $\mathbf{x}$ as shown in **Figure** 1.9.1. Thus one possible description of the motion of the material body can be given by the vector equation

$$\mathbf{x} = \mathbf{x}(\mathbf{X}, t) \tag{1.9.3}$$

where, for each time t, $\mathbf{x}(\mathbf{X}, t)$ is a continuously differentiable function if no discontinuities of the shock-wave type occur. Obviously, one has $\mathbf{x}(\mathbf{X}, 0)=\mathbf{X}$. Eqn.(1.9.3) defines the material (or Lagrangian) description of the motion of a continuum.

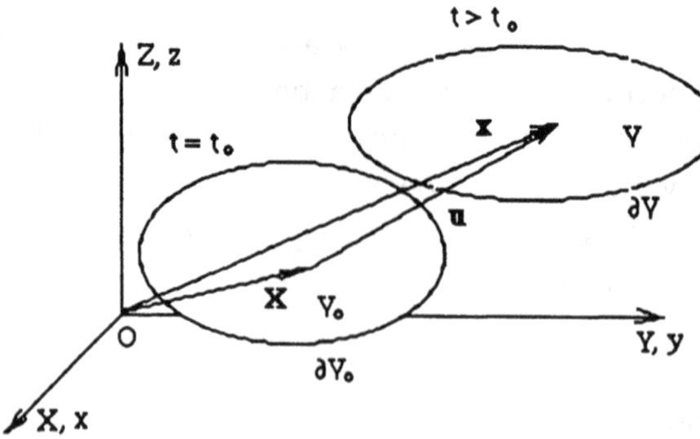

Figure 1.9.1 Configurations of a moving material body.

The velocity and acceleration of the material particle labelled **X** in the material description are given respectively by

$$\mathbf{V}(\mathbf{X}, t) = \frac{\partial \mathbf{x}(\mathbf{X}, t)}{\partial t} \tag{1.9.4}$$

$$\mathbf{a}(\mathbf{X}, t) = \frac{\partial^2 \mathbf{x}(\mathbf{X}, t)}{\partial t^2}. \tag{1.9.5}$$

It is important to notice that the spatial arguments of **V** and **a** do not denote the current position of the material particle (which is of course **x**), but the initial position. For many purposes it is more convenient to express velocity and acceleration in terms of **x** rather than **X**. This can be accomplished by using a spatial (Eulerian) description of motion.

We shall assume that the particle located initially at **X** moves to one and only one point **x**, and conversely that no two separated particles in the initial configuration V_0 arrive at the same point **x**. Thus, the mapping of **x** -> **X** is one to one and eqn.(1.9.3) can be solved to obtain **X** as a function of **x** and t

$$\mathbf{X} = \mathbf{X}(\mathbf{x}, t). \tag{1.9.6}$$

The necessary condition for the existence of the unique solution (1.9.6) of eqn.(1.9.3) is the non-vanishing of the Jacobian J defined by

$$J = \det\left(\frac{\partial x_i}{\partial X_K}\right) \equiv | x_{i,K} | \tag{1.9.7}$$

where one has used the denotation $x_{i,K} \equiv \partial x_i / \partial X_K$.

In a spatial description, the instantaneous motion of the material body is described by a velocity vector field **v**(**x**, t) associated with the instantaneous location of each material particle

$$\mathbf{v}(\mathbf{x}, t) = \mathbf{V}(\mathbf{X}(\mathbf{x}, t), t). \tag{1.9.8}$$

It is important to recognize that **v** as a function of **x** and t is a totally different function from **V** as a function of **X** and t. The velocity vector field **v**(**x**, t) may be interpreted as follows. Consider some fixed point **x** in space. At this location suppose that an apparatus is installed to record the velocity of the particles passing through $\mathbf{x}$ as a function of time. The measurements yield $\mathbf{v}(\mathbf{x}, t)$.

The acceleration of the particle currently passing through $\mathbf{x}$ is given by the material time derivative of $\mathbf{v}(\mathbf{x}, t)$

$$\frac{d\mathbf{v}(\mathbf{x}, t)}{dt} = \frac{\partial \mathbf{v}}{\partial t} + (\mathbf{v}\cdot\nabla)\mathbf{v}. \tag{1.9.9}$$

The first term in eqn.(1.9.9) is interpreted as the local part of the acceleration arising from the time dependence of the velocity field while the second term represents the convective part of the acceleration as the contribution of the motion of the material particle in the instantaneous velocity field.

1.9.3 **Conservation law of mass**

Mathematically, the conservation law of mass, which states that mass can be neither created nor destroyed, may now be expressed by

$$\int_{V_0} \rho_0(\mathbf{X})dV_0 = \int_V \rho(\mathbf{x}, t)dV = \int_{V_0} \rho(\mathbf{x}(\mathbf{X},t), t)J\,dV_0\,. \tag{1.9.10}$$

Since this relation must hold for any volume element of a continuous material body, we then have

$$\rho_0(\mathbf{X}) = \rho(\mathbf{x}(\mathbf{X}, t), t)J \tag{1.9.11}$$

which expresses the mass density in the initial configuration in terms of the mass density in the current configuration. Since physically ρ and ρ_0 are both positive, the Jacobian J is then also a positive quantity.

With the use of the material time derivative, the conservation law of mass may also be written alternatively by the following mathematical expression

$$\frac{d}{dt}\int_V \rho(\mathbf{x}, t)dV = 0 \tag{1.9.12}$$

where V is a moving volume of space occupied by particles of the body.

The integral form of the conservation of mass (1.9.12) can be applied to arbitrary motions of a material medium, in which the field of the density, velocity, and other mechanical characteristics can be discontinuous.

For continuous motion described by smooth functions, eqn.(1.9.12) can be replaced by a partial differential equation from the material time derivative of the volume integral

$$\frac{d}{dt}\int_V \rho(\mathbf{x}, t)dV = \int_V \left(\frac{d\rho}{dt} + \rho\nabla\cdot\mathbf{v}\right)dV = 0 \qquad (1.9.13)$$

from which one may get the local equation of conservation of mass by

$$\frac{d\rho}{dt} + \rho\nabla\cdot\mathbf{v} = 0. \qquad (1.9.14)$$

This equation is also called the equation of continuity.

1.10 Continuum deformation and strain analysis

1.10.1 Deformation and strain tensor

In this section, a brief description of the geometric aspects of the deformation of a continuum and the strain analysis will be given. The nature of the deformation of material in the immediate vicinity of a particle labelled $\mathbf{X}$ is made by examining the behavior of all line vector elements connecting the material particle $\mathbf{X}$ with an arbitrary neighboring material particle $\mathbf{X}+d\mathbf{X}$. In the reference configuration the line vector element is $d\mathbf{X}$. At time t, the line vector element connecting the same particles now located at $\mathbf{x}(\mathbf{X}, t)$ and $\mathbf{x}(\mathbf{X}+d\mathbf{X}, t)$ is

$$d\mathbf{x} = \mathbf{x}(\mathbf{X}+d\mathbf{X}, t) - \mathbf{x}(\mathbf{X}, t). \qquad (1.10.1)$$

The effect of the motion is to carry the line vector element from $\mathbf{X}$ to $\mathbf{x}$ and concomitantly to transform the line vector element from $d\mathbf{X}$ to $d\mathbf{x}$. With the use of the assumption that $\mathbf{x}$ is a differentiable function, eqn.(1.10.1) may be written to the first order approximation as

$$dx_i = x_{i,K}dX_K \equiv F_{iK}dX_K \tag{1.10.2}$$

where F_{iK} is called the deformation gradient tensor defined by $F_{iK} \equiv x_{i,K}$. The squares of the length of the line vector elements $\mathbf{dX}$ and $\mathbf{dx}$ are respectively

$$dS^2 = d\mathbf{X} \cdot d\mathbf{X} = \delta_{KL}dX_K dX_L \tag{1.10.3}$$

$$ds^2 = d\mathbf{x} \cdot d\mathbf{x} = \delta_{ij}dx_i dx_j \tag{1.10.4}$$

where δ_{KL} and δ_{ij} denote the Euclidean metric tensors for their respective coordinate systems, which, in the case of choosing the same Cartesian coordinate system, both become the Kronecker delta.
The difference between the squares of the length of line elements may now be expressed as

$$ds^2 - dS^2 = (\delta_{ij}F_{iK}F_{jL} - \delta_{KL})dX_K dX_L \tag{1.10.5a}$$

or

$$ds^2 - dS^2 = 2E_{KL}dX_K dX_L \tag{1.10.5b}$$

in which one has introduced the Lagrangian strain tensor E_{KL} by

$$E_{KL} \equiv \frac{1}{2}(C_{KL} - \delta_{KL}) \tag{1.10.6}$$

where $C_{KL} \equiv F_{iK}F_{iL}$ is called the Cauchy deformation tensor. The tensors C_{KL} and E_{KL} are obviously symmetric. From eqn.(1.10.5), one finds a fundamental result that a necessary and sufficient condition for the motion of a body being a rigid-body motion is that all the components of the strain tensor E_{KL} are zero throughout the body.

By introducing further a displacement vector $\mathbf{u}$ defined by (**Fig.** 1.9.1)

$$\mathbf{u} = \mathbf{x} - \mathbf{X} \qquad (\text{or } u_K = \delta_{Ki}x_i - X_K) \tag{1.10.7}$$

with δ_{Ki} being the so-called shifter which plays a similar role as the Kronecker delta, eqn.(1.10.6) can then be expressed as

$$E_{KL} = \frac{1}{2}(u_{K,L} + u_{L,K} + u_{I,K}u_{I,L}). \tag{1.10.8a}$$

For infinitesimal strains, eqn.(1.10.8a) is reduced to

$$E_{KL} = \varepsilon_{kl} = \frac{1}{2}(u_{k,l} + u_{l,k}) \tag{1.10.8b}$$

where ε_{ij} is called the infinitesimal strain tensor. The physical meaning of the Lagrangian strain tensor (and the infinitesimal strain tensor) can be illustrated by considering a simple stretch of a line vector element $d\mathbf{X}$ = (0, 0, dX) along the X-axis, the result of which gives

$$E_{XX} = \frac{1}{2}\frac{(ds - dS)(ds + dS)}{dS^2} . \tag{1.10.9}$$

In the case of an infinitesimal strain, eqn.(1.10.9) is reduced to

$$E_{XX} \approx \varepsilon_{xx} = \frac{ds - dS}{dS} \tag{1.10.10}$$

which shows that ε_{xx} represents the extension, or change of length per unit length of a line vector element parallel to the x-axis.

1.10.2 **Rate of deformation and rigid body rotation**

Based on the above knowledges, one can now derive the following useful relations for the study of the time-rate of change of deformation fields:

$$\frac{d}{dt}(x_{i,K}) = v_{i,j}x_{j,K} \tag{1.10.11}$$

$$\frac{d}{dt}|x_{i,K}| = (\nabla\cdot v)|x_{i,K}| \tag{1.10.12}$$

$$\frac{d}{dt}(E_{KL}) = D_{ij}x_{i,K}x_{j,L} \tag{1.10.13}$$

in which $\mathbf{v}$ is the velocity field defined in eqn.(1.9.8) and $v_{i,j}$ is called the velocity gradient which can be decomposed into a symmetric part and an antisymmetric part

$$v_{i,j} = D_{ij} + \Omega_{ij} \tag{1.10.14}$$

where D_{ij} is called the rate of strain tensor defined by

$$D_{ij} = \frac{1}{2}(v_{i,j} + v_{j,i}) \tag{1.10.15}$$

which, in the case of infinitesimal strains, becomes $D_{ij} = d(\varepsilon_{ij})/dt$. Ω_{ij} is called the spin tensor defined by

$$\Omega_{ij} = \frac{1}{2}(v_{i,j} - v_{j,i}) \tag{1.10.16}$$

which can be associated uniquely with an axial vector Ω, called the vorticity vector defined by

$$\Omega_i = -\frac{1}{2}\varepsilon_{ijk}\Omega_{jk} \qquad \text{(i.e. } \Omega = \frac{1}{2}\nabla \times \mathbf{v}\text{).} \tag{1.10.17}$$

This vorticity vector Ω measures the instantaneous angular velocity of a rigid body rotation. The infinitesimal rigid body rotation experienced by a continuum at an infinitesimal interval of time Δt may thus be expressed as

$$\omega_i = \Omega_i \Delta t \tag{1.10.18}$$

where ω_i is called the rotation vector defined by

$$\omega = \frac{1}{2}\nabla \times \mathbf{u} \qquad \text{(or } \omega_k = \frac{1}{2}\varepsilon_{kij}\omega_{ij}\text{)} \tag{1.10.19}$$

in which ω_{ij} is called the rotation tensor of the displacement field $\mathbf{u}$, defined by

$$\omega_{ij} = \frac{1}{2}(u_{j,i} - u_{i,j})\,. \tag{1.10.20}$$

By noting eqn.(1.10.13), the rigid body motion of a continuum body may now be stated as the motion of the body with $\mathbf{D}(\mathbf{x}, t) = 0$

throughout the body for all time t. In addition, motions in which $\Omega = 0$ throughout the body for all time t are said to be irrotational.

1.10.3 Compatibility conditions

For some problems, the question of how to determine the displacement field **u** may arise when the strain field has been given. Since the strain tensor is symmetric, we then have six equations for the determination of three unknown functions u_i (i=1, 2, 3). Thus, to have a single-valued continuous displacement field, the so-called compatibility conditions have to be introduced, which in the case of infinitesimal strains, are

$$\varepsilon_{ij,kl} + \varepsilon_{kl,ij} - \varepsilon_{ik,jl} - \varepsilon_{jl,ik} = 0. \tag{1.10.21}$$

A proof of the compatibility conditions (1.10.21) may be found in the work of Fung (1965).

Of the 81 equations represented by eqn.(1.10.21), only six are essential and they can be written in unabridged notation as

$$\frac{\partial^2 \varepsilon_{xx}}{\partial y \partial z} = \frac{\partial}{\partial x}\left(-\frac{\partial \varepsilon_{yz}}{\partial x} + \frac{\partial \varepsilon_{zx}}{\partial y} + \frac{\partial \varepsilon_{xy}}{\partial z}\right) \tag{1.10.22}$$

$$\frac{\partial^2 \varepsilon_{yy}}{\partial z \partial x} = \frac{\partial}{\partial y}\left(-\frac{\partial \varepsilon_{zx}}{\partial y} + \frac{\partial \varepsilon_{xy}}{\partial z} + \frac{\partial \varepsilon_{yz}}{\partial x}\right) \tag{1.10.23}$$

$$\frac{\partial^2 \varepsilon_{zz}}{\partial x \partial y} = \frac{\partial}{\partial z}\left(-\frac{\partial \varepsilon_{xy}}{\partial z} + \frac{\partial \varepsilon_{yz}}{\partial x} + \frac{\partial \varepsilon_{zx}}{\partial y}\right) \tag{1.10.24}$$

$$2\frac{\partial^2 \varepsilon_{xy}}{\partial x \partial y} = \frac{\partial^2 \varepsilon_{xx}}{\partial y^2} + \frac{\partial^2 \varepsilon_{yy}}{\partial x^2} \tag{1.10.25}$$

$$2\frac{\partial^2 \varepsilon_{yz}}{\partial y \partial z} = \frac{\partial^2 \varepsilon_{yy}}{\partial z^2} + \frac{\partial^2 \varepsilon_{zz}}{\partial y^2} \tag{1.10.26}$$

$$2\frac{\partial^2 \varepsilon_{zx}}{\partial z \partial x} = \frac{\partial^2 \varepsilon_{zz}}{\partial x^2} + \frac{\partial^2 \varepsilon_{xx}}{\partial z^2}. \tag{1.10.27}$$

These conditions are derived for infinitesimal strains referred to rectangular Cartesian coordinates. In general cases, one requires that the Riemannian-Christoffel curvature tensor formed from a metric tensor for the Euclidean material manifold vanishes (Eringen (1971)).

1.11 The laws of motion and stress hypothesis

1.11.1 Objective tensors

In the formulation of physical laws, it is often desirable to employ quantities that are independent of the motion of the observer. Such quantities are called objective. In classical mechanics, a tensorial quantity is said to be objective or frame-indifferent if it obeys an appropriate tensor transformation law for all times in any two objectively equivalent (rigid body) motions $\mathbf{x}(\mathbf{X}, t)$ and $\mathbf{x}'(\mathbf{X}, t')$, i.e.

$$x'_k(\mathbf{X}, t') = Q_{kl}(t)x_l(\mathbf{X}, t) + b_k(t), \qquad t' = t - \alpha \tag{1.11.1}$$

with $b_k(t)$ being a time-dependent vector, α a constant characterizing a shift of the origin of time on the clocks in reference frames, and $Q_{kl}(t)$ a time-dependent orthogonal tensor satisfying the proper condition

$$Q_{kl}Q_{ml} = Q_{lk}Q_{lm} = \delta_{km}. \tag{1.11.2}$$

For example, if a vector A_k and a second-order tensor S_{kl} are objective, then they must obey the following transformation law appropriate to two objectively equivalent motions:

$$A'_k(\mathbf{X}, t') = Q_{kl}(t)A_l(\mathbf{X}, t) \tag{1.11.3}$$

$$S'_{kl}(\mathbf{X}, t') = Q_{km}(t)Q_{ln}(t)S_{mn}(\mathbf{X}, t). \tag{1.11.4}$$

For vectors and tensors that are independent of time, objectivity readily applies. For time-dependent quantities, this is not always the case. For instance, the velocity vector v_k and the spin tensor Ω_{kl} are not objective while the rate of strain tensor D_{kl} is objective (see Eringen (1967) for more detail discussions).

The transformation relation (1.11.1) between $\mathbf{x}'$ and $\mathbf{x}$ is also called the Euclidean transformation. In particular, a Euclidean transformation with Q_{kn} being time-independent (for which one may take particular coordinate systems such that one has $Q_{kn}=\delta_{kn}$), $\alpha=0$ and $\mathbf{b} = -\mathbf{v}t$ is called the Galilean transformation (see eqn.(1.8.1)).

1.11.2 The laws of motion

The motion of a continuum in space is governed by the laws of continuum mechanics based on Newton's laws of motion and laws of thermodynamics. Newton's laws of motion state that, in an inertial frame of reference, the material rate of change of the linear momentum of a body is equal to the resultant of applied forces acting on the body, and that the material rate of change of the angular momentum of the body with respect to the coordinate origin is equal to the resultant moment of applied forces and couples about the same origin.

Mathematically, the balance law of linear momentum for a continuum medium with volume V bounded by ∂V may be written as

$$\frac{d}{dt}\int_V \rho \mathbf{v} dV = \int_{\partial V} \mathbf{t}^{(n)} dS + \int_V \mathbf{f} dV \tag{1.11.5}$$

where $\mathbf{t}^{(n)}$ denotes the surface traction, which are, for instance, aerodynamic pressure, mechanical contact pressure and magnetic pressure. $\mathbf{f}$ is the body force, such as earth's gravitational force and electromagnetic body forces.

The balance law of angular momentum is

$$\frac{d}{dt}\int_V \mathbf{x} \times \rho \mathbf{v}\, dV = \int_{\partial V} \mathbf{x} \times \mathbf{t}^{(n)} dS + \int_V (\mathbf{x} \times \mathbf{f} + \mathbf{c})\, dV \tag{1.11.6}$$

in which $\mathbf{c}$ denotes the body couple which may result from the presence of electromagnetic fields in magnetized or electrically polarized materials. No surface couple effect is being taken into account here (Eringen (1964)). The integral form of the balance laws of motion (1.11.5) and (1.11.6) are valid for any motion including discontinuous motion, in which the distributions of the characteristics of the motion and state within the volume V can be step functions of the time t (shock processes).

From a continuum point of view, the forces and couples acting on the material body may be classified into three categories, i.e. the extrinsic body loads, the extrinsic surface loads, and the internal loads. The extrinsic body loads are the forces and couples that arise from the external effects, such as the force of gravity. They act on the mass point of the body. In continuum mechanics, a load density per unit mass is assumed to exist. The extrinsic surface loads may arise from the action of one body on another through the bounding surface. The surface density of these loads is also assumed to exist, and the extrinsic surface force per unit area is called the surface traction. The internal loads are the result of the mutual action of pairs of particles that are located in the interior of the body. The effect of interparticle forces in a continuum appears in the form of a resultant effect of one part of the body on another part through the latter's bounding surface. This concept gives rise to the stress hypothesis as follows.

1.11.3 **Stress tensor**

The internal loads and their connection to surface loads may be understood by applying the balance law of momenta on a region υ with its surface $\partial\upsilon$ fully or partially contained in the body. In particular, we may consider a small tetrahedron with its vertex in the interior of υ and having three of its faces on the coordinate surfaces and the fourth face on $\partial\upsilon$. We then introduce the stress vector $-\mathbf{t}^k$ acting on the coordinate surface x_k=const. (k = 1, 2, 3) and apply eqn.(1.11.5) to this tetrahedron. We obtain, by using the mean value theorem,

$$\frac{d}{dt}(\rho^*\mathbf{v}^*\Delta\upsilon) = \mathbf{t}^{(n)*}\Delta S - \mathbf{t}^{k*}\Delta S_k + \rho\mathbf{f}^*\Delta\upsilon \tag{1.11.7}$$

where ρ^*, $\mathbf{v}^*$ and $\mathbf{f}^*$ denote respectively the values of ρ, $\mathbf{v}$ and $\mathbf{f}$ at some interior points of υ and $\mathbf{t}^{(n)*}$ and $\mathbf{t}^{k*}$ are the values of $\mathbf{t}^{(n)}$ and $\mathbf{t}^k$ at some point of ΔS and ΔS_k respectively. Using the principle of conservation of mass, and letting ΔS -> 0, we can find, by noting that ρ^*, $\dot{\mathbf{v}}^*$ and $\mathbf{f}^*$ are bounded quantities,

$$\mathbf{t}^{(n)}\,dS = \mathbf{t}^k dS_k \tag{1.11.8}$$

and

$$dS_k = n_k\,dS \tag{1.11.9}$$

where **n** is the unit outward normal vector of the surface ΔS.

Substituting eqn.(1.11.9) into (1.11.8), we have

$$\mathbf{t}^{(\mathbf{n})} = \mathbf{t}^k n_k \tag{1.11.10}$$

from which, since the stress vectors $\mathbf{t}^k$ are, by definition, independent of **n**, we can get immediately the following *Cauchy's lemma:* The stress vector $\mathbf{t}^{(\mathbf{n})}$ acting upon opposite sides of the same surface at a given point are equal in magnitude and opposite in direction, i.e.

$$\mathbf{t}^{(\mathbf{n})} = -\mathbf{t}^{(-\mathbf{n})} . \tag{1.11.11}$$

We may now introduce the stress tensor with its component t_{kj} defined as the jth component of the stress vector $\mathbf{t}^k$ acting on the positive side of the kth coordinate surface, i.e.

$$\mathbf{t}^k = t_{kj}\,\mathbf{i}_j \qquad (k=1, 2, 3) \tag{1.11.12}$$

with $\mathbf{i}_j$ (j = 1,2,3) being the unit base vectors of a rectangular Cartesian coordinate system. For example, t_{13} is the x_3-component of the stress vector $\mathbf{t}^1$ acting on the the surface x_1=const.

By eqns.(1.11.10) and (1.11.12), we may further get the following *Cauchy's theorem:* From the stress vectors acting across three mutually perpendicular planes at a point, all stress vectors at that point are determinate; they are given by

$$t_k^{(n)} = t_{jk} n_j \tag{1.11.13}$$

as linear functions of the stress tensor t_{jk}.

In particular, at the boundary surface of a body with the prescribed surface loading $\mathbf{t}^{(n)o}$, we have the following stress boundary condition

$$t_{jk} n_j = t_k^{(n)o} \tag{1.11.14}$$

with **n** being the unit outward normal vector of the boundary surface.

It can be shown that the Cauchy stress tensor is objective since it obeys the appropriate tensor transformation rule

$$t'_{jk}(\mathbf{X}, t') = Q_{jm}(t)Q_{kn}(t)t_{mn}(\mathbf{X}, t) \tag{1.11.15}$$

in any two objectively equivalent motions described by eqn.(1.11.1). The objectivity of the Cauchy stress tensor has important consequences on constitutive equations which will be discussed later.

With the aid of the stress tensor t_{jk}, one may get the following local balance equation of linear momentum

$$\rho \frac{dv_i}{dt} = t_{ji,j} + f_i \tag{1.11.16}$$

and the local balance equation of angular momentum

$$c_i + \varepsilon_{ijk}t_{jk} = 0 \tag{1.11.17}$$

in the region of continuous smooth motions, where eqns.(1.11.5) and (1.11.6) are valid for every volume element. It can be seen from eqn.(1.11.17) that, in the absence of body couples, Cauchy's stress tensor t_{jk} is symmetric, i.e. $t_{[jk]} = t_{jk} - t_{kj} = 0$. Eqn.(1.11.16) is often called the Eulerian equation of motion of a continuum.

1.12 The laws of continuum thermodynamics

The actual motion and physical states of material bodies in contact with their surroundings at a certain temperature have to be subject to the restrictions of classical laws of thermodynamics. In the classical theory of thermodynamics, thermodynamic systems are characterized by a set of independent macroscopic state variables, such as the temperature, the strain tensor, the electro-magnetic fields, etc. The thermodynamic system may be closed or open, depending on whether or not it exchanges matter with its surroundings. Here, we shall only deal with the closed thermodynamic system, which does not exchange matter with its surroundings. The closed system of a continuous body may, however, interact with its surrounding by exchanging heat and work performed by body forces and surface forces on it. A system having no interaction with its surrounding is called an isolated system. It is worth mentioning that where electric conduction in materials is concerned, we shall only consider the electric conduction in metallic solids so that the elèctric current is due to the flow of free electrons in the metallic solids, and, thus, the effect of matter transfer of electrons may be ignored since the mass of the electron is very small compared with the mass of the atoms composing the material. In contrast to the metallic solids, the electric current in electrolytic media is due to the flow of ions which form the substance, and, therefore, a passage of an electric current due to the migration of ions under electric field is always associated with transport of matter. In addition, chemical reactions are always present during the passage of an electric current from a metallic to an electrolytic conductor or vice versa (Kortum (1965)).

In each problem of a thermodynamic system, it is important to select a particular set of independent state variables. However, the choice is to a certain extent arbitrary. We shall see this point more clearly in later discussion of concrete problems. If a certain state variable can be expressed as a single-valued function of a set of other state variables, it is called a state function, and the functional relationship is said to be an equation of state. For a given system, if the state variables does not depend on space coordinates, the system is said to be homogeneous. If the values of the state variables are independent of time, the system is said to be in thermodynamic equilibrium. Otherwise, if the state variables vary with time, the system is said to undergo a thermodynamic process. The thermodynamic process can be either irreversible or reversible, depending on whether the energy of the system is dissipative or not during the process.

In the theory of thermodynamics for continuous media, the first law of thermodynamics is the law of energy conservation. The mathematical expression of the law of energy conservation for an electromagnetic continuum body in electromagnetic fields may be written as

$$\frac{d}{dt}\int_V \rho\left(\frac{1}{2}\mathbf{v}\cdot\mathbf{v} + u\right)dV = \int_{\partial V} (\mathbf{t}^{me(n)}\cdot\mathbf{v} - \mathbf{q}\cdot\mathbf{n})dS + \int_V (\mathbf{f}^{me}\cdot\mathbf{v})dV + W^{em} \quad (1.12.1)$$

where u denotes the internal stored energy per unit mass, $\mathbf{q}$ the heat flux, $\mathbf{t}^{me(n)}\cdot\mathbf{v}$ the power per unit volume by mechanical surface traction and $\mathbf{f}^{me}\cdot\mathbf{v}$ the power per unit volume by mechanical body force. W^{em} is the power supply of electromagnetic origin, which, according to specific materials, will be given explicitly in later sections.

The first law of thermodynamics only states that energy is conserved in any process. We may naturally ask the question what kind of process is physically possible for a given thermodynamic system with a certain amount of energy. For instance, our common experience tells us that heat flow is always from a body with high temperature to one at lower temperature. What determines the direction of the process? The second law of thermodynamics is such a law which gives a restriction on the possible thermodynamic processes by using the concepts of absolute temperature and entropy. In thermodynamics of continuum systems, the entropy is postulated to be a function of state in irreversible processes as well as in reversible processes, and the balance equation of entropy can be expressed by

$$\frac{d}{dt}\int_V \rho\eta dV = -\int_{\partial V} \mathbf{S}\cdot\mathbf{n}\, dS + \int_V \rho\sigma_s dV \tag{1.12.2}$$

where η denotes the specific entropy (entropy per unit mass), $\mathbf{S}$ the entropy flux, and σ_s the rate of internal entropy production per unit mass. It is important to realize that entropy is an attribute of a material body, just as its mass or its electric charges. Thus, the entropy of a system is equal to the sum of entropies of its parts.

The second law of thermodynamics can now be stated by the simple mathematical expression

$$\int_V \rho\sigma_s dV \geq 0 \qquad \text{for all possible thermodynamic processes} \tag{1.12.3}$$

where the equality sign is true only for reversible processes, and the inequality sign is valid for irreversible processes.

When eqn.(1.12.2) is valid for any part of a continuous body, we get the local production of entropy

$$\rho\sigma_s = \rho\frac{d\eta}{dt} + \nabla\cdot\mathbf{S} \geq 0 \tag{1.12.4}$$

and the interface condition

$$[\,\mathbf{S}\,]\cdot\mathbf{n} \geq 0 \tag{1.12.5}$$

in the absence of surface heat sources.

The local expression (1.12.4) of the second law of thermodynamics is called the local Clausius-Duhem inequality. We may now introduce the definition for a thermodynamically admissible process by stating that a process is called thermodynamically admissible if and only if it obeys the local Clausius-Duhem inequality and possesses a nonnegative finite temperature.

In addition, we may also introduce the definition that a thermodynamic process is called mechanically admissible if it obeys the laws of motion, namely, the conservation law of mass, the balance laws of momenta, and the conservation law of energy.

For a simple thermodynamic process, in which there are no chemical reactions and matter transfers, the entropy flux reads

$$\mathbf{S} = \frac{\mathbf{q}}{T}\,. \tag{1.12.6}$$

where $\mathbf{q}$ is the heat flux and T the local absolute temperature being a strictly positive scalar with the unit °K. The scale of the absolute temperature is fixed by defining the temperature at equilibrium between liquid water and ice at a pressure of 1 atm at 273.16 °K, i.e. 273.16 degrees on Kelvin's scale.

It can been seen from eqn.(1.12.4) that, in an isolated system, the second law of thermodynamics tells us that the possible spontaneous processes occur only in the direction of increasing entropy. Further development of irreversible thermodynamics for various thermodynamic systems requires a detailed description of the entropy production σ_s. We shall give further discussion later in the following relevant sections.

1.13 Electromagnetic forces, couples and powers

1.13.1 On formulations of electrodynamics for deformable media

During the past three decades, the macroscopic continuum theory of electromagnetism for moving and deformable solids has been studied extensively because of the rapid development of technology and applied physical sciences. Many formulations of the macroscopic Maxwell's equations were proposed, but none of them has, so far, been accepted as a universal law owing to the complication of the electromagnetic phenomena for moving deformable solids, especially for high velocity and highly accelerated motion of the solids. The difficulties in formulating the electromagneto-mechanical interaction phenomena are mainly due to the following facts:

(a) *The invariance properties of electromagnetic field equations and mechanical equations.*

In classical electrodynamics and classical mechanics, it is well-known that Maxwell's equations are invariant under Lorentz transformation but the mechanical balance laws are invariant under Galilean transformation. Therefore a consistent formulation of electromagneto-mechanical interaction should be treated from the relativity point of view, especially, for high velocity and accelerated motion of deformable solids, where the formulation has to be based on the general principle of relativity. So far, the formulations with less controversy are those for electromagneto-mechanical interaction phenomena in the electro-quasistatic and the magneto-quasistatic approximations as we have shown in section 1.8. In these cases, with the neglect of relativistic effects, the electromagneto-mechanical interaction may be treated consistently in the classical Galilean relativity.

(b) *The definitions of electromagnetic body force, body couple, and the energy supply.*

This difficulty mainly lies in the separation of the electromagneto-mechanical interactions in near and far field effects. As we know, in the classical continuum theory of elasticity, the introduction of a stress vector implies that the range of action of interparticle forces may be taken to be effectively zero; i.e. the interparticle forces, which are dealt with by the stress vector, are considered as short-range forces acting only between neighboring points. The mechanical body forces are, for instance, the gravitational force between the body and the earth, which are long-range force. The gravitational force also acts at short distances but no two parts of the body and the earth respectively are separated by a distance comparable with the microscopic dimensions of the elastic body. Therefore, in the classical elasticity, the surface (stress) force and the body force are uniquely well defined quantities.

However, in an electromagnetic body, the electromagnetic inter-particle forces are partly of short-range character (e.g. exchange forces) and partly of long-range character (e.g. electric and magnetic dipole-dipole forces). In addition, the long-range part of the electromagnetic interparticle forces also act at short distances and these forces act on some of the particles which are separated from each other by short distances comparable to the microscopic dimensions of the solid. So far, no unique separation between the electromagnetic stress forces and the electromagnetic body forces seems to be available.

(c) *The forms of Maxwell's equations for deformable solids, especially, the definitions of the macroscopic quantities of the electric polarization and the magnetization.*

It has been shown that, in the theory of electrodynamics for free space, the electric and magnetic fields may be well defined from measurable forces acting on testing charges in free space. However, the electromagnetic fields inside materials cannot be measured directly. They have to be inferred from the measurable fields adjacent to the material in free space. Different forms of the polarization and the magnetization vectors then appear as a consequence of different choices of fundamental field variables in the material, which are somewhat abstract quantities. The differences between these formulations are mainly in their different expressions for electromagnetic body forces, body couples and power supply.

At present, there are essentially four types of formulations, i.e. the Maxwell-Minkowski formulation, the Chu formulation (two-dipole model), the Lorentz-Ampére formulation (dipole-current circuit model), and the statistical formulation. A comparison of these four types of formulations has been made by Pao (1978) and Hutter and van de Ven (1978).

In this section, we shall present a set of commonly used expressions (Brown (1966), Alblas (1979), Moon (1984) and Maugin (1985)) for the electromagnetic body forces, body couples and powers, based on the dipole model being consistent with its use in our derivation of the macroscopic Maxwell's equations for materials as shown in section 1.5. The influence of the motion of a deformable solid on electromagnetic field is taken into account here by introducing locally and instantaneously a Galilean frame of reference $S' = S_c(\mathbf{x}, t)$ co-moving with the element of the continuous matter with velocity $\mathbf{v}(\mathbf{x},t)$. Thus, the transformation relations for electromagnetic fields given in section 1.8 can be used here for the deformable electromagnetic solid with the local velocity $\mathbf{v}(\mathbf{x}, t)$.

1.13.2 **Force, couple and power for non-magnetizable conductors**

For a non-magnetizable conductor in the magneto-quasistatic approximation, the electromagnetic body force in the Lorentz force form is

$$\mathbf{f}^{em} = \mathbf{J} \times \mathbf{B} \tag{1.13.1}$$

where $\mathbf{J}$ is the macroscopic current density vector. Here, the free charge density ρ_e has been assumed to be zero in the magneto-quasistatic approximation.

By introducing an electromagnetic stress tensor t^{em}_{ji}, defined by

$$t^{em}_{ji} = \frac{1}{\mu_o}[B_j B_i - \frac{1}{2}(B_k B_k)\delta_{ji}], \tag{1.13.2}$$

we may write eqn.(1.13.1) in the following form

$$\mathbf{f}^{em} = \mathbf{J} \times \mathbf{B} = \nabla\cdot\mathbf{t}^{em} \tag{1.13.3}$$

where one has introduced the notation $(\nabla\cdot\mathbf{t}^{em})_i = (t^{em}_{ji})_{,j}$.

The electromagnetic body couple vanishes for the non-polarizable and non-magnetizable conductors, and the electromagnetic power reads

$$W^{em} = \int_V (\mathbf{J} \times \mathbf{B})\cdot\mathbf{v}dV + \int_V \mathbf{J}\cdot\mathbf{E}'dV \tag{1.13.4}$$

with $\mathbf{E}' = \mathbf{E} + \mathbf{v} \times \mathbf{B}$. Here, $\mathbf{J}\cdot\mathbf{E}'$ represents the Joule heating of the conductor.

1.13.3 **Force, couple and power for magnetizable insulators**

For a magnetizable insulator in the magneto-quasistatic approximation, the total macroscopic magnetic force acting on the magnetizable body in an external magnetic induction field $\mathbf{B}^{\circ}$ may be found in the dipole model from eqn.(1.4.33) as

$$\mathbf{F}^{m} = \int_{V} (\mathbf{M}\cdot\nabla)\mathbf{B}^{\circ} dV \tag{1.13.5}$$

which is in consistent with the result given by Brown (1966). Here, we have noticed the fact that internal forces in the material body always balance each other so that they do not contribute to the total magnetic force on the body.

Eqn.(1.13.5) can also be transformed into the following expression (see Brown (1966) and Alblas (1979))

$$\mathbf{F}^{m} = \int_{V} \mathbf{f}^{m}\, dV + \int_{\partial V} \frac{1}{2}\mu_{o}(\mathbf{M}\cdot\mathbf{n})^{2}\mathbf{n} dS \tag{1.13.6}$$

where one has introduced a magnetic body force defined by

$$\mathbf{f}^{m} = \mu_{o}(\mathbf{M}\cdot\nabla)\mathbf{H} \tag{1.13.7}$$

with $\mathbf{M}$ being the magnetization of the material.

The magnetic surface traction at the outer boundary surface of the magnetized body is

$$\mathbf{t}^{m(n)} = \frac{1}{2}\mu_{o}(\mathbf{M}\cdot\mathbf{n})^{2}\mathbf{n}\,. \tag{1.13.8}$$

By introducing a magnetic stress tensor t^{em}_{ji}, defined by

$$t^{em}_{ji} = B_{j}H_{i} - \frac{1}{2}\mu_{o}(H_{k}H_{k})\delta_{ji} \tag{1.13.9}$$

eqn.(1.13.7) may also be expressed as

$$\mathbf{f}^{m} = \mu_0(\mathbf{M}\cdot\nabla)\mathbf{H} = \nabla\cdot\mathbf{t}^{em} . \tag{1.13.10}$$

It is shown in eqn.(1.13.6) that the presence of the magnetic surface traction means that the long-range magnetic force cannot be expressed simply as a form per unit volume; it depends also on the shape of the volume considered. However, one also notices that, for practical magnetic materials, the magnetic surface traction is usually much smaller than the yield stress traction of the material (see Moon (1984)) so that, for some engineering problems, this magnetic surface traction may be ignored at the first-order approximation.

The total magnetic torque on the magnetic body may be found in the dipole model (see eqn.(1.4.34)) as

$$\mathbf{L}^{m} = \int_V [\mathbf{M} \times \mathbf{B}^{o} + \mathbf{x} \times (\mathbf{M}\cdot\nabla)\mathbf{B}^{o}]dV. \tag{1.13.11}$$

This expression for magnetic torque may also be transformed into the following form (Brown (1966))

$$\mathbf{L}^{m} = \int_V [\mathbf{c}^{m} + \mathbf{x} \times (\mu_0\mathbf{M}\cdot\nabla)\mathbf{H}]dV + \int_{\partial V} \frac{1}{2}[\mathbf{x} \times (\mu_0\mathbf{M}\cdot\mathbf{n})^2\mathbf{n}]dS \tag{1.13.12}$$

where one has introduced a magnetic body couple $\mathbf{c}^{m}$ defined by

$$\mathbf{c} = \mathbf{c}^{m} = \mu_0\mathbf{M} \times \mathbf{H} = \mathbf{M} \times \mathbf{B} \tag{1.13.13}$$

which can be related to the magnetic stress from eqn.(1.13.9) by $t^{em}_{[ji]} = \varepsilon_{jik}c^{m}_{k}$. The electromagnetic power for the insulating magnetizable solid in the dipole model reads

$$W^{em} = \int_V \mu_0[(\mathbf{M}\cdot\nabla)\mathbf{H}]\cdot\mathbf{v}dV + \int_{\partial V} \frac{1}{2}\mu_0(\mathbf{M}\cdot\mathbf{n})^2\mathbf{n}\cdot\mathbf{v}dS + \int_V \rho\mu_0\mathbf{H}\cdot\frac{d}{dt}\left(\frac{\mathbf{M}}{\rho}\right)dV \tag{1.13.14}$$

in which the last term on the right-hand side of eqn.(1.13.14) represents the rate of work done by magnetic couples. Note that no surface magnetic torques are being considered in this model.

1.13.4 **Force, couple and power for magnetizable conductors**

For a magnetizable conductor in the magneto-quasistatic approximation, the total electromagnetic force on the conducting magnetizable solid in the absence of surface currents can be expressed (Alblas (1979)) in the dipole model as

$$\mathbf{F}^{em} = \int_V \{\mathbf{J} \times \mathbf{B}^o + (\mathbf{M}\cdot\nabla)\mathbf{B}^o\}dV \tag{1.13.15}$$

which can be further transformed into the following form

$$\mathbf{F}^{em} = \int_V \mathbf{f}^{em}\, dV + \int_{\partial V} \tfrac{1}{2}\mu_o(\mathbf{M}\cdot\mathbf{n})^2\mathbf{n}dS \tag{1.13.16}$$

where one has introduced an electromagnetic body force defined by

$$\mathbf{f}^{em} = \mathbf{J} \times \mu_o\mathbf{H} + \mu_o(\mathbf{M}\cdot\nabla)\mathbf{H} = \mathbf{J} \times \mathbf{B} + (\nabla\mathbf{H})\cdot\mu_o\mathbf{M}\,. \tag{1.13.17}$$

By introducing the electromagnetic stress tensor t^{em}_{ji}, defined in the same form given by eqn.(1.13.9), eqn.(1.13.17) can also be written as

$$\mathbf{f}^{em} = \mathbf{J} \times \mu_o\mathbf{H} + \mu_o(\mathbf{M}\cdot\nabla)\mathbf{H} = \nabla\cdot\mathbf{t}^{em}\,. \tag{1.13.18}$$

The electromagnetic body couple acting on the magnetizable conductor can still be given by eqn.(1.13.13). The total magnetic torque on the conducting magnetic body in the presence of the external magnetic field $\mathbf{B}^o$ may be expressed in the dipole model as

$$\mathbf{L}^{m} = \int_V \{\mathbf{M} \times \mathbf{B}^o + \mathbf{x} \times [(\mathbf{M}\cdot\nabla)\mathbf{B}^o + \mathbf{J} \times \mathbf{B}^o]\}\, dV\,. \tag{1.13.19}$$

The electromagnetic power on the magnetizable conductor may be written in the dipole model as

$$W^{em} = \int_V \mathbf{J}\cdot\mathbf{E}'dV + \int_V (\mathbf{J}\times\mu_0\mathbf{H})\cdot\mathbf{v}dV + \int_V \mu_0[(\mathbf{M}\cdot\nabla)\mathbf{H}]\cdot\mathbf{v}dV$$

$$+ \int_{\partial V} \frac{1}{2}\mu_0(\mathbf{M}\cdot\mathbf{n})^2\mathbf{n}\cdot\mathbf{v}dS + \int_V \rho\mu_0\mathbf{H}\cdot\frac{d}{dt}\left(\frac{\mathbf{M}}{\rho}\right)dV \tag{1.13.20}$$

in which the first term on the right-hand side of eqn.(1.13.20) is the Joule heating of the conductor. Here we emphasize that the electromagnetic field quantities **E**, **H**, and **B**, defined here, are still governed by the set of Maxwell's equations at the magneto-quasistatic approximation given by eqns.(1.8.2)-(1.8.5) with the transformation relations by eqns.(1.8.7)-(1.8.10) rather than Chu's form of Maxwell's equations (see Hutter and van de Ven (1978)). The present formulation can, however, be used consistently only at the magneto-quasistatic approximation.

1.13.5 **Force, couple and power for dielectric solids**

For a dielectric solid in the absence of free volume and surface charges, the total electric force acting on an electrically polarized body in the presence of an external electric field $\mathbf{E}^0$ may be expressed in the dipole model and in the electro-quasistatic approximation (see eqn.(1.4.15)) by

$$\mathbf{F}^e = \int_V (\mathbf{P}\cdot\nabla)\mathbf{E}^0 dV\ . \tag{1.13.21}$$

This force expression can also be transformed into the following form (see Alblas (1979))

$$\mathbf{F}^e = \int_V \mathbf{f}^e\, dV + \int_{\partial V} \frac{1}{2\varepsilon_0}(\mathbf{P}\cdot\mathbf{n})^2\mathbf{n}dS \tag{1.13.22}$$

where one has introduced an electric body force defined by

$$\mathbf{f}^e = (\mathbf{P}\cdot\nabla)\mathbf{E} \tag{1.13.23}$$

with **P** being the polarization of the material.

The electric surface traction at the outer boundary surface of the electrically polarized body is

$$\mathbf{t}^{e(n)} = \frac{1}{2\varepsilon_o}(\mathbf{P}\cdot\mathbf{n})^2\mathbf{n} \ . \tag{1.13.24}$$

By introducing an electric stress tensor t^e_{ji}, defined by

$$t^e_{ji} = D_jE_i - \frac{1}{2}\varepsilon_o(E_kE_k)\delta_{ji} \tag{1.13.25}$$

eqn.(1.13.23) may also be expressed as

$$\mathbf{f}^e = (\mathbf{P}\cdot\nabla)\mathbf{E} = \nabla\cdot\mathbf{t}^e \ . \tag{1.13.26}$$

Similar to the magnetic case, the total electric torque on the dielectric body in the dipole model (see eqn.(1.4.16)) can be found by

$$\mathbf{L}^e = \int_V [\mathbf{x}\times(\mathbf{P}\cdot\nabla)\mathbf{E}^o + \mathbf{P}\times\mathbf{E}^o]dV \tag{1.13.27a}$$

$$= \int_V [\mathbf{c}^e + \mathbf{x}\times(\mathbf{P}\cdot\nabla)\mathbf{E}]dV + \int_{\partial V}\frac{1}{2\varepsilon_o}[\mathbf{x}\times(\mathbf{P}\cdot\mathbf{n})^2\mathbf{n}]dS \tag{1.13.27b}$$

where one has introduced the electric body couple $\mathbf{c}^e$ defined by

$$\mathbf{c} = \mathbf{c}^e = \mathbf{P}\times\mathbf{E} \tag{1.13.28}$$

which can be related to the electric stress from eqn.(1.13.25) by $t^e_{[ji]} = \varepsilon_{jik}c^e_k$. The electromagnetic power on the dielectric solid may be written in the dipole model as

$$W^{em} = \int_V [(\mathbf{P}\cdot\nabla)\mathbf{E}]\cdot\mathbf{v}dV + \int_{\partial V}\frac{1}{2\varepsilon_o}(\mathbf{P}\cdot\mathbf{n})^2\mathbf{n}\cdot\mathbf{v}dS + \int_V \rho\mathbf{E}\cdot\frac{d}{dt}\left(\frac{\mathbf{P}}{\rho}\right)dV \ . \tag{1.13.29}$$

where the last term on the right-hand side of this equation represents the rate of work done by electric couples. No surface electric torques are being considered in this model.

We have presented here some expressions of electromagnetic forces, couples and electromagnetic powers for some specified materials based on the dipole model at the magneto- and electro-quasistatic approximations. From those expressions, we are now ready to further discuss constitutive equations which characterize phenomenologically the electromagneto-mechanical interactions in the materials. As a final comment of this section, we emphasize that any meaningful phenomenological formulation (or model) can only be justified eventually by its consequences in agreement with experimental results.

1.14 Constitutive equations of electromagneto-thermoelastic solids

1.14.1 Axioms of constitutive theory

In above sections, we have presented the basic laws of electromagneto-mechanical phenomena, which are valid for all types of material media irrespective of their internal constitutions. To complete the study of material responses under mechanical and electromagnetic loadings, material models characterizing the internal constitution of the matter are, therefore, required. In continuum mechanics, we are not concerned with the atomic or molecular structures of the material, rather, we are interested in the global macroscopic behavior of the material, which can be modelled phenomenologically by a set of equations called the constitutive equations. To formulate a constitutive theory which can represent a material adequately, certain physical and mathematical rules (axioms) have to be satisfied according to classical continuum mechanics (see, for instance, Eringen (1967)). Here, we shall present several axioms which are relevant to our material modelling.

The *axiom of determinism* states that past and present "causes" (independent constitutive (state) variables) determine the present "effects" (dependent constitutive (state) variables).

The *axiom of memory* states that the values of the independent constitutive variables at distant past from the present do not affect appreciably the values of the constitutive functions (dependent constitutive variables). In particular, when hereditary phenomena, such as viscoelasticity, are not of interest, we may adopt a simpler statement that the present "causes" determine the present "effects", which means that both independent and dependent constitutive variables are now defined for the same event ($\mathbf{x}$, t). Here, the independent constitutive variables are, for instance, the motion of the material body, the temperature, the electric and magnetic fields. The dependent constitutive variables are, for instance, the stress tensor, the heat flux vector, the electric polarization, the magnetization, and the internal energy, etc.

The *axiom of equipresence* states that at the outset all constitutive functionals should be expressed in terms of the same list of independent constitutive variables until the contrary is deduced. This axiom is a precautionary measure. It helps us not to forget or be prejudiced against a certain class of variables and favor others in the expression of constitutive functionals, though some of the variables may be eliminated finally by the basic laws of continuum mechanics and various approximations.

The *axiom of neighborhood* states that the values of the independent constitutive variables at distant material points from $\mathbf{X}$ do not affect appreciably the values of the dependent constitutive variables at $\mathbf{X}$. In the present work, we are not going to discuss the general nonlocal theory. Only local first-order gradient theories will be considered.

The *axiom of objectivity* states that constitutive equations must be form-invariant with respect to rigid motions of the spatial frame of reference, defined by eqn.(1.11.1). We have seen that Cauchy's stress tensor is an objective tensor, which will impose the restriction on its functional dependence on dependent constitutive variables. To illustrate further the objectivity, a simple example cited by Truesdall (1961) may be of interest to be included here. Consider a spring mounted on a rotating table with one end attached to the table and the other fixed to a mass. As the table is rotating the spring extends. From the viewpoint of an external observer, the force in the spring is merely that required to satisfy Newton's law for the accelerated motion of the mass. For an observer utilizing a reference frame fixed in the rotating table the spring force would derive seemingly from a counter balancing body force exerted on the mass by an external field. Nevertheless for both observers the force in the spring is the same, and hence so are the contact forces exerted on both the table and mass by the spring, which means that the contact forces are frame-independent (objective) quantities.

Other self-evident axioms are the axiom of material invariance and the axiom of admissibility. The *axiom of material invariance* states that constitutive equations must be form-invariant with respect to a group of orthogonal transformations {**S**} and translations {**C**} of the material coordinates, defined by

$$\overline{\mathbf{X}} = \mathbf{S}\cdot\mathbf{X} + \mathbf{C} \,. \tag{1.14.1}$$

These restrictions express the geometric symmetries, represented by {**S**}, and inhomogeneities represented by {**C**}, at **X**, in the physical properties of the material body. When {**S**} is the proper orthogonal group (det **S** = 1), the material is called hemitropic. If {**S**} is the full orthogonal group (det **S** = ± 1), the material is called isotropic. A material that is not hemitropic is called anisotropic. When the constitutive response functions do not depend on the translations {**C**} of the origin of material coordinates, the material is said to be homogeneous. Otherwise, the material is called inhomogeneous. Obviously, constitutive equations of a homogeneous material are independent of the material coordinate **X** explicitly. It should be noticed that a material may posses different types of material symmetries for its different physical and mechanical properties. For example, a material isotropic with respect to mechanical property may not be isotropic with respect to other physical properties, such as magnetization.

The *axiom of admissibility* states that all constitutive equations must be consistent with the basic laws of the conservation of mass, the balance of momenta, the conservation of energy and the second laws of thermodynamics.

1.14.2 Constitutive equations for thermoelastic conductors

With the aid of the above knowledge, we are now ready to formulate the constitutive equations for electromagnetic materials. At first, we consider a soft-ferromagnetic conducting thermoelastic solid, for which no hysteresis is exhibited, and gyromagnetic and exchange effects are neglected. The energy balance equation for such a material body in electromagnetic fields may be expressed from eqns.(1.12.1) and (1.13.20) at the magneto-quasistatic approximation by

$$\frac{d}{dt}\int_V \rho(\frac{1}{2}\mathbf{v}\cdot\mathbf{v} + u)dV = \int_{\partial V}(\mathbf{t}^{(n)}\cdot\mathbf{v} - \mathbf{q}\cdot\mathbf{n})dS + \int_V \mathbf{f}\cdot\mathbf{v}dV$$

$$+ \int_V \mathbf{J}\cdot\mathbf{E}'dV + \int_V \rho\mu_0\mathbf{H}\cdot\frac{d\mu}{dt}dV \qquad (1.14.2)$$

in which $\mu = \mathbf{M}/\rho$ denotes the magnetization per unit mass. Here $\mathbf{f}$ is the sum of the mechanical body force $\mathbf{f}^{me}$ and the electromagnetic body force $\mathbf{f}^{em}$ given by eqn.(1.13.17). The stress traction $\mathbf{t}^{(n)}$ is related to Cauchy's stress tensor by eqn.(1.11.13).

Eqn.(1.14.2) represents the global conservation of energy, valid for any motion of the material body, including discontinuous distribution of material characteristics. In the region of continuous smooth motions of a continuous body, where eqns.(1.14.2) is valid for every volume element, one can then get the following local balance equation of energy

$$\rho\dot{u} = \mathbf{t} : (\nabla\mathbf{v}) - \nabla\cdot\mathbf{q} + \mathbf{J}\cdot\mathbf{E}' + \rho\mu_0\mathbf{H}\cdot\dot{\mu} \qquad (1.14.3)$$

with the aid of the balance law of linear momentum (1.11.5) and the conservation law of mass (1.9.14). Here, the notations of $\mathbf{t} : (\nabla\mathbf{v}) = t_{ji}v_{i,j}$ and $\dot{u} = du/dt$ have been used.

With the use of the second law of thermodynamics (1.12.4) and eqn.(1.14.3), one finds

$$\rho T\sigma_s = -\rho(\dot{\Psi} + \eta\dot{T}) + \mathbf{t} : (\nabla\mathbf{v}) + \rho\mu_0\mathbf{H}\cdot\dot{\mu} - \frac{1}{T}\mathbf{q}\cdot\nabla T + \mathbf{J}\cdot\mathbf{E}' \geq 0 \qquad (1.14.4)$$

where Ψ is a thermodynamic free energy function defined by

$$\Psi = u - \eta T. \qquad (1.14.5)$$

Since no hereditary phenomena and nonlocality are being considered for the soft ferromagnetic conducting thermoelastic solid, Ψ may be taken to be of the following form

$$\Psi = \Psi(E_{KL}, G_L, T) \tag{1.14.6}$$

with the quantities G_L defined by $G_L = \mu_i x_{i,L}$ due to the fact that Ψ must be an objective scalar functional under the full Euclidean transformation group (1.11.1) according to the axiom of objectivity. The free energy function Ψ may also depend on the position vector **X** if the material is inhomogeneous.

To satisfy the inequality expressed by eqn.(1.14.4) according to the axiom of admissibility, one must have

$$\eta = -\frac{\partial \Psi}{\partial T} \tag{1.14.7}$$

$$t_{ji} = \rho\frac{\partial \Psi}{\partial E_{KL}} x_{i,K}x_{j,L} + \rho\frac{\partial \Psi}{\partial G_L}\mu_i x_{j,L} = \rho\frac{\partial \Psi}{\partial E_{KL}} x_{i,K}x_{j,L} + \mu_o M_i H_j \tag{1.14.8}$$

$$t_{[ji]} = \rho\frac{\partial \Psi}{\partial G_L}\mu_{[i}x_{j],L} = \mu_o H_{[j}M_{i]} = \mu_o(H_j M_i - H_i M_j) \tag{1.14.9}$$

$$\mu_o H_i = \frac{\partial \Psi}{\partial G_L} x_{i,L} \tag{1.14.10}$$

and

$$-\frac{1}{T}\mathbf{q}\cdot\nabla T + \mathbf{J}\cdot\mathbf{E}' \geq 0 \tag{1.14.11}$$

where the heat flux **q** and the current density **J** can be, in general, functions of the Lagrangian strain E_{KL}, the magnetization G_L, the temperature T, the temperature gradient $(\nabla_R T)_L = T_{,i}x_{i,L}$ and the electric field $W'_L = E'_i x_{i,L}$ according to the axiom of equipresence. It is noticed that the disappearance of the dependence of the thermodynamic free energy function Ψ on the temperature gradient $\nabla_R T$ and the electric field W'_L is due to the fact that Ψ is independent of the rate of the independent constitutive variables in the non-hereditary material model.

It can be seen that eqn.(1.14.9) implies that the balance law of angular momentum (1.11.17) is identically satisfied by noting eqn.(1.13.13).

As to the inequality (1.14.11) which characterizes the irreversibility of thermodynamic processes of the system, the following set of phenomenological equations may be introduced to describe the irreversible thermodynamic process of the system which is supposed to be not far from its equilibrium state by

$$q_i = -\kappa_{ij}T_{,j} + \Gamma^Q_{ij}E'_j \qquad (1.14.12)$$

$$J_i = \sigma_{ij}E'_j - \Gamma^J_{ij}T_{,j} \qquad (1.14.13)$$

with $\mathbf{E}' = \mathbf{E} + \mathbf{v} \times \mathbf{B}$.

This set of phenomenological relations is usually called the generalized Fourier-Ohm's law. Here, κ_{ij}, σ_{ij}, Γ^Q_{ij} and Γ^J_{ij} are respectively the thermal conductivity, the electric conductivity and the thermoelectric coefficient tensors. In general, these coefficient tensors may still be functions of the elastic strain, the magnetization as well as the temperature according to the axiom of equipresence. To make the Clausius-Duhem inequality (1.14.11) be satisfied for all possible thermodynamic processes, the thermal conductivity tensor κ_{ij} and the electric conductivity tensor σ_{ij} have to be both positive definite. For most practical engineering applications, we may simply take these coefficient tensors be constant tensors provided that those neglected effects, such as Hall effect, thermogalvanomagnetic effect etc., are not important in the problems considered (Landau et al. (1984)). In such a case, the following relations of the Onsager symmetry apply

$$\kappa_{ij} = \kappa_{ji}, \quad \sigma_{ij} = \sigma_{ji} \quad \text{and} \quad \Gamma^Q_{ij} = T_o\Gamma^J_{ji} \qquad (1.14.14)$$

where T_o is a constant uniform reference temperature.

It must be emphasized that, in an anisotropic solid, the coefficient tensors Γ^Q_{ij} and Γ^J_{ij} are in general not symmetrical. For isotropic solids, according to the axiom of material invariance, we may simply have

$$\kappa_{ij} = \kappa\delta_{ij}, \quad \sigma_{ij} = \sigma\delta_{ij}, \quad \Gamma^Q_{ij} = T_o\Gamma\delta_{ij} \quad \text{and} \quad \Gamma^J_{ij} = \Gamma\delta_{ij} \qquad (1.14.15)$$

with the thermal conductivity κ and the electric conductivity σ being positive ($\kappa > 0$ and $\sigma > 0$).

The local balance equation of entropy (1.12.4) now reads

$$\rho T\dot{\eta} = -\nabla\cdot\mathbf{q} + \mathbf{J}\cdot\mathbf{E}' \tag{1.14.16}$$

which is the equation of heat conduction with Joule heating being the heat source.

The derived equations of (1.14.7), (1.14.8), (1.14.10), (1.14.12) and (1.14.13) form the complete set of constitutive equations for the soft ferromagnetic thermoelastic conductors.

It is sometimes useful to make an inversion of the independent constitutive (state) variable of the magnetization μ to the magnetic intensity field $\mathbf{H}$ as the independent constitutive (state) variable so that the formulation may be more convenient to be used than those presented in eqns.(1.14.7)-(1.14.10) for practical applications, especially for studying nonlinear magnetic behaviors. To do so we can introduce a thermodynamic Gibbs function Φ defined by

$$\Phi = \Psi - \mu_o\mathbf{H}\cdot\mu\ . \tag{1.14.17}$$

Thus, eqn.(1.14.4) may now be expressed as

$$\rho T\sigma_s = -\rho(\dot{\Phi} + \eta\dot{T}) + \mathbf{t} : (\nabla\mathbf{v}) - \mu_o\mathbf{M}\cdot\dot{\mathbf{H}} - \frac{1}{T}\mathbf{q}\cdot\nabla T + \mathbf{J}\cdot\mathbf{E}' \geq 0 \tag{1.14.18}$$

from which one can arrive at the following form of constitutive equations:

$$\eta = -\frac{\partial\Phi}{\partial T} \tag{1.14.19}$$

$$t_{ji} = \rho\frac{\partial\Phi}{\partial E_{KL}}x_{i,K}x_{j,L} + \rho\frac{\partial\Phi}{\partial H_L}H_i x_{j,L} = \rho\frac{\partial\Phi}{\partial E_{KL}}x_{i,K}x_{j,L} - \mu_o M_j H_i \tag{1.14.20}$$

$$t_{[ji]} = \rho\frac{\partial\Phi}{\partial H_L}H_{[i}x_{j],L} = -\mu_o M_{[j}H_{i]} \tag{1.14.21}$$

$$\mu_0 M_i = -\rho \frac{\partial \Phi}{\partial H_L} x_{i,L} \tag{1.14.22}$$

with $H_L = H_i x_{i,L}$. A consequence of such an inversion may now be seen from eqn.(1.14.10) and eqn.(1.14.22).

1.14.3 **Constitutive equations for thermoelastic dielectric solids**

Here, we shall derive the constitutive equations for thermoelastic dielectric solids in an electro-quasistatic approximation. We start also from the energy balance equation for the thermoelastic dielectrics by

$$\frac{d}{dt}\int_V \rho(\frac{1}{2}\mathbf{v}\cdot\mathbf{v}+u)dV = \int_{\partial V}(\mathbf{t}^{(n)}\cdot\mathbf{v} - \mathbf{q}\cdot\mathbf{n})dS + \int_V \mathbf{f}\cdot\mathbf{v}dV + \int_V \rho\mathbf{E}\cdot\frac{d}{dt}(\frac{\mathbf{P}}{\rho})dV \tag{1.14.23}$$

in which the total body force $\mathbf{f}$ is the sum of the mechanical body force $\mathbf{f}^{me}$ and the electric body force $\mathbf{f}^{e}$ given in eqn.(1.13.23).

Eqn.(1.14.23) represents the global conservation of energy, valid for any motion of the thermoelastic dielectric body, including discontinuous distribution of material characteristics. In the region of continuous smooth motions of a continuous body, where eqn.(1.14.23) is valid for every volume element, we can get the following local balance equation of energy

$$\rho\dot{u} = \mathbf{t} : (\nabla\mathbf{v}) - \nabla\cdot\mathbf{q} + \rho\mathbf{E}\cdot\frac{d}{dt}(\frac{\mathbf{P}}{\rho}) \tag{1.14.24}$$

with the aid of the balance law of linear momentum (1.11.5) and the conservation law of mass (1.9.14).

By eqn.(1.14.24) and the second law of thermodynamics (1.12.4), we find

$$\rho T\sigma_s = -\rho(\dot{\Phi}^e + \eta\dot{T}) + \mathbf{t} : (\nabla\mathbf{v}) + \mathbf{P}\cdot\dot{\mathbf{E}} - \frac{1}{T}\mathbf{q}\cdot\nabla T \geq 0 \tag{1.14.25}$$

in which the thermodynamical function Φ^e is defined by

$$\Phi^e = u - \eta T - (\mathbf{E}\cdot\mathbf{P})/\rho\,. \tag{1.14.26}$$

Since no hereditary phenomena and nonlocality are being considered for the thermoelastic dielectrics, Φ^e may be taken to be of the following form

$$\Phi^e = \Phi^e(E_{KL}, W_L, T) \tag{1.14.27}$$

with the quantities W_L defined by $W_L = E_i x_{i,L}$ due to the fact that Φ^e must be an objective scalar functional under the full Euclidean transformation group (1.11.1) according to the axiom of objectivity. The free energy function Φ^e may also depend on the position vector $\mathbf{X}$ if the material is inhomogeneous.

To satisfy the inequality expressed by eqn.(1.14.25) according to the axiom of admissibility, one must have

$$\eta = -\frac{\partial \Phi^e}{\partial T} \tag{1.14.28}$$

$$t_{ji} = \rho\frac{\partial \Phi^e}{\partial E_{KL}} x_{i,K} x_{j,L} + \rho\frac{\partial \Phi^e}{\partial W_L} E_i x_{j,L} = \rho\frac{\partial \Phi^e}{\partial E_{KL}} x_{i,K} x_{j,L} - P_j E_i \tag{1.14.29}$$

$$t_{[ji]} = \rho\frac{\partial \Phi^e}{\partial W_L} E_{[i} x_{j],L} = -P_{[j} E_{i]} \tag{1.14.30}$$

$$P_i = -\rho\frac{\partial \Phi^e}{\partial W_L} x_{i,L} \tag{1.14.31}$$

and

$$-\frac{1}{T}\mathbf{q}\cdot\nabla T \geq 0. \tag{1.14.32}$$

It can be seen that eqn.(1.14.30) implies that the balance law of angular momentum (1.11.17) is identically satisfied by noting eqn.(1.13.28). It can also be seen that the Clausius-Duhem inequality (1.14.32) implies simply the fact that the heat can only flow from the place of high temperature to the place of lower temperature consistent with our common experience.

The Fourier law for heat flux reads here

$$q_i = -\kappa_{ij}T_{,j} \tag{1.14.33}$$

with κ_{ij} being the thermal conductivity tensor, which must be non-negative definite for making the Clausius-Duhem inequality be satisfied generally. In general, the thermal conductivity tensor κ_{ij} is a function of the elastic strain, the electric field and the temperature field. In many practical cases, we may simply let it be a constant tensor provided that those neglected effects are not important in the problems considered. In such a case, the thermal conductivity tensor is a constant tensor being positive definite and satisfying the symmetric relation $\kappa_{ij} = \kappa_{ji}$. In particular, for isotropic thermoelastic dielectrics, we simply have

$$\kappa_{ij} = \kappa\delta_{ij} \qquad (\kappa > 0) \tag{1.14.34}$$

according to the axiom of material invariance.
The local balance equation of entropy (1.12.4) now reads

$$\rho T\dot{\eta} = -\nabla\cdot\mathbf{q} \tag{1.14.35}$$

which is the equation of heat conduction in the dielectrics in the absence of heat sources. The derived equations of (1.14.28), (1.14.29), (1.14.31) and (1.14.33) form the complete set of constitutive equations for thermoelastic dielectric solids.

1.15 Simplified theories of electromagneto-thermoelastic solids

The general dynamic equations presented in above sections for studying electromagneto-mechanical interactions in thermoelastic solids are shown to be highly nonlinear. Thus, solutions of such a general set of equations are hardly possible to be found exactly. Some further simplifications are, therefore, often be made in order to practically deal with engineering problems for various application purposes. In this section, we shall, first, summarize the field equations and boundary conditions in a general form for electromagneto-thermoelastic solids, and then introduce some simplified theories which are particularly convenient for studying some practical problems with sufficient accuracy.

1.15.1 **Field and boundary equations**

In general, the set of the mechanical motion equation, the heat conduction equation, Maxwell's equations and boundary conditions for electromagneto-thermoelastic solids can be summarized as follows.
The mechanical motion equation is

$$\nabla \cdot \mathbf{t} + \mathbf{f} = \rho \frac{d\mathbf{v}}{dt} \tag{1.15.1}$$

which may also be written as

$$\nabla \cdot (\mathbf{t} + \mathbf{t}^{em}) + \mathbf{f}^{me} = \rho \frac{d\mathbf{v}}{dt} \tag{1.15.2}$$

where $\mathbf{f}^{me}$ denotes the mechanical body force, and $\mathbf{t}^{em}$ the electromagnetic stress tensor defined by eqn.(1.13.2), (1.13.9) and (1.13.25) for the conducting nonmagnetic solid, the magnetic solid, and the dielectric solid respectively.

The stress interface condition at the interface S between two material media is

$$\mathbf{n} \cdot [\![\mathbf{t} + \mathbf{t}^{em}]\!] = 0 \qquad \text{on } S \tag{1.15.3}$$

where $[\![A]\!] = A^+ - A^-$ denotes the jump of the quantity A across the interface S.

On the boundary ∂V ($\equiv \partial V_t + \partial V_u$) of the material body, the mechanical boundary conditions can be given from eqn.(1.15.3) as

$$\mathbf{n} \cdot \mathbf{t} = \mathbf{t}^{o(n)} + \mathbf{n} \cdot [\![\mathbf{t}^{em}]\!] \qquad \text{on } \partial V_t \tag{1.15.4}$$

$$\mathbf{u} = \mathbf{u}^o \qquad \text{on } \partial V_u \tag{1.15.5}$$

where $\mathbf{t}^{o(n)}$ are the prescribed surface traction on ∂V_t and $\mathbf{u}^o$ the prescribed displacement on ∂V_u. Here, we have simply replaced $\mathbf{t}^-$ by $\mathbf{t}$ and $\mathbf{u}^-$ by $\mathbf{u}$ on the boundary of the material body, where the quantity with superscript "-" means its value being taken on the material side of the body at the boundary surface.

The heat conduction equation with Joule heating as the heat source reads

$$\rho T\dot{\eta} = - \nabla\cdot\mathbf{q} + \mathbf{J}\cdot\mathbf{E}' \qquad (1.15.6)$$

with the energy interface condition at the interface S between material media

$$\mathbf{n}\cdot[\![\,\mathbf{q} + \mathbf{E}'\times\mathbf{H}' - \mathbf{v}\cdot(\mathbf{t} + \mathbf{t}^{em})\,]\!] = 0 \quad \text{on S} \qquad (1.15.7)$$

which states that the jump in the energy of traction across a material interface is balanced by that of the sum of the normal components of the heat flux vector and of Poynting's vector. This interface condition is derived from the global balance equation of energy (1.12.1) by noting the possible discontinuity of the electromagnetic traction and of Poynting's vector across the material interface.

With the use of eqn.(1.15.3), eqn.(1.15.7) may be reduced to

$$\mathbf{n}\cdot[\![\,\mathbf{q} + \mathbf{E}'\times\mathbf{H}'\,]\!] = 0 \qquad \text{on S} \qquad (1.15.8)$$

if the material velocity $\mathbf{v}$ is continuous across the material interface, which may be the case where two material media are supposed to be perfectly bonded at the interface.

Furthermore, if no surface currents exist at the interface of the material media which are also supposed to be non-magnetic, one then has

$$\mathbf{n}\cdot[\![\,\mathbf{q}\,]\!] = 0 \qquad \text{on S} \qquad (1.15.9)$$

since, in such a case, the magnetic intensity field $\mathbf{H}'$ and the tangential component of the electric field $\mathbf{E}'$ are both continuous across the material interface. Eqn.(1.15.9) simply states that the normal component of the heat flux vector is continuous across the material interface.

Maxwell's equations at magneto-quasistatic approximation are given by

$$\nabla \times \mathbf{H} = \mathbf{J} \tag{1.15.10}$$

$$\nabla \cdot \mathbf{B} = 0 \tag{1.15.11}$$

$$\nabla \cdot \mathbf{J} = 0 \tag{1.15.12}$$

$$\nabla \times \mathbf{E} + \frac{\partial \mathbf{B}}{\partial t} = 0 \tag{1.15.13}$$

with $\mathbf{B} = \mu_o(\mathbf{H} + \mathbf{M})$, and with the electromagnetic interface conditions

$$\mathbf{n} \times [\, \mathbf{E} + \mathbf{v} \times \mathbf{B} \,] = 0 \tag{1.15.14}$$

$$\mathbf{n} \cdot [\, \mathbf{B} \,] = 0 \tag{1.15.15}$$

$$\mathbf{n} \times [\, \mathbf{H} \,] = \mathbf{K}_f \tag{1.15.16}$$

$$\mathbf{n} \cdot [\, \mathbf{J} \,] + \nabla_T \cdot \mathbf{K}_f = 0 \tag{1.15.17}$$

where $\mathbf{K}_f$ is the surface current density, and $\nabla_T \cdot \mathbf{K}_f$ denotes a tangential (two-dimensional) surface divergence of $\mathbf{K}_f$. At an interface between two conductors of finite conductivities, the surface current density $\mathbf{K}_f$ vanishes. In such a case, the normal component of current density and the tangential component of magnetic intensity field are continuous across the material interface.

Maxwell's equations for dielectric solids in the absence of free electric charges and at the electro-quasistatic approximation are

$$\nabla \times \mathbf{E} = 0 \tag{1.15.18}$$

$$\nabla \cdot \mathbf{D} = 0 \tag{1.15.19}$$

with $\mathbf{D} = \varepsilon_o\mathbf{E} + \mathbf{P}$, and with the electric interface conditions

$$\mathbf{n} \times [\, \mathbf{E} \,] = 0 \tag{1.15.20}$$

$$\mathbf{n} \cdot [\, \mathbf{D} \,] = 0 \tag{1.15.21}$$

where no macroscopic free currents are considered in the dielectric solid.

In addition to the interface conditions, initial conditions have to be given for general dynamic cases. For instance, initial values of the displacement and velocity fields, the temperature field and the electromagnetic fields may be specified according to concrete problems. No need, however, arises for the initial conditions in the case of steady-state problems.

In what follows, we shall introduce some simplified theories of solids with infinitesimal deformation and small temperature variation with respect to a certain reference configuration of the solids. The whole "linearization" (with respect to the infinitesimal deformation) procedure starting from the general equations given in the above sections will, however, not be derived here since careful readers can do it without much conceptual difficulty by choosing a properly defined reference (or intermediate) configuration. Such a procedure may be necessary if we are interested in the "exact" meanings of those phenomenological material coefficients, and in the consistent treatment of higher order effects. We shall come back to this point later when we deal with the problem of superconducting-mechanical interactions.

1.15.2 Non-magnetic thermoelastic conductors

The first simplified theory to be presented is concerned with non-magnetic conducting thermoelastic solids, for which a set of linearized constitutive equations may be given by

$$t_{ij} = C_{ijkl}\varepsilon_{kl} - \beta_{ij}\theta \tag{1.15.22}$$

$$\eta = \eta_o + \frac{C_v}{T_o}\theta + \frac{1}{\rho_o}\beta_{ij}\varepsilon_{ij} \tag{1.15.23}$$

$$q_i = -\kappa_{ij}\theta_{,j} + T_o\Gamma_{ij}E'_j \tag{1.15.24}$$

$$J_i = \sigma_{ij}E'_j - \Gamma_{ij}\theta_{,j} \tag{1.15.25}$$

with $\mathbf{E}' = \mathbf{E} + \mathbf{v} \times \mathbf{B}$, and ε_{kl} being the infinitesimal strain defined in eqn.(1.10.8b). θ is the temperature deviation defined by $\theta = T - T_o$ with the condition $|\theta| << T_o$, where T_o (>0) is the reference temperature. η_o is the specific entropy (per unit mass) of the material at a stress-free reference configuration with the reference temperature T_o. ρ_o is the mass density of the material, which is supposed to be a constant in the infinitesimal deformation approximation, and C_v is the specific heat per unit mass. C_{ijkl} and β_{ij} are respectively the elastic moduli and the thermal moduli. All material coefficients defined here are only functions of the position vector **x** if the material is inhomogeneous. Otherwise they are all constants. In addition, the material coefficients C_{ijkl}, β_{ij}, κ_{ij} and σ_{ij} obviously satisfy their symmetric conditions:

$$C_{ijkl} = C_{klij} = C_{jikl} = C_{ijlk}, \quad \beta_{ij} = \beta_{ji}, \quad \kappa_{ij} = \kappa_{ji}, \quad \sigma_{ij} = \sigma_{ji}. \tag{1.15.26}$$

In particular, in the case of isotropy, we have

$$C_{ijkl} = \lambda\delta_{ij}\delta_{kl} + G(\delta_{ik}\delta_{jl} + \delta_{il}\delta_{jk}) \tag{1.15.27a}$$

$$\beta_{ij} = \beta\delta_{ij}, \quad \kappa_{ij} = \kappa\delta_{ij}, \quad \Gamma_{ij} = \Gamma\delta_{ij}, \quad \sigma_{ij} = \sigma\delta_{ij} \tag{1.15.27b}$$

where λ and G are also called Lame's constants.
The mechanical motion equation (1.15.1) becomes

$$\nabla\cdot\mathbf{t} + \mathbf{J}\times\mathbf{B} = \rho_o\frac{\partial^2\mathbf{u}}{dt^2} \tag{1.15.28}$$

in the absence of mechanical body forces. The stress interface condition at the interface S between two material media is

$$\mathbf{n}\cdot[\,\mathbf{t} + \mathbf{t}^{em}\,] = 0 \qquad \text{on S} \tag{1.15.29}$$

in which the electromagnetic stress $\mathbf{t}^{em}$ is defined by eqn.(1.13.2). The stress interface condition (1.15.29) may be reduced to

$$\mathbf{n}\cdot[\,\mathbf{t}\,] = 0 \qquad \text{on S} \tag{1.15.30}$$

if no free surface currents are present at the material interface since, in such a case, the electromagnetic traction is continuous across the material interface (including the boundary surface between the material body and free space).

On the boundary ∂V ($\equiv \partial V_t + \partial V_u$) of the material body, the mechanical boundary conditions are

$$\mathbf{n} \cdot \mathbf{t} = \mathbf{t}^{o(n)} + \mathbf{n} \cdot [\, \mathbf{t}^{em} \,] \qquad \text{on } \partial V_t \tag{1.15.31}$$

$$\mathbf{u} = \mathbf{u}^{o} \qquad \text{on } \partial V_u \tag{1.15.32}$$

where $\mathbf{t}^{o(n)}$ and $\mathbf{u}^{o}$ are respectively the prescribed surface traction on ∂V_t and the prescribed displacement on ∂V_u. The second term on the r.h.s. of eqn.(1.15.31) vanishes in the absence of free surface currents at the boundary of the material body. We shall discuss further the problems concerning surface currents when electrodynamics of superconductors are studied.

The heat conduction equation (1.15.6) may now be expressed as

$$\rho_o C_v \frac{\partial \theta}{\partial t} = (\kappa_{ij}\theta_{,j})_{,i} - T_o(\Gamma_{ij}E'_j)_{,i} - T_o\beta_{ij}\frac{\partial \varepsilon_{ij}}{\partial t} + J_i E'_i \tag{1.15.33}$$

with the interface condition for heat flux

$$\mathbf{n} \cdot [\, \mathbf{q} + \mathbf{E}' \times \mathbf{H} \,] = 0 \qquad \text{on S} \tag{1.15.34}$$

where we have assumed that the material velocity is continuous across the material interface. In the absence of free surface currents at the material interface, eqn.(1.15.34) is reduced to

$$\mathbf{n} \cdot [\, \mathbf{q} \,] = 0 \qquad \text{on S.} \tag{1.15.35}$$

On the boundary ∂V ($\equiv \partial V_q + \partial V_\theta$) of the material body where no free surface currents present, the boundary conditions for heat flux and temperature may be written as

$$\mathbf{q} \cdot \mathbf{n} = q^{(n)} \qquad \text{on } \partial V_q \tag{1.15.36}$$

$$\theta = \theta^{o} \qquad \text{on } \partial V_\theta \tag{1.15.37}$$

where $q^{(n)}$ and θ^o are both prescribed on the part of the boundary ∂V_q and ∂V_θ respectively.

Another important condition, replacing any one of (1.15.36) and (1.15.37), is concerned with heat convection from a part ∂V_r of the boundary surface ∂V, which may be expressed by

$$\mathbf{q}\cdot\mathbf{n} + h(\theta - \theta_1) = 0 \qquad \text{on } \partial V_r \tag{1.15.38}$$

where h (≥ 0) is termed the boundary conductance (or the surface heat transfer coefficient) being an appropriate function of the surface coordinates and θ_1 the known temperature outside of the body near ∂V_r. Maxwell's equations for studying the non-magnetic conducting thermoelastic solids at the magneto-quasistatic approximation are given by eqns.(1.15.10)-(1.15.13) with $\mathbf{B} = \mu_o\mathbf{H}$ and with the electromagnetic interface conditions given by eqn.(1.15.14)-(1.15.17).

15.3 Soft ferromagnetic elastic insulators

The next simplified theory to be introduced here is concerned with soft ferromagnetic elastic insulators. For insulating soft ferromagnetic elastic solids, by neglecting thermal effects by assuming that the solid is kept at a certain constant and uniform temperature, the constitutive equations taking into account of magnetostrictive effect may be given by

$$t_{ij} = C_{ijkl}\varepsilon_{kl} + \Pi_{ijkl}M_kM_l + \mu_oM_jH_i \tag{1.15.39}$$

$$\mu_oH_i = \alpha_{ij}M_j + 2\Pi_{klij}\varepsilon_{kl}M_j \tag{1.15.40}$$

where $\mathbf{M}$ is the magnetization vector per unit volume. α_{ij} and Π_{klij} are respectively the magnetic anisotropic coefficient tensor and the magnetostrictive coefficient tensor, which obviously satisfy the following symmetrical conditions

$$\alpha_{ij} = \alpha_{ji} \qquad \text{and} \qquad \Pi_{klij} = \Pi_{lkij} = \Pi_{klji}\,. \tag{1.15.41}$$

These material coefficients are all constants if the material is homogeneous at the constant temperature.

The mechanical motion equation may now be written as

$$\nabla \cdot \mathbf{t} + \mu_0 (\mathbf{M} \cdot \nabla) \mathbf{H} = \rho_0 \frac{\partial^2 \mathbf{u}}{dt^2} \tag{1.15.42}$$

in the absence of mechanical body forces, and the stress interface condition reads

$$\mathbf{n} \cdot [\, \mathbf{t} + \mathbf{t}^{em} \,] = 0 \qquad \text{on } S \tag{1.15.43}$$

where the electromagnetic stress $\mathbf{t}^{em}$ is defined by eqn.(1.13.9).

On the boundary ∂V ($\equiv \partial V_t + \partial V_u$) of the material body, the mechanical boundary conditions can be expressed by

$$\mathbf{n} \cdot \mathbf{t} = \mathbf{t}^{o(n)} + \frac{1}{2} \mu_0 (\mathbf{M} \cdot \mathbf{n})^2 \, \mathbf{n} \qquad \text{on } \partial V_t \tag{1.15.44}$$

$$\mathbf{u} = \mathbf{u}^o \qquad \text{on } \partial V_u \tag{1.15.45}$$

where $\mathbf{t}^{o(n)}$ and $\mathbf{u}^o$ are respectively the prescribed surface traction on ∂V_t and the prescribed displacement on ∂V_u.

Maxwell's equations for the study of the soft ferromagnetic elastic insulators at the magneto-quasistatic approximation are

$$\nabla \times \mathbf{H} = 0 \tag{1.15.46}$$

$$\nabla \cdot \mathbf{B} = 0 \tag{1.15.47}$$

with $\mathbf{B} = \mu_0 (\mathbf{H} + \mathbf{M})$, and with the interface conditions

$$\mathbf{n} \cdot [\, \mathbf{B} \,] = 0 \tag{1.15.48}$$

$$\mathbf{n} \times [\, \mathbf{H} \,] = 0. \tag{1.15.49}$$

1.15.4 **Soft ferromagnetic thermoelastic conductors**

In the case of conducting soft ferromagnetic thermoelastic solids neglecting the magnetostrictive effect, the linearized constitutive equations are given from eqns.(1.14.19)-(1.14.22) by

$$t^s_{ij} = C_{ijkl}\varepsilon_{kl} - \beta_{ij}\theta \tag{1.15.50}$$

$$M_i = \chi^m_{ij} H_j + \frac{1}{\mu_o}\pi^m_i \theta \tag{1.15.51}$$

$$\eta = \eta_o + \frac{C_v}{T_o}\theta + \frac{1}{\rho_o}\beta_{ij}\varepsilon_{ij} + \frac{1}{\rho_o}\pi^m_i H_i \tag{1.15.52}$$

together with the generalized Fourier-Ohm's law by

$$q_i = - \kappa_{ij}\theta_{,j} + T_o\Gamma_{ij}E'_j \tag{1.15.53}$$

$$J_i = \sigma_{ij}E'_j - \Gamma_{ij}\theta_{,j}\,. \tag{1.15.54}$$

Here, π^m_i denote the pyromagnetic coefficients, and χ^m_{ij} the magnetic susceptibilities which obviously satisfy the symmetrical condition $\chi^m_{ij}=\chi^m_{ji}$. All other coefficients have already been defined. These material coefficients are all constant if the material is homogeneous. By eqn.(1.15.50), we have introduced a symmetric stress tensor t^s_{ij} which is related to Cauchy's stress tensor t_{ij} by $t^s_{ij} \equiv t_{ij} + \mu_o M_i H_j$. As to the generalized Fourier-Ohm's law, in some cases, one may simply ignore the thermoelectric effect by setting $\Gamma_{ij} = 0$.

The mechanical motion equation is

$$\nabla\cdot(\mathbf{t}^s + \mathbf{t}^M) = \rho_o\frac{\partial^2\mathbf{u}}{dt^2} \tag{1.15.55}$$

in the absence of mechanical body forces, and the stress interface condition reads

$$\mathbf{n}\cdot[\,\mathbf{t}^s + \mathbf{t}^M\,] = 0 \qquad \text{on } S \tag{1.15.56}$$

where the electromagnetic stress $\mathbf{t}^M$ is defined here by

$$t_{ij}^{M} = \mu_o[H_iH_j - \frac{1}{2}(H_kH_k)\delta_{ij}] \tag{1.15.57}$$

which is also symmetric and has the similar form to the well-known Maxwell's (magnetic) stress tensor in free space.

On the boundary ∂V ($\equiv \partial V_t + \partial V_u$) of the material body, the mechanical boundary conditions can be expressed by

$$\mathbf{n} \cdot \mathbf{t}^s = \mathbf{t}^{o(n)} + \mu_o(\mathbf{M}\cdot\mathbf{n})\mathbf{H}^- + \frac{1}{2}\mu_o(\mathbf{M}\cdot\mathbf{n})^2\,\mathbf{n} \qquad \text{on } \partial V_t \tag{1.15.58}$$

$$\mathbf{u} = \mathbf{u}^o \qquad \text{on } \partial V_u \tag{1.15.59}$$

where $\mathbf{t}^{o(n)}$ and $\mathbf{u}^o$ are respectively the prescribed surface traction on ∂V_t and the prescribed displacement on ∂V_u. Note that free surface currents are supposed to be absent in the material considered.

The heat conduction equation is

$$\rho_o C_v \frac{\partial \theta}{\partial t} = (\kappa_{ij}\theta_{,j})_{,i} - T_o(\Gamma_{ij}E'_j)_{,i} - T_o\beta_{ij}\frac{\partial \varepsilon_{ij}}{\partial t} - T_o\pi_i^m\frac{\partial H_i}{\partial t} + J_iE'_i \tag{1.15.60}$$

with the interface condition for heat flux

$$\mathbf{n} \cdot [\,\mathbf{q} + \mathbf{E}' \times \mathbf{H}\,] = 0 \qquad \text{on } S \tag{1.15.61}$$

where we have assumed that the material velocity is continuous across the material interface.

On the boundary ∂V ($\equiv \partial V_q + \partial V_\theta$) of the material body, the boundary conditions for heat flux and temperature may be written as

$$\mathbf{q}\cdot\mathbf{n} = q^{(n)} + \mathbf{n} \cdot [\,\mathbf{E}' \times \mathbf{H}\,] \qquad \text{on } \partial V_q \tag{1.15.62}$$

$$\theta = \theta^o \qquad \text{on } \partial V_\theta \tag{1.15.63}$$

where $q^{(n)}$ and θ^o are both prescribed on the part of the boundary ∂V_q and ∂V_θ respectively. The second term on the r.h.s. of eqn.(1.15.62) will vanish if one is considering an isotropic solid neglecting the thermoelectric effect by noting the fact that the tangential component of the magnetic intensity field **H** and the electric field vector **E**' are now both continuous across the boundary of the material body due to $\mathbf{n}\times[\![\mathbf{E}']\!] = 0$ and $\mathbf{E}'\cdot\mathbf{n} = 0$ at the boundary.

Maxwell's equations for the study of the conducting soft ferromagnetic thermoelastic solids at the magneto-quasistatic approximation are given by eqns.(1.15.10)-(1.15.13) together with the electromagnetic interface conditions by eqn.(1.15.14)-(1.15.17) where the free surface current $\mathbf{K}_f$ is here set to zero.

1.15.5 **Linear thermopiezoelectric solids**

The last simplified theory to be presented here is concerned with the linear thermopiezoelectric solids, for which the linearized constitutive equations may be expressed by

$$t_{ij} = C_{ijkl}\varepsilon_{kl} - e_{k\cdot ij}E_k - \beta_{ij}\theta \qquad (1.15.64)$$

$$P_i = \chi^e_{ij} E_j + e_{i\cdot kl}\varepsilon_{kl} + \pi^e_i\theta \qquad (1.15.65)$$

$$\eta = \eta_o + \frac{C_v}{T_o}\theta + \frac{1}{\rho_o}\beta_{ij}\varepsilon_{ij} + \frac{1}{\rho_o}\pi^e_i E_i \qquad (1.15.66)$$

together with the Fourier law by

$$q_i = -\kappa_{ij}\theta_{,j} \qquad (1.15.67)$$

where π^e_i denote the pyroelectric coefficients. χ^e_{ij} and $e_{i\cdot kl}$ are respectively the electric susceptibilities and the piezoelectric coefficients, which obviously satisfy the following symmetrical conditions

$$\chi^e_{ij} = \chi^e_{ji} \qquad \text{and} \qquad e_{i\cdot kl} = e_{i\cdot lk}\ . \qquad (1.15.68)$$

All other coefficients have been already defined above. These material coefficients are all constants if the material is homogeneous. Obviously, the piezoelectric phenomena occur only in dielectric solids presenting no center of symmetry. In contrast to piezoelectricity, piezo-magnetism is seldom encountered in nature since in such materials certain types of magnetic structures have to exist and adequate magnetic symmetric conditions must be met, which is rarely the case (see Landau et al. (1984)).

The linearized mechanical motion equation for the thermopiezoelectric solid is

$$(C_{ijkl}u_{k,l})_{,j} - (e_{k \cdot ij}E_k)_{,j} - (\beta_{ij}\theta)_{,j} = \rho_0 \frac{\partial^2 u_i}{dt^2} \qquad (1.15.69)$$

in the absence of mechanical body forces, and the stress interface condition is

$$\mathbf{n} \cdot [\,\mathbf{t}\,] = 0 \qquad \text{on } S\,. \qquad (1.15.70)$$

On the boundary ∂V ($\equiv \partial V_t + \partial V_u$) of the thermopiezoelectric body, the mechanical boundary conditions can be expressed by

$$\mathbf{n} \cdot \mathbf{t} = \mathbf{t}^{o(n)} \qquad \text{on } \partial V_t \qquad (1.15.71)$$

$$\mathbf{u} = \mathbf{u}^o \qquad \text{on } \partial V_u\,. \qquad (1.15.72)$$

The linearized heat conduction equation is

$$\rho_0 C_v \frac{\partial \theta}{\partial t} = (\kappa_{ij}\theta_{,j})_{,i} - T_0 \beta_{ij} \frac{\partial \varepsilon_{ij}}{\partial t} - T_0 \pi_i^e \frac{\partial E_i}{\partial t} \qquad (1.15.73)$$

in the absence of internal heat sources.

The interface condition for heat flux is

$$\mathbf{n} \cdot [\,\mathbf{q}\,] = 0 \qquad \text{on } S \qquad (1.15.74)$$

where we have assumed that the material velocity is continuous across the material interface.

On the boundary ∂V ($\equiv \partial V_q + \partial V_\theta$) of the material body, the heat flux boundary conditions read

$$\mathbf{q} \cdot \mathbf{n} = q^{(n)} \qquad \text{on } \partial V_q \qquad (1.15.75)$$

$$\theta = \theta^{o} \qquad \text{on } \partial V_\theta \tag{1.15.76}$$

where $q^{(n)}$ and θ^{o} are both prescribed on the part of the boundary ∂V_q and ∂V_θ respectively. Similarly we may also have the boundary condition of heat convection as shown by eqn.(1.15.38).

Maxwell's equations at the electro-quasistatic approximation are

$$\nabla \times \mathbf{E} = 0 \tag{1.15.77}$$

$$\nabla \cdot \mathbf{D} = 0 \tag{1.15.78}$$

with $\mathbf{D} = \varepsilon_0\mathbf{E} + \mathbf{P}$, and with the electric interface conditions

$$\mathbf{n} \cdot [\,\mathbf{D}\,] = 0 \tag{1.15.79}$$

$$\mathbf{n} \times [\,\mathbf{E}\,] = 0 \tag{1.15.80}$$

where we have assumed that free charges are absent in the thermopiezoelectric body considered. Furthermore, we may introduce a scalar electric potential function ϕ defined by

$$\mathbf{E} = -\nabla\phi \tag{1.15.81}$$

so that eqn.(1.15.77) is identically satisfied, and eqn.(1.15.78) may now be expressed by using the constitutive equation (1.15.65) as

$$(\varepsilon^{e}_{ij}\,\phi,_{j} - e_{i\cdot kl}u_{k,l} - \pi^{e}_{i}\theta),_{i} = 0 \tag{1.15.82}$$

where $\varepsilon^{e}_{ij} = \varepsilon_0\delta_{ij} + \chi^{e}_{ij}$ are called the permittivities of the dielectric solid.

The set of five linearized equations (1.15.69), (1.15.73) and (1.15.82), with the aid of eqn.(1.15.81), constitute the complete set of field equations of thermopiezoelectricity for the determination of the five unknowns u_i, θ and ϕ. Of course, in all above theories, initial conditions have to be specified if dynamic problems are involved.

We shall now conclude the theories for studying electromagneto-thermoelastic interaction in classical solid materials. Other types of approximate material models of possible interest in engineering applications may be worked out systematically by the reader from those general equations given in the above sections. The problems concerning hard ferromagnetic materials for which the exchange interaction and the saturation effect etc. are important in theoretical modelling will be discussed later in the study of ferromagnetic superconductors.

Chapter 2

Electrodynamics of superconductors

The discovery of superconductivity and its recent advances in producing high temperature superconducting materials have foretold that superconductivity technology will have an enormous impact on industry and our society. At present, the greatly increased interest of world industry in the engineering applications of superconductivity technology are requiring basic researches on the applied electromagnetics and mechanics of superconducting materials in order to develop the new technology and to support its applications. In particular, a complete appreciation of the macroscopic electromagnetic behavior of superconductors is needed before one can undertake the most basic engineering calculations for any concrete problem concerning the applications.

The purpose of this chapter is, therefore, to introduce systematically the basic concepts of superconductivity and electrodynamic models for the study of superconductors. Emphasis will be put on the presentation of phenomenological theories, such as the London theory, Ginzburg-Landau theory and the electrodynamic theory of Josephson junctions. Problems concerning two-fluid model accounting for the effect of normal electrons in a.c electrodynamics of superconductors, the G-L theory of superconducting thin films, the Abrikosov's flux structures, and the London-Bean model for a.c. losses are studied. Examples illustrating these theoretical models are chosen and their mathematical calculations will be carried out carefully. Finally, a brief introduction is given on the micromechanism of superconductivity and BCS theory. In the text, SI units will be used throughout, which are thought to be particularly convenient for electrical engineers and applied physicists.

2.1 Introduction to superconducting phenomena

2.2.1 **Zero resistance and transition temperature**

The discovery of superconductivity started from the finding of Kamerlingh Onnes in 1911 that the resistance of mercury has an abrupt drop at a temperature of 4.2 °K and has practically a zero dc-resistance value at temperatures below 4.2 °K (see **Figure** 2.1.1) This new phenomenon of zero-resistance at low temperature was soon found in many other metals and alloys. An important characteristic of the loss of dc-resistance observed is the sharpness of the transition. Careful experiments by de Hass and Voogd (1931) suggested that in "ideal' conditions the transition from the normal resistance state of the testing material to its superconducting state would be practically discontinuous. The temperature at which superconductivity first occurs in a material is thus termed the critical (or transition) temperature of the material and is denoted by T_c. The transition temperatures of some important materials are, for instance, 9.4 °K for Niobium, 7.19 °K for Lead, 3.72 °K for Tin and 1.20 °K for Aluminum, etc. The highest transition temperature of superconductors did not exceed 23 °K found in the compound Nb_3Ge until 1986 when two scientists, J.G. Bednorz and K.A. Müller (1986) reported possible superconductivity in a mixture of La and Ba copper oxides at temperature of 30 °K.

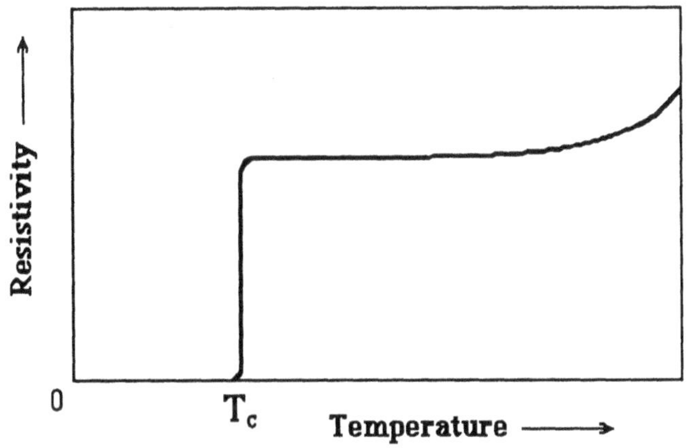

Figure 2.1.1 Loss of resistance of a superconductor at low temperature.

Following their discovery, which led them to receive the 1987 Nobel prize in Physics, the high-T_c superconductors have, so far, been found in various ceramic oxides having transition temperatures as high as 125 °K. The importance of these discoveries is based on the fact that the superconductivity in such ceramics can be maintained in cryostats with relatively cheap liquid nitrogen (boiling point 77.4 °K) rather than liquid helium so that large-scale applications of superconductivity become economically viable.

Practical measurement of the transition temperature T_c of a superconductor is usually made with using the following three basic methods. The first one is to measure the change in resistance of the material. For pure samples containing few metallurgical defects, the transition is sharp, with a typical width less than 0.01 °K, and changes little with current level. Inhomogeneous samples, however, have broad transitions. In this case, T_c is usually defined as the point where the resistivity of the testing material decreases to half its normal value. The primary problem with the resistance method is the current-dependent shift of T_c for the inhomogeneous samples. Therefore, the resistance method is best suited to homogeneous samples. The second method is to measure the change in magnetic permeability of the material, which is based on the large decrease in magnetic permeability of the material as it passes into the superconducting state. The advantage of this method is that it requires no direct electrical contacts with the sample, and consequently, it can be used for samples which are small, irregular or in powder form. However, this method has its disadvantage in that the highest T_c component of an inhomogeneous sample can shield lower T_c components, thus dominating the measurement. The third method utilizes the fact that the specific heat of a superconductor shows a discontinuity at the transition temperature T_c.

Though the transition temperature seems to be unaffected by frequencies, it has been found that the behavior of zero dc-resistance of a superconductor at temperatures below its critical temperature is modified significantly at very high frequencies of alternating current up to 100 MHz, as shown by London (1940). When the frequency increased to the infra-red frequencies of the order of 10^{13} Hz, the resistance of the superconductors would be the same and independent of temperature both in the normal and superconducting state (Shoenberg (1952)).

2.1.2 **Critical magnetic field**

Besides the discovery of zero dc-resistance of superconductors, Kamerlingh Onnes (1914) also found another important property of superconductors, namely, that when the superconductor is placed in a sufficiently strong magnetic field, superconductivity can be destroyed. The superconductivity, however, reappears when the applied magnetic field is removed. The minimum magnetic field required to destroy the superconductivity depends on the shape and orientation of the specimen at a given temperature. If the specimen is in the shape of a long cylinder and its axis is placed parallel to the applied magnetic field, the transition is sharp and the minimum magnetic field required to destroy the superconductivity is called the critical (magnetic) field and denoted by H_c.

Within only small deviations, the temperature dependence of H_c was found to be well represented by a parabolic relation (see **Figure** 2.1.2)

$$H_c(T) = H_0[1 - (\frac{T}{T_c})^2] . \qquad (2.1.1)$$

where H_0 denotes the critical field at zero temperature and T_c the transition temperature, which are properties of superconductors. For most superconducting elements, B_0 $(=\mu_0 H_0)$ is of the order of 10^{-2} Tesla. Some values are, for instance, $B_0 = 10^{-2}$ Tesla and T_c=1.2 °K

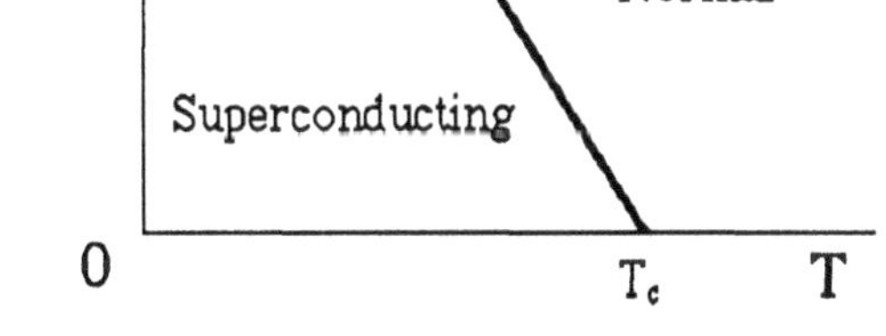

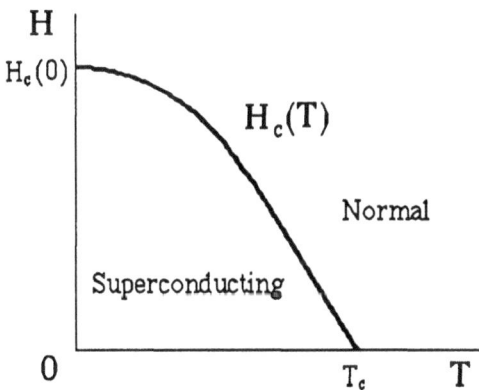

Figure 2.1.2 The critical magnetic field as a function of temperature.

for Aluminium, $B_0 = 8 \times 10^{-2}$ Tesla and $T_c = 7.2$ °K for Lead, and $B_0=2.7\times10^{-2}$ Tesla and $T_c = 3.4$ °K for Indium, etc. (Shoenberg (1952) and Rose-Innes and Rhoderick (1969)). However, in some metallic compounds and alloys, for example, Nb_3Sn, the magnetic field required to destroy the superconductivity can be as high as 10 Tesla. In these materials the transition in the field is usually not very sharp. The practical determination of the critical field is usually from the measurements of the superconducting magnetization curves, and the critical field is simply the field at which the magnetization becomes zero.

2.1.3 **Meissner effect**

For many years after the discovery of superconductivity, it was tacitly assumed that the electromagnetic behavior of a superconductor could be deduced from its infinite conductivity (i.e. zero resistivity). According to classical theory of electrodynamics, the infinite conductivity implies that the electric field **E** is zero in a superconductor due to Ohm's law. It follows that $\nabla \times \mathbf{E} = 0$ and hence by one of Maxwell's equations (1.5.18) that $\partial \mathbf{B}/\partial t = 0$, which means that the magnetic induction **B** inside is constant in time, and would be, therefore, dependent on the past history of the state of the superconductor. For instance, the final state of a perfect conductor depends on whether the material is cooled first and then the magnetic field is applied, or vice versa. This may be understood by the following an example. We suppose that a specimen becomes a perfect conductor at low temperature in the absence of any magnetic field, and that a magnetic field is then applied. Thus, since the magnetic flux cannot change in the perfect conductor, the magnetic field inside the specimen must remain zero even after the application of the magnetic field. Next, we consider that the specimen is at first supposed to be embedded in a magnetic field, and that the specimen is then cooled to a low temperature, at which it becomes a perfect conductor. Thus, because the magnetic flux cannot change in the perfect conductor, the magnetic field inside the specimen must remain as it was even when the applied magnetic field is removed. This means that a magnetic field may exist inside a perfect conductor.

Such a property of superconductors and its consequences were taken for granted until 1933 when Meissner and Ochsenfeld (1933) measured the field surrounding a superconductor and concluded that the magnetic induction field **B** inside a macroscopic specimen of pure superconductor would always be zero (strictly speaking, experiment showed that it was very small compared with its value in the normal state), independent of initial conditions (see **Figure** 2.1.3). Here, the macroscopic specimen means that its size is large enough so that size effects are not important since, otherwise, the magnetic behavior would be seriously modified if the specimen size becomes comparable to the penetration depth ($\sim 10^{-7}$ m), which we shall discuss later. The absence of any magnetic flux in a pure superconductor independent of the initial conditions now known as the Meissner effect is an additional fundamental property of the superconductors since it cannot be deduced from the perfect conductivity. This means that the superconductor behaves, in practice, not simply like a perfect conductor, which would imply only $\partial \mathbf{B}/\partial t = 0$, but also like a perfect diamagnetic ($\mathbf{B}=0$). In the thermodynamic sense, the Meissner effect indicates that the superconductive state in a given external magnetic field is a single stable state to which the laws of thermodynamics apply, and the magnetically induced transition between the superconducting and normal states of a superconductor is, in principle, a reversible phase transition.

The fascinating phenomena of superconductivity and their potential applications had attracted not only many experimentalists trying to find new superconductors, but also many theoreticians aiming to understand the physical phenomena of superconductivity and to describe the electromagnetic and thermodynamic behavior of superconductors. Extensive theoretical studies have been made along the two

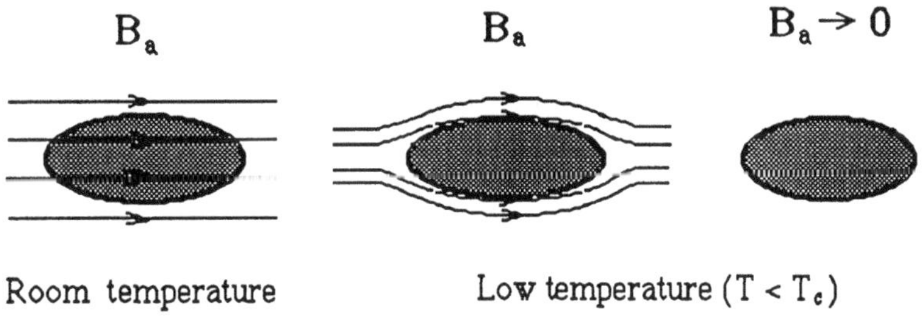

Figure 2.1.3 The Meissner effect of a superconductor.

main lines: the microscopic theories were developed to study the possible mechanism of superconductivity and its properties while the macroscopic theories were introduced to describe phenomenologically the macroscopic behaviors of the superconductors. In the following section, we shall start with the introduction of a well-known macroscopic theory of superconductors, developed by London (1935, 1950).

2.2 London theory for superconductors in weak fields

Although the development of quantum mechanics in the 1920's led to the understanding of the normal process of electrical conduction in metals, the origins of superconductivity were to remain obscure until 1956. This situation encouraged the development of phenomenological theories, which had established that most superconducting phenomena can be derived from a small number of empirical postulates. The task of a microscopic theory of superconductivity is thus reduced to that of explaining these postulates. An early step towards a phenomenological theory of superconductivity was made by Gorter and Casimir (1934) in their two-phase fluid model. Later, the discovery of the Meissner effect led to the appearance of London's theory (1935), which is now being introduced as follows. It should be noted that, in this chapter, we shall only deal with non-magnetic superconductors on the assumption that their local magnetic moment does not exist or is negligible.

2.2.1 Free superelectron model

Before the discovery of Meissner effect, Becker et al. (1933) analyzed the electrodynamic behavior of a perfect conductor with the use of a simple free superelectron model, in which the superelectrons accelerate without any resistance under the electromagnetic fields **E** and **B**

$$m^* \frac{d\mathbf{v}_s}{dt} = e^*\mathbf{E} + e^*\mathbf{v}_s \times \mathbf{B} \qquad (2.2.1)$$

where m* and e* are respectively the effective mass and charge of the superelectrons (It was later found in BCS theory that the superelectrons are formed in Cooper pairs so that one has $e^* \approx 2e$, with e being the charge of one electron, and $m^* \approx 2m$ with m being the effective mass of one electron).

If there are n_s^* numbers of Cooper pairs of superelectrons per unit volume moving with the local velocity $\mathbf{v}_s$, one may introduce a supercurrent density $\mathbf{J}_s$ by

$$\mathbf{J}_s = n_s^* e^* \mathbf{v}_s = - n_s^* |e^*| \mathbf{v}_s \tag{2.2.2}$$

where the minus sign indicates that the supercurrent flows in a direction opposite to the direction of the motion of the superelectrons. Substituting this into eqn.(2.2.1) and taking a linear approximation, one can find

$$\frac{\partial \mathbf{J}_s}{\partial t} = \frac{n_s^* e^{*2}}{m^*} \mathbf{E} \, . \tag{2.2.3}$$

By taking the curl of both sides of eqn.(2.2.3) and using Maxwell's equations (1.5.18) for the superconducting body, one may then write

$$\frac{\partial \mathbf{B}}{\partial t} = - \frac{m^*}{n_s^* e^{*2}} \nabla \times \left(\frac{\partial \mathbf{J}_s}{\partial t}\right) \, . \tag{2.2.4}$$

inside the superconductor.

By noting that for non-magnetic superconductors, any flux density in the material body must be due to the currents, one can get from eqns.(2.2.4) and (1.5.16), neglecting displacement current

$$\frac{\partial \mathbf{B}}{\partial t} = - \frac{m^*}{\mu_0 n_s^* e^{*2}} \nabla \times \left(\nabla \times \frac{\partial \mathbf{B}}{\partial t}\right) \tag{2.2.5}$$

which may also be written, by noting the fact that the divergence of $\mathbf{B}$ is zero, as

$$\frac{\partial \mathbf{B}}{\partial t} = \lambda_L^2 \, \nabla^2 \left(\frac{\partial \mathbf{B}}{\partial t}\right) \tag{2.2.6}$$

where λ_L is a parameter of length dimension defined by

$$\lambda_L = \sqrt{\frac{m^*}{\mu_o n_s^* e^{*2}}} \qquad \text{(m).} \tag{2.2.7}$$

It can be seen that eqn.(2.2.6) predicts that the external field would fall off exponentially inside a perfect conductor to a "trapped" field $\mathbf{B}^o$, depending on the initial state of the superconductor. This prediction is clearly contradicted by the Meissner effect which states that the internal magnetic induction field in an ideal superconductor is always zero, independent of the initial states. Thus, H. and F. London (1935) suggested that the magnetic behavior of a superconducting metal might be correctly described according to the Meissner effect if eqn.(2.2.6) applied not only to $\partial \mathbf{B}/\partial t$ but to $\mathbf{B}$ itself, i.e.

$$\mathbf{B} = \lambda_L^2 \nabla^2 \mathbf{B} \, . \tag{2.2.8}$$

Equivalently, from eqn.(2.2.4), one may also write

$$\mathbf{B} = -\mu_o \lambda_L^2 \nabla \times \mathbf{J}_s \tag{2.2.9}$$

which is now called the second London equation, describing the Meissner effect of the superconductor. It is noticed that in the London theory the parameter λ_L, characterizing the superconductor, is assumed to be constant in space. In particular, λ_L is assumed to be independent of the strength of the applied magnetic field and also of the dimensions of the specimen though its value may depend on temperatures.

By introducing a magnetic vector potential $\mathbf{A}$ defined by $\mathbf{B} = \nabla \times \mathbf{A}$ and by choosing a proper gauge (London gauge (London (1950))), one may also write from eqn.(2.2.9)

$$\mathbf{J}_s = -\frac{1}{\mu_o \lambda_L^2} \mathbf{A} \tag{2.2.10}$$

which gives a special form of the second London equation, being somewhat reminiscent of Ohm's law. It should, however, be noticed that the relation (2.2.10) is valid only for problems involving simply connected superconductors due to the specially chosen gauge ($\nabla\cdot\mathbf{A} = 0$ and $\mathbf{A}\cdot\mathbf{n} = 0$ on the boundary surface of the superconductor with $\mathbf{n}$ being the unit normal vector).

Eqn.(2.2.3), namely

$$\frac{\partial \mathbf{J}_s}{\partial t} = \frac{1}{\mu_o \lambda_L^2}\mathbf{E} \tag{2.2.11}$$

is called the first London equation, describing the resistanceless property of a superconductor. This equation shows that there is no electric field in the superconductor unless the current is changing. Eqns.(2.2.3) and (2.2.9) together describe the electrodynamics of the superconductor in the London theory, and they are known as the London equations.

2.2.2 **Boson-gas model**

The second London equation (2.2.9) may also be derived by using a Boson-gas model (Duzer and Turner (1981)). A fundamental assumption used in this model is that a macroscopic (ensemble average) wave function with a well-defined amplitude and phase angle can be used to represent the condensed electron-pair fluid. The macroscopic wave function is expressed by

$$\psi = |\psi| e^{i\theta} \tag{2.2.12}$$

where θ is a scalar function of position, and $|\psi|$ the amplitude of the wave function related to the density n_s^* of Cooper pairs of superelectrons by

$$|\psi|^2 = n_s^* \ . \tag{2.2.13}$$

The fraction of the superelectrons n_s^* is assumed to be independent of position. Problems concerning the spatial variation of distribution of superelectrons will be studied in later sections.

The classical canonical momentum for a particle of charge e^* and mass m^* with velocity $\mathbf{v}_s$ in the presence of a magnetic field can be given by

$$\mathbf{p} = m^*\mathbf{v}_s + e^*\mathbf{A} . \tag{2.2.14}$$

Thus, for a distribution of the Cooper pairs with the density n_s^*, all having the same momentum, we may write the momentum density as

$$n_s^*\mathbf{p} = n_s^*(m^*\mathbf{v}_s + e^*\mathbf{A}) . \tag{2.2.15}$$

Using the quantum-mechanical description in which $n_s^*\mathbf{p}$ is the expected value of the canonical-momentum operator $-i\hbar\nabla$, operating on the pair-fluid wave function (2.2.12), we can get the following relation

$$\hbar\nabla\theta = m^*\mathbf{v}_s + e^*\mathbf{A} \tag{2.2.16}$$

which may also be written, by eqn.(2.2.2), as

$$\frac{\hbar}{e^*}\nabla\theta = \mu_0\lambda_L^2\,\mathbf{J}_s + \mathbf{A} \tag{2.2.17}$$

where $\hbar = 1.054 \times 10^{-34}$ joule-sec is called Plank's constant (Dirac's notation). The second London equation (2.2.9) then follows immediately by taking the curl of both sides of eqn.(2.2.17) and noting $\mathbf{B} = \nabla \times \mathbf{A}$.

We may recall that in electromagnetic theory the magnetic vector potential $\mathbf{A}$ as well as the electric scalar potential ϕ are not unique quantities, but are only defined in terms of their derivatives relating to the physical quantities of electric and magnetic fields by

$$\mathbf{B} = \nabla \times \mathbf{A} \qquad \text{and} \qquad \mathbf{E} = -\nabla\phi - \frac{\partial \mathbf{A}}{\partial t} . \tag{2.2.18}$$

Thus, any transformation (called a gauge transformation) of the form

$$\mathbf{A}' = \mathbf{A} + \nabla\chi \qquad \text{and} \qquad \phi' = \phi - \frac{\partial \chi}{\partial t} \tag{2.2.19}$$

leaves the field quantities unchanged. Here, χ is a mathematical function that allows the gauge for the electric and magnetic potentials to be changed such that the electromagnetic equations may be transformed into the most convenient form to be analyzed.

Since the gauge transformation must leave all measurable quantities unchanged, the gauge transformation of (2.2.19) will result in a necessary change of the local phase of the wave function (2.2.12) by

$$\theta' = \theta + \frac{e^*}{\hbar}\chi \tag{2.2.20}$$

in order to leave the velocity $\mathbf{v}_s$ (a measurable quantity) unchanged.

2.2.3 London theory of superconductors in alternating fields

The derived London equations do not replace Maxwell's equations, which, of course, still apply to all currents and the fields they produce. The London equations are additional conditions obeyed by the supercurrent. In the most general case the total current density $\mathbf{J}$ is the sum of a normal current density $\mathbf{J}_n$ and a supercurrent density $\mathbf{J}_s$

$$\mathbf{J} = \mathbf{J}_n + \mathbf{J}_s\,. \tag{2.2.21}$$

The normal current need only obey Maxwell's equations and Ohm's law, which for a stationary rigid isotropic conductive body, reads

$$\mathbf{J}_n = \sigma\mathbf{E}\,. \tag{2.2.22}$$

In the steady state, when fields and currents are not changing with time, the only current is the supercurrent, i.e. $\mathbf{J}_n = 0$, and we need only employ the London equations (2.2.9) and (2.2.11). For alternating fields, an electric field must be present to accelerate the electrons which have a small inertial mass and so the supercurrent does not rise instantaneously but only at the rate at which the electrons accelerate in the electric field, i.e. the supercurrent will lag behind the field because of the inertia of the superelectrons. Because there now is an electric field present, some of the current will be carried by the normal electrons. However, the inertia of electrons is very small and so, unless we go to extremely high frequencies, only a tiny fraction of the current is carried by the normal electrons and there is a correspondingly minute dissipation of power. If the frequency of an applied field is sufficiently high, however, a superconducting metal responds in the same way as a normal metal. The behavior of a superconductor at optical frequencies is, therefore, essentially no different from that of a normal metal.

To discuss quantitatively the time-varying cases, we may transform the set of electrodynamic equations for the superconductors into a convenient form to be analyzed. At first, we shall suppose that the superconductor is isotropic and homogeneous, and has the normal conductivity σ, permittivity ε and permeability μ_0. Thus, by means of eqns.(2.2.9), (2.2.11), (2.2.21), (2.2.22) and (1.5.18), we may write

$$-\mu_0\lambda_L^2\,\nabla\times\mathbf{J}_s = \mathbf{B} + \sigma\mu_0\lambda_L^2\frac{\partial\mathbf{B}}{\partial t} \tag{2.2.23}$$

and

$$\mu_0\lambda_L^2\frac{\partial\mathbf{J}}{\partial t} = \mathbf{E} + \sigma\mu_0\lambda_L^2\frac{\partial\mathbf{E}}{\partial t}\,. \tag{2.2.24}$$

From these two equations and Maxwell's equations (1.5.16)-(1.5.19), we may obtain the following four equations for the four fields **B**, **E**, **J** and ρ_e:

$$\nabla\times(\nabla\times\mathbf{B}) + \frac{1}{\lambda_L^2}\mathbf{B} + \sigma\mu_0\frac{\partial\mathbf{B}}{\partial t} + \varepsilon\mu_0\frac{\partial^2\mathbf{B}}{\partial t^2} = 0 \tag{2.2.25}$$

$$\nabla\times(\nabla\times\mathbf{E}) + \frac{1}{\lambda_L^2}\mathbf{E} + \sigma\mu_0\frac{\partial\mathbf{E}}{\partial t} + \varepsilon\mu_0\frac{\partial^2\mathbf{E}}{\partial t^2} = 0 \tag{2.2.26}$$

$$\nabla\times(\nabla\times\mathbf{J}) + \frac{1}{\lambda_L^2}\mathbf{J} + \sigma\mu_0\frac{\partial\mathbf{J}}{\partial t} + \varepsilon\mu_0\frac{\partial^2\mathbf{J}}{\partial t^2} = 0 \tag{2.2.27}$$

$$\frac{1}{\lambda_L^2}\rho_e + \sigma\mu_0\frac{\partial\rho_e}{\partial t} + \varepsilon\mu_0\frac{\partial^2\rho_e}{\partial t^2} = 0\,. \tag{2.2.28}$$

The equation for the total free charge density ρ_e is an ordinary differential equation in t alone with constant coefficients at a given temperature, and can thus be solved immediately in full generality by

$$\rho_e = A_1 e^{-\gamma_1 t} + A_2 e^{-\gamma_2 t} \tag{2.2.29}$$

where A_1 and A_2 are integration constants independent of time but arbitrary functions of space. γ_1 and γ_2 are given by

$$\gamma_1 = \frac{\sigma}{2\varepsilon}(1 + \sqrt{1 - \frac{4\varepsilon}{\sigma^2\mu_0\lambda_L^2}}) \approx \frac{\sigma}{2\varepsilon} \tag{2.2.30a}$$

$$\gamma_2 = \frac{\sigma}{2\varepsilon}(1 - \sqrt{1 - \frac{4\varepsilon}{\sigma^2\mu_0\lambda_L^2}}) \approx \frac{1}{\sigma\mu_0\lambda_L^2} . \tag{2.2.30b}$$

A relaxation time τ may be introduced by the slower of the two exponentials

$$\tau = \frac{1}{\gamma_2} \approx \sigma\mu_0\lambda_L^2 \tag{2.2.31}$$

which is usually on the order of 10^{-12} sec.

Thus, any free charges which might occur in the superconductor would disappear within this extremely short time. A quasistatic approximation may, therefore, be introduced to treat superconducting phenomena of low frequencies, less than 10^9 Hz. In the quasistatic case, eqns.(2.2.25)-(2.2.27) may be reduced to

$$\nabla^2\mathbf{B} = \frac{1}{\lambda_L^2}\mathbf{B} \tag{2.2.32}$$

$$\nabla^2\mathbf{E} = \frac{1}{\lambda_L^2}\mathbf{E} \tag{2.2.33}$$

$$\nabla^2\mathbf{J} = \frac{1}{\lambda_L^2}\mathbf{J} \tag{2.2.34}$$

together with the quasistatic form of the continuity equation of charges

$$\nabla \cdot \mathbf{J} = 0 \tag{2.2.35}$$

where $\mathbf{J}$ is the total current density, being the sum of the supercurrent density $\mathbf{J}_s$ and the normal current density $\mathbf{J}_n$ which satisfies the classical Ohm's law. Here, we have also used the fact that the divergence of electric field in the superconductor is zero, i.e. $\nabla \cdot \mathbf{E} = 0$ at the quasistatic approximation.

For problems of superconductors involving higher frequency with, however, still negligible displacement currents, we may have the following set of field equations for the superconductor

$$\nabla^2 \mathbf{B} = \frac{1}{\lambda_L^2} \mathbf{B} + \sigma \mu_o \frac{\partial \mathbf{B}}{\partial t} \tag{2.2.36}$$

$$\nabla^2 \mathbf{E} = \frac{1}{\lambda_L^2} \mathbf{E} + \sigma \mu_o \frac{\partial \mathbf{E}}{\partial t} \tag{2.2.37}$$

$$\nabla^2 \mathbf{J} = \frac{1}{\lambda_L^2} \mathbf{J} + \sigma \mu_o \frac{\partial \mathbf{J}}{\partial t} \tag{2.2.38}$$

with $\nabla \cdot \mathbf{E} = 0$.

To complete the theory, the interface conditions at the material interface between a superconductor and a normal conductor (or a dielectric, including free space) may be given by

$$\mathbf{n} \times [\,\mathbf{E}\,] = 0 \qquad \text{and} \qquad \mathbf{n} \cdot [\,\varepsilon \mathbf{E}\,] = \alpha_f \tag{2.2.39}$$

$$\mathbf{n} \cdot [\,\mathbf{B}\,] = 0 \qquad \text{and} \qquad \mathbf{n} \times [\,\mathbf{B}\,] = 0 \tag{2.2.40}$$

$$\mathbf{n} \cdot [\,\mathbf{J}\,] = 0\,. \tag{2.2.41}$$

Furthermore, at the material interface of two different superconductors, one has also the interface condition

$$\mathbf{n} \times [\,\lambda_L^2 \mathbf{J}\,] = 0\,. \tag{2.2.42}$$

It is worth mention that there really exist two ways of describing the electromagnetic behavior of the superconductor in the literature. To avoid any possible confusion, some discussion is necessary. In the first way as we have done in the presentation of the London theory, the two field quantities **E** and **B** are sufficient for the study of the electrodynamics of the superconductors since in the London theory we only deal with non-magnetic superconductors (note that magnetic superconducting materials have been recently found, which will, however, be studied in the next chapter). The Meissner effect is due to the induced superconducting currents, which are, in effect, screening currents serving to cancel the magnetic flux deep inside the superconductor.

In the second way, we regard the superconductor as perfect diamagnetic materials (if the penetration of the magnetic field is neglected) with, however, a modified current density expression. In this description, two more field quantities **H** and **M** are introduced. More specifically, the dc field equations for the superconductor are written in the second description by

$$\nabla \times \mathbf{H} = \mathbf{J} \tag{2.2.43a}$$

$$\nabla \times \mathbf{B} = 0 \tag{2.2.43b}$$

with $\mathbf{B} = \mu_0(\mathbf{H} + \mathbf{M})$ and $\mathbf{E} = 0$ in the superconductor, and with the interface conditions

$$\mathbf{n} \cdot [\![\mathbf{B}]\!] = 0 \tag{2.2.44a}$$

$$\mathbf{n} \times [\![\mathbf{H}]\!] = \mathbf{K}_f \tag{2.2.44b}$$

where $\mathbf{K}_f$ and $\mathbf{J}$ denote the external surface and volume current densities, such as the applied external current densities and the persistent current densities which may exist in multiply-connected superconductors. The magnetization $\mathbf{M} = -\mathbf{H}$ is used to describe the perfect diamagnetic effect (Meissner effect) in the superconductor if one may neglect the penetration of the magnetic field into its very thin layer near the surface of the superconductor. In particular, one may easily find that the magnetic intensity field **H** inside a long superconducting cylinder with the negligible demagnetizing effect is equal to the external magnetic intensity field $\mathbf{H}^e = \mathbf{B}^e/\mu_0$, applied in the direction parallel to the axis of the cylinder.

It can be seen that the difference between the two ways arises from what is meant by the symbol **J** for the current density in the relevant field equations. In the first way the current density **J** stands for all types of currents, the externally applied currents as well as the screening currents, while, in the second way, it stands only for the externally applied current densities (including the persistent currents such as the circulating current in a superconducting ring). The second description is particularly convenient to the study, for instance, of thermodynamic properties of the superconductors as we shall see in later sections. Finally it should be noticed that the London theory of superconductors is only valid for studying superconductors in weak magnetic fields so that the Meissner effect and the assumption of the independence of the penetration depth in applied magnetic fields applies.

2.3 Electrodynamic solutions of superconductors

2.3.1 Superconducting half-space in a static magnetic field

To illustrate the use of the London theory, we shall present here some simple analytical electrodynamic solutions of superconductors in weak fields. The first problem which we shall consider is a superconducting half-space in a uniform static magnetic field $\mathbf{B}^e$ applied in the y-direction at its surface x=0, as shown in **Figure** 2.3.1. To find the distribution of the magnetic field inside the superconducting half space, we may use eqn.(2.2.32), which reads

$$\frac{d^2B_y}{dx^2} = \frac{1}{\lambda_L^2} B_y \, . \tag{2.3.1}$$

This equation obviously has the solution

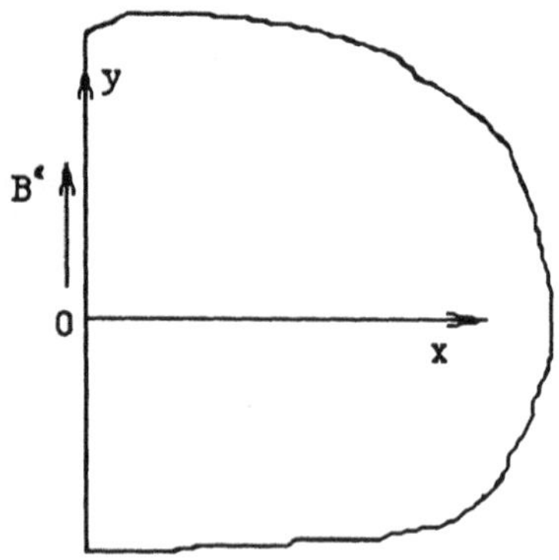

Figure 2.3.1 Magnetic field applied parallel to a superconducting half-space.

$$B_y(x) = B^e \exp(-x/\lambda_L) \tag{2.3.2}$$

which shows that the magnetic field in the superconducting half space decays exponentially with the increase of x, and that the measure of the penetration of the external field is by λ_L which accordingly is called the London penetration depth. This penetration depth is usually of the order of 10^{-7} m so that the exponential decay of the magnetic field is extremely rapid.

Because the penetration depth is so small, one usually does not notice the flux penetration in magnetic measurements on ordinary sized specimens, and these appear to be perfectly diamagnetic with $\mathbf{B} = 0$. This nearly complete flux exclusion is known as the Meissner effect.

The penetration of the magnetic flux becomes noticeable, however, if one makes measurements on small samples, such as thin films whose dimensions are not much greater than the penetration depth. In these cases, there is an appreciable magnetic flux density through the material, which means that there is no longer perfect diamagnetism and consequently the properties are rather different from those of the bulk superconductor. The special features of these thin films will be discussed later in section 2.10.

From Maxwell's equation $\mu_0 \mathbf{J}_s = \nabla \times \mathbf{B}$, we may also find the induced superconducting current density in the superconducting medium by

$$J_{sz} = \frac{1}{\mu_0}\frac{dB_y}{dx} = -\frac{B^e}{\mu_0 \lambda_L}\exp(-x/\lambda_L)\,. \tag{2.3.4}$$

This equation shows that the superconducting screen current flows also within a thin layer near the surface of the superconducting half-space.

2.3.2 **Superconducting cylinder carrying dc current**

The second problem which we shall consider is to find the superconducting current density distribution in an infinite cylinder with a radius R and carrying a total amount of dc current I flowing along its axis as shown in **Figure** 2.3.2. In this case, we can use eqn.(2.2.34), which in a cylindrical coordinate system is

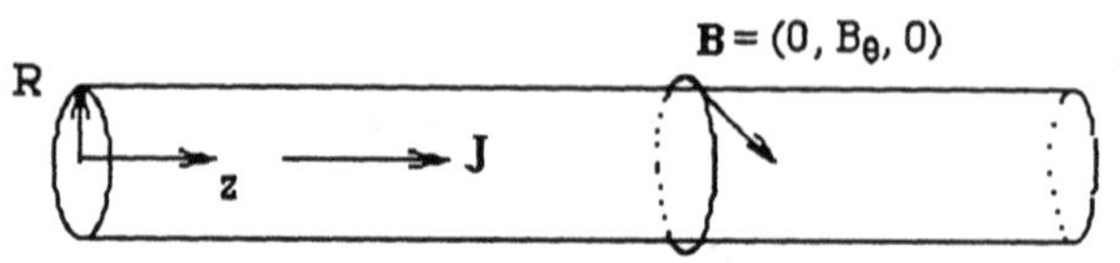

Figure 2.3.2 Current passed through an infinite cylindrical superconductor.

$$\frac{d^2J_z}{dr^2} + \frac{1}{r}\frac{dJ_z}{dr} - \frac{1}{\lambda_L^2}J_z = 0 . \tag{2.3.5}$$

A general solution of this equation can be found as

$$J_z(r) = CI_0(r/\lambda_L) \tag{2.3.6}$$

where C is a constant determined by the condition of total current

$$I = 2\pi \int_0^R J_z(r)rdr . \tag{2.3.7}$$

The result gives

$$J_z(r) = \frac{I}{2\pi R\lambda_L}\frac{I_0(r/\lambda_L)}{I_1(R/\lambda_L)} \qquad \text{for } 0 \le r \le R \tag{2.3.8}$$

where I_0 and I_1 are the modified Bessel functions of the first kind of the zeroth-order and of the first-order respectively.

Two limiting cases may be of interest, which are

$$J_z(r) \approx \frac{I}{2\pi R\lambda_L}\sqrt{\frac{R}{r}}\exp(-\frac{R-r}{\lambda_L}) \qquad \text{for } \lambda_L << r \le R \tag{2.3.9}$$

and

$$J_z(r) = \frac{I}{\pi R^2} \qquad \text{for } r \le R << \lambda_L \tag{2.3.10}$$

which shows a uniform current density distribution in the case of $R \ll \lambda_L$. The maximum current density in eqn.(2.3.9) is found to be $J_{zmax}=I/(2pR\lambda_L)$ at $r = R$. The magnetic field in the superconductor may then be found from the second London equation (2.2.9). In particular, at the surface of the superconducting cylinder, the magnetic field may be found simply by Ampere's circuital law as

$$B_\theta(r=R) = \frac{\mu_0 I}{2\pi R} . \tag{2.3.11}$$

2.3.3 **Superconducting cylinder in a static magnetic field**

The third problem to be solved is the study of the field distribution in a superconducting cylinder of infinite length and of radius R in a uniform static magnetic field B^e parallel to its axis (oz). To solve this problem, one may use the field equation (2.2.32) and can find

$$B_z(r) = B^e \frac{I_0(r/\lambda_L)}{I_0(R/\lambda_L)} \qquad \text{for } 0 \leq r \leq R . \tag{2.3.12}$$

The average induced magnetization in the superconducting cylinder may then be written as

$$M_z = \frac{2\pi B^e}{\mu_0 \pi R^2} \int_0^R \{\frac{I_0(r/\lambda_L)}{I_0(R/\lambda_L)} - 1\} r dr$$

$$= -\frac{B^e}{\mu_0} \{1 - \frac{2\lambda_L}{R} \frac{I_1(R/\lambda_L)}{I_0(R/\lambda_L)}\} \tag{2.3.13}$$

which, in the limiting cases of $R \gg \lambda_L$ and $R \ll \lambda_L$, may be reduced respectively to

$$M_z \approx -\frac{B^e}{\mu_0} \{1 - \frac{2\lambda_L}{R}\} \qquad \text{for } R \gg \lambda_L \tag{2.3.14}$$

and

$$M_z \approx -\frac{B^e R^2}{8\mu_o \lambda_L^2} \qquad \text{for} \quad R << \lambda_L . \tag{2.3.15}$$

Eqn.(2.3.14) shows that the superconducting cylinder presents an almost perfect diamagnetic effect provided that its radius R is much larger than its penetration depth λ_L.

In general, we have shown that a superconductor in an applied magnetic field does not show a perfect diamagnetic effect since screening currents cannot be confined entirely to the surface. The penetration of the magnetic flux into the thin surface layer of the superconductor is characterized by the penetration depth λ_L. Though this layer is so thin, it plays a very important part in determining the properties of superconductors. Experimentally it has been found that the penetration depth λ_L is temperature-dependent, which fits very closely the following relation

$$\lambda_L(T) = \frac{\lambda_o}{\sqrt{1 - (\frac{T}{T_c})^4}} \tag{2.3.16}$$

with λ_o being defined as the value of λ_L at zero temperature. This relation shows that at lower temperatures, the penetration depth is nearly independent of temperature while, at temperatures above about 0.8 of the transition temperature, the penetration depth increases rapidly and approaches infinity as the temperature approaches T_c. The penetration depth λ_o was studied experimentally and some experimental data were reported, which are, for instance, 3.9×10^{-8} m for Pb, 5.1×10^{-8} m for Sn, 3.9×10^{-8} m for In (Lock (1951)) and 3.9×10^{-8} m for Al (Tedrow et al. (1971)).

2.3.4 **Two-fluid model**

In order to study the behavior of superconductors in AC fields, we shall first introduce here a two-fluid model for superconductors proposed by Gorter and Casimir (1934). Though the original intention of the two-fluid model is to study thermodynamic properties of the superconductors, the combination of this model with London theory has turned out to be of much wider applications.

It is postulated in the two-fluid model that conduction electrons in superconductors below T_c can be divided into two distinct groups. A fraction n_s (= $2n_s^*$) of the conduction electrons is condensed into a superconducting aggregate while the remainder n_n is in the normal state. The total electron density n is thus the sum of the densities of the superelectrons n_s and the normal electrons n_n. Gorter and Casimir found that the best agreement with the thermal properties of superconductors could be obtained if the fraction was chosen in the form

$$\frac{n_s}{n} = 1 - (\frac{T}{T_c})^4 \qquad (2.3.17)$$

which shows that the density of superelectrons rises rapidly when the temperature falls below T_c. By noting eqn.(2.2.7), one finds that this model predicts the temperature-dependent relation for the London penetration depth given by eqn.(2.3.16). The agreement of this model with experiment is hardly surprising due to the somewhat arbitrary nature of this theory. However, its successes in dealing with time-varying field problems of superconductors with the combination of London theory are not trivial.

In non-stationary processes, the presence of normal conduction electrons requires a modification of London electrodynamic equations. In the simple model, we may assume that the dynamics of the conduction electron fluid in the presence of an electric field may be described by the following equations

$$m^* \frac{d\mathbf{v}_s}{dt} = - e^*\mathbf{E} \qquad (2.3.18)$$

$$m\frac{d\mathbf{v}_n}{dt} + m\frac{\mathbf{v}_n}{\tau} = - e\mathbf{E} \qquad (2.3.19)$$

where $\mathbf{v}_s$ and $\mathbf{v}_n$ are respectively the average velocities of the pair superelectrons and the normal electrons. τ is the momentum relaxation time accounting for the effect of collisions of normal electrons. The corresponding current densities are

$$\mathbf{J}_s = - n_s e\mathbf{v}_s \qquad (2.3.20)$$

$$\mathbf{J}_n = - n_n e \mathbf{v}_n \ . \tag{2.3.21}$$

Ignoring nonlinear effects, eqns.(2.3.18) and (2.3.19) may be expressed in terms of the current densities as

$$\frac{\partial \mathbf{J}_s}{\partial t} = \frac{n_s e^2}{m}\mathbf{E} \tag{2.3.22}$$

$$\tau\frac{\partial \mathbf{J}_n}{\partial t} + \mathbf{J}_n = \frac{n_n \tau e^2}{m}\mathbf{E} \ . \tag{2.3.23}$$

For steady-state alternating electric fields of the form $\mathbf{E}e^{i\omega t}$, the total current density $\mathbf{J}$ may be written as

$$\mathbf{J} = \mathbf{J}_s + \mathbf{J}_n = \sigma_{eff}\mathbf{E} \tag{2.3.24}$$

where σ_{eff} denotes an effective complex conductivity, which in the low-frequency approximation (typically $2\pi\omega < 10^{11}$ Hz so that $\omega^2\tau^2 << 1$) reads

$$\sigma_{eff} = \sigma_n\frac{n_n}{n} - \frac{i}{\omega\mu_o\lambda_L^2} \tag{2.3.25}$$

with $i = \sqrt{-1}$, where $\sigma_n = n\tau e^2/m$ is the conductivity of the material in the normal state. By eqn.(2.3.17), the effective complex conductivity can be further expressed in its temperature-dependent form

$$\sigma_{eff} = [1 - (\frac{T}{T_c})^4][\sigma_n\eta(T) - \frac{i}{\omega\mu_o\lambda_o^2}] \tag{2.3.26}$$

with

$$\eta(T) = (\frac{T}{T_c})^4[1 - (\frac{T}{T_c})^4]^{-1} \ . \tag{2.3.27}$$

2.3.5 Superconducting half-space in a time-harmonic magnetic field

We shall now consider an AC electrodynamic problem of superconductors in which the effect due to the presence of normal conduction electrons will be taken into account with the use of the two-fluid model and London theory. The geometry of the superconductor is taken to be the same as the example shown in section 2.3.1 (a half-space of superconductor) with, however, now a time-harmonic applied uniform magnetic field $\mathbf{B}^e e^{i\omega t}$ in the y-direction as shown in **Figure** 2.3.1.

By using eqn.(2.2.36), we can obtain

$$B_y(x, t) = B^e \exp(i\omega t - kx) \tag{2.3.28}$$

with

$$k = \frac{1}{\lambda_L}\sqrt{1 + i\omega\sigma\mu_o\lambda_L^2} \tag{2.3.29}$$

where σ denotes the temperature-dependent normal conductivity, which in the two-fluid model becomes $\sigma = \sigma_n n_n/n$.

The surface impedance (which is a measurable quantity for bulk sample) can be found by

$$Z_s = \frac{E_z(x=0^+)}{\int_0^\infty J_z dx} = \frac{k}{\sigma_{eff}} = \frac{i\omega\mu_o\lambda_L\sqrt{1 - i\sigma\omega\mu_o\lambda_L^2}}{\sqrt{1 + \sigma^2\omega^2\mu_o^2\lambda_L^4}} . \tag{2.3.30}$$

Separating it into real and imaginary parts, we have

$$Z_s = R_s + i X_s \tag{2.3.31}$$

with R_s being the surface resistance given by

$$R_s = \frac{1}{\sqrt{2}}\{\frac{\omega^2\mu_o^2\lambda_L^2}{1 + \sigma^2\omega^2\mu_o^2\lambda_L^4}[\sqrt{1 + \sigma^2\omega^2\mu_o^2\lambda_L^4} - 1]\}^{1/2} \tag{2.3.32}$$

and X_s being the surface inductive reactance by

$$X_s = \frac{1}{\sqrt{2}}\{\frac{\omega^2\mu_o^2\lambda_L^2}{1 + \sigma^2\omega^2\mu_o^2\lambda_L^4}[\sqrt{1 + \sigma^2\omega^2\mu_o^2\lambda_L^4} + 1]\}^{1/2} . \tag{2.3.33}$$

In particular, for the case of $\sigma\omega\mu_o\lambda_L^2 << 1$ (with, for instance, frequencies less than 10^9 Hz), eqns.(2.3.32) and (2.3.33) can be reduced respectively to

$$R_s = \frac{1}{2}\sigma\omega^2\mu_o^2\lambda_L^3 = \frac{\sigma_n\omega^2\mu_o^2\lambda_o^3}{2}\left(\frac{T}{T_c}\right)^4\left[1 - \left(\frac{T}{T_c}\right)^4\right]^{-3/2} \tag{2.3.34}$$

and

$$X_s = \omega\mu_o\lambda_L = \omega\mu_o\lambda_L\left[1 - \left(\frac{T}{T_c}\right)^4\right]^{-1/2}. \tag{2.3.35}$$

It is shown that the loss (R_s) predicted by this model for the superconductor increases as the square of the frequency, whereas for normal conductors the loss increases only as the square root of the frequency.

It has been recognized that the frequency and temperature dependence of the surface impedance of superconductors predicted by the two-fluid model are close to observed experimental behavior. As a result, many circuit and waveguide calculations are so far based on the two-fluid model. However, at higher microwave frequencies (above 10 GHz), appreciable deviations occur and further theoretical efforts are needed (Hartwig and Passow (1975)) and Duzer and Turner (1981)). Nevertheless, in practical applications, the simple two-fluid model is often used in many circuit and waveguide calculations.

2.4 Magnetic shielding using superconductors

This section is concerned with the problem of magnetic shielding by using superconducting materials. In practice, the magnetic shielding is of considerable importance since essentially field-free regions are often necessary or desirable for experimental purposes or for the reliable working of some electronic devices. Here we shall analyze the field distribution around a superconducting shell of spherical shape with an outer radius R and an inner radius a, embedded in a uniform external static magnetic field $\mathbf{B}^e$ with its direction along the z-axis as shown in **Figure** 2.4.1.

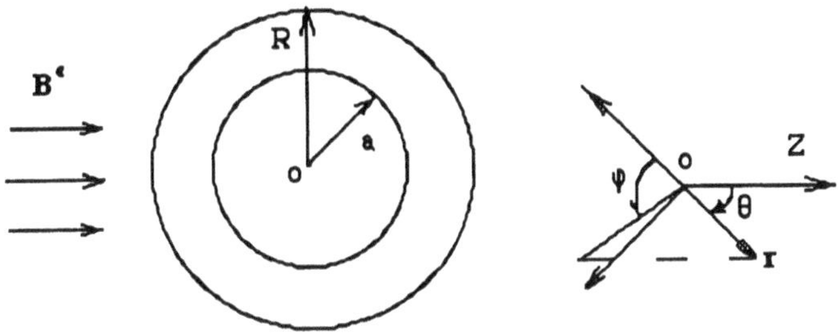

Figure 2.4.1 A superconducting shell is embedded in an external magnetic field.

In this problem, the field equations for the superconducting region are

$$\nabla^2 \mathbf{J} = \frac{1}{\lambda_L^2}\mathbf{J} \qquad (a < r < R) \qquad (2.4.1)$$

and

$$\mathbf{B} = -\mu_0 \lambda_L^2 \nabla \times \mathbf{J} \qquad (a < r < R). \qquad (2.4.2)$$

By noting the symmetry of the system and of the external field assumed, we can choose a spherical coordinate system (r, θ, φ) so that we may write $\mathbf{J} = (0, 0, J_\varphi(r, \theta))$. Furthermore, we may let $J_\varphi = f(r)\sin\theta$ and substitute it into the eqn.(2.4.1) for J_φ, which leads to an ordinary differential equation for f(r)

$$\frac{d^2f}{dr^2} + \frac{2}{r}\frac{df}{dr} - (\frac{2}{r^2} + \beta^2)f = 0 \qquad (a < r < R). \qquad (2.4.3)$$

Eqn.(2.4.3) has a general solution for f(r), which reads

$$f(r) = \frac{A}{r^2}(\sinh\beta r - \beta r \cosh\beta r) + \frac{C}{r^2}(\cosh\beta r - \beta r \sinh\beta r) \qquad (2.4.4)$$

where A and C are two integration constants, and β is a material parameter defined by $\beta = 1/\lambda_L$.

The magnetic field in the superconducting shell may, thus, be found from eqn.(2.4.2) as

$$B_r = -\frac{2\mu_0\cos\theta}{\beta^2 r^3}\{A(\sinh\beta r - \beta r\cosh\beta r) + C(\cosh\beta r - \beta r\sinh\beta r)\} \tag{2.4.5a}$$

$$B_\theta = -\frac{\mu_0\sin\theta}{\beta^2 r^3}\{A[(1+\beta^2r^2)\sinh\beta r - \beta r\cosh\beta r] + C[(1+\beta^2r^2)\cosh\beta r - \beta r\sinh\beta r]\} \tag{2.4.5b}$$

$$B_\varphi = 0 \tag{2.4.5c}$$

where functions $\cosh\beta r$ and $\sinh\beta r$ are defined respectively by

$$\cosh x = \frac{e^x + e^{-x}}{2} \qquad \text{and} \qquad \sinh x = \frac{e^x - e^{-x}}{2} \; . \tag{2.4.6}$$

The field equations in the free space are simply the static magnetic equations

$$\nabla\times\mathbf{B} = 0 \qquad \text{and} \qquad \nabla\cdot\mathbf{B} = 0 \qquad (\text{for } R < r < +\infty \text{ or } r < a). \tag{2.4.7}$$

The symmetry of the problem also suggests trying the solution from a superposition of the uniform applied magnetic field $\mathbf{B}^e$ and the field of a magnetic dipole at the origin with dipole moment m, oriented in the direction of the applied field, i.e.

$$B_r = (B^e + \frac{2\mu_0 m}{4\pi r^3})\cos\theta \tag{2.4.8a}$$

$$B_\theta = (-B^e + \frac{\mu_0 m}{4\pi r^3})\sin\theta \tag{2.4.8b}$$

$$B_\varphi = 0 \tag{2.4.8c}$$

for $R < r < +\infty$.

In addition, the magnetic field in the cavity ($r < a$) may be tried by the form

$$B_r = B^c\cos\theta\,, \qquad B_\theta = -B^c\sin\theta\,, \qquad B_\varphi = 0 \qquad \text{for } r < a\,. \tag{2.4.9}$$

Here, the magnetic dipole moment m and the magnitude of the magnetic field B^c in the cavity are determined together with the constants A and C by the interface conditions that the magnetic field **B** is continuous across the surfaces at r = a and r = R.
After some manipulation, we can derive the following results

$$A = \frac{3RB^e}{2\mu_o} \frac{3 + \beta^2 a^2 - 3\beta a \, th\beta a}{\{(3+\beta^2a^2-3\beta a \, th\beta a)\sinh\beta R + [3\beta a - (3+\beta^2a^2)th\beta a]\cosh\beta R\}} \qquad (2.4.10)$$

$$C = \frac{3RB^e}{2\mu_o} \frac{3\beta a - (3+\beta^2a^2)th\beta a}{\{(3+\beta^2a^2-3\beta a \, th\beta a)\sinh\beta R + [3\beta a - (3+\beta^2a^2)th\beta a]\cosh\beta R\}} \qquad (2.4.11)$$

and

$$m = -\frac{2\pi R^3 B^e}{\mu_o}\{1 + \frac{3}{\beta^2R^2} - \frac{3}{\beta R}\alpha_m\} \qquad (2.4.12)$$

with the constant α_m given by

$$\alpha_m = \frac{(3+\beta^2a^2-3\beta a \, th\beta a)\cosh\beta R + [3\beta a - (3+\beta^2a^2)th\beta a]\sinh\beta R}{(3+\beta^2a^2-3\beta a \, th\beta a)\sinh\beta R + [3\beta a - (3+\beta^2a^2)th\beta a]\cosh\beta R}, \qquad (2.4.13)$$

and

$$B^c = \frac{3\beta RB^e\cosh\beta a[1 - (th\beta a)^2]}{(3+\beta^2a^2-3\beta a \, th\beta a)\sinh\beta R + [3\beta a - (3+\beta^2a^2)th\beta a]\cosh\beta R} \qquad (2.4.14)$$

with thx = sinhx/coshx.
It is shown that inside the cavity there is a uniform magnetic field parallel to $\mathbf{B}^e$ and its magnitude B^c is given by eqn.(2.4.14). The uniqueness of the solution is due to the uniqueness theorem provided by London (1950).

Let us now consider some special cases. In the case of a=0, eqn.(2.4.12) is reduced to

$$m = -\frac{2\pi R^3 B^e}{\mu_o}\{1 + \frac{3}{\beta^2R^2} - \frac{3}{\beta R}\coth\beta R\} \qquad (2.4.15)$$

which reproduces the result (see London (1950) for a superconducting sphere of radius R in a uniform external magnetic field $\mathbf{B}^e$. In particular, if the radius R of the superconducting sphere is much larger than its penetration depth λ_L, one has approximately

$$m \approx -\frac{2\pi B^e}{\mu_o}(R - \lambda_L)^3 \qquad \text{for} \quad R >> \lambda_L . \tag{2.4.16}$$

Furthermore we may define an average induced magnetization in the superconducting sphere by

$$M = \frac{m}{\frac{4\pi}{3}R^3} = -\frac{3B^e}{2\mu_o}\frac{(R - \lambda_L)^3}{R^3} \approx -\frac{3B^e}{2\mu_o} \qquad \text{for} \quad R >> \lambda_L \tag{2.4.17}$$

which shows that the superconducting sphere presents a perfect diamagnetic effect if we ignore the penetration of the field.

Next we consider the case where a and R are both much larger than the penetration depth of the superconducting shell, i.e. a, R >> λ_L, which is usually the case of practical interest. The magnetic field inside the cavity can then be found from eqn.(2.4.14) as approximately

$$B^c \approx \frac{B^e R \lambda_L}{a^2}\exp\{-\frac{R - a}{\lambda_L}\} \qquad \text{for} \quad a, R >> \lambda_L \tag{2.4.18}$$

which shows a perfect magnetic shielding if λ_L -> 0.

We may now make a comparison between the magnetic shielding by superconducting materials and the magnetic shielding by permeable materials with high relative permeability, the result of which is given by (see Jackson (1975))

$$B^c \approx \frac{9B^e}{2\mu_r(1 - \frac{a^3}{R^3})} \qquad \text{for} \quad \mu_r >> 1 . \tag{2.4.19}$$

Quantitatively, by taking the numerical values of a = 0.1 m, R = 0.11 m (R - a = 10 mm) for the shield shell made of magnetically permeable material with $\mu_r \approx 10^6$ (for instance, supermalloy), one finds from eqn.(2.4.19) that $B^c/B^e \approx 1.8 \times 10^{-5}$. However, if one takes the numerical values of a = 0.1 m, R = 0.1001 m (R - a = 0.1 mm) for the shield shell (though which is very thin) made of superconducting materials with, for instance, $\lambda_L \approx 10^{-7}$ m, one can find from eqn.(2.4.18) that $B^c/B^e \approx 10^{-439}$, which is extremely small. Thus, it is shown that a shield made of superconducting material can cause a much greater reduction of the field inside it than the one made of high-permeability materials even with the highest permeability available. A theoretical proposal may be made according to this analysis that one may use a composite of a normal (for instance, dielectric) matrix covered with a thin superconducting film at its surface as the magnetic shielding material, where the normal matrix material is mainly used to stand for necessary mechanical loadings and other technical requirements.

2.5 Energy-momentum theorems for superconductors

2.5.1 Energy theorem

The energy theorem for superconducting solids may follow from Maxwell's equations in the usual way as shown in section 1.6. The energy integral (1.6.29), for a continuous isotropic non-magnetic superconductor, may be written into its local form

$$\nabla \cdot \mathbf{S} + \frac{\partial}{\partial t}\left[\frac{1}{2}\left(\varepsilon E^2 + \frac{1}{\mu_0} B^2\right)\right] = - \mathbf{J} \cdot \mathbf{E} \tag{2.5.1}$$

where **S** is the Poynting vector, and **J**·**E** is the work done by the field on the moving electric charges. Unlike normal conductors, a superconducting solid, however, does not waste this work completely by transformation into irreversible Joule heat. This can be seen by writing

$$\mathbf{J} \cdot \mathbf{E} = (\mathbf{J}_s + \mathbf{J}_n) \cdot \mathbf{E} = \frac{\partial}{\partial t}\left(\frac{1}{2}\mu_0 \lambda_L^2 J_s^2\right) + \mathbf{J}_n \cdot \mathbf{E} \tag{2.5.2}$$

with the aid of eqn.(2.2.11) for the supercurrent $\mathbf{J}_s$.

The energy equation (2.5.1) may then be expressed by

$$\nabla \cdot (\mathbf{E} \times \mathbf{H}) + \frac{\partial}{\partial t}\left[\frac{1}{2}\left(\varepsilon E^2 + \frac{1}{\mu_0} B^2\right) + \left(\frac{1}{2}\mu_0 \lambda_L^2 J_s^2\right)\right] = - \mathbf{J}_n \cdot \mathbf{E} \tag{2.5.3}$$

where the new term $\frac{\partial}{\partial t}(\frac{1}{2}\mu_o\lambda_L^2 J_s^2)$, is evidently positive when the supercurrent is generated, negative when it is switched off. Hence the energy $\frac{1}{2}\mu_o\lambda_L^2 J_s^2$ represents reversible work which can be entirely recovered. It is the kinetic-energy density of the supercurrent. In fact, the normal current also has kinetic energy, which, however, is usually neglected since Ohm's law does not take into account the inertia of the normal conduction electrons. The term on the right-hand side of eqn.(2.5.3) is always negative or zero. This term is the energy dissipation by Joule heat due to the normal current. For stationary conditions, where there is no normal current, it is exactly zero, and for quasistationary conditions, it is extremely small. Hence, there is almost no energy dissipation in a superconductor except in the case of very rapidly alternating fields.

2.5.2 **Momentum theorem**

In Maxwell's theory, the conservation law of linear momentum for a homogeneous linear electromagnetic body, subject to only the Lorentz body force $(\rho_e\mathbf{E} + \mathbf{J} \times \mathbf{B})$ with ρ_e being the free electric charge density, may be expressed by

$$\frac{\partial}{\partial t}(\mathbf{P}_{mech} + \mathbf{P}_{field}) = \int_{\partial V} \mathbf{t}^{em}\cdot\mathbf{n}dS \tag{2.5.4}$$

in which $\mathbf{P}_{field}$ is the total electromagnetic momentum defined by

$$\mathbf{P}_{field} = \int_V \varepsilon\mathbf{E} \times \mathbf{B}dV \tag{2.5.5}$$

and $\mathbf{t}^{em}$ is the Maxwell stress tensor defined by

$$t_{ij}^{em} = \varepsilon(E_iE_j - \frac{1}{2}E_kE_k\delta_{ij}) + \frac{1}{\mu}(B_iB_j - \frac{1}{2}B_kB_k\delta_{ij}) \tag{2.5.6}$$

where μ is the magnetic permeability, which equals μ_o for a non-magnetic material.

The total mechanical kinetic-momentum $\mathbf{P}_{mech}$ is defined by the relation

$$\frac{\partial \mathbf{P}_{mech}}{\partial t} = \int_V (\rho_e \mathbf{E} + \mathbf{J} \times \mathbf{B})\, dV \ . \tag{2.5.7}$$

The local differential form of the conservation law of linear momentum can then be written as

$$\frac{\partial}{\partial t}(\mathbf{p}_{mech} + \mathbf{p}_{field}) = \nabla \cdot \mathbf{t}^{em} \tag{2.5.8}$$

in which the electromagnetic field momentum density is defined by

$$\mathbf{p}_{field} = \varepsilon \mathbf{E} \times \mathbf{B} \tag{2.5.9}$$

and the mechanical kinetic-momentum density $\mathbf{p}_{mech}$ is defined by

$$\frac{\partial \mathbf{p}_{mech}}{\partial t} = \rho_e \mathbf{E} + \mathbf{J} \times \mathbf{B} \ . \tag{2.5.10}$$

In the case of a superconductor, we may write

$$\rho_s \mathbf{E} + \mathbf{J}_s \times \mathbf{B} = \mu_o \lambda_L^2 [\, \rho_s \frac{\partial \mathbf{J}_s}{\partial t} - \mathbf{J}_s \times (\nabla \times \mathbf{J}_s)] = \mu_o \lambda_L^2 [\rho_s \frac{\partial \mathbf{J}_s}{\partial t} - \mathbf{J}_s (\nabla \, . \, \mathbf{J}_s)] - \nabla \cdot \mathbf{t}^{su} \tag{2.5.11}$$

in which $\mathbf{t}^{su}$ is a stress tensor defined by

$$t_{ij}^{su} = - \mu_o \lambda_L^2 \, [J_{si} J_{sj} - \frac{1}{2} (\mathbf{J}_s \cdot \mathbf{J}_s) \delta_{ij}] \ . \tag{2.5.12}$$

By now writing the total current $\mathbf{J}=\mathbf{J}_s+\mathbf{J}_n$ and the total free charge density $\rho_e=\rho_s+\rho_n$, and by assuming the conservation relation of the superelectrons, $\nabla \cdot \mathbf{J}_s = - \partial \rho_s / \partial t$, the momentum theorem for the superconductor may then be written in the following form:

$$\frac{\partial}{\partial t}(\mathbf{p}_{field} + \mathbf{p}_{n\text{-}mech} + \mathbf{p}_{super}) = \nabla \cdot (\, \mathbf{t}^{em} + \mathbf{t}^{su} \,) \tag{2.5.13}$$

where the kinetic-momentum density of supercurrent is defined by

$$\mathbf{p}_{super} = \mu_o \lambda_L^2 \, \rho_s \, \mathbf{J}_s \tag{2.5.14}$$

and the mechanical kinetic-momentum density $\mathbf{p}_{n\text{-mech}}$ is defined by

$$\frac{\partial \mathbf{p}_{n\text{-mech}}}{\partial t} = \rho_n \mathbf{E} + \mathbf{J}_n \times \mathbf{B} . \tag{2.5.15}$$

Under stationary conditions one has $\mathbf{E} = 0$ and consequently $\mathbf{J}_n = 0$. Hence in this case one has

$$\nabla \cdot (\mathbf{t}^{em} + \mathbf{t}^{su}) = 0 \tag{2.5.16}$$

which means that under stationary conditions a superconductor is not subject to any volume forces. However, at the surface of the superconductor the momentum current tensor has a discontinuity. It jumps from its value inside, $\mathbf{t}^{em} + \mathbf{t}^{su}$, to its ordinary value, $\mathbf{t}^{em}$, outside, because the tensor $\mathbf{t}^{su}$ is defined only within the superconductor.

2.5.3 **Superconducting pressure**

We shall now calculate the total electromagnetic force acting on the superconductor at the quasistatic approximation where the electromagnetic stress tensor $\mathbf{t}^{em}$ is given by eqn.(1.13.2) (or the second term on the r.h.s. of eqn.(2.5.6) with $\mu = \mu_0$). Thus the total electromagnetic force on a superconductor may be found as

$$\mathbf{F}^{em} = \frac{1}{\mu_0} \int_{\partial V} [(\mathbf{B}\cdot\mathbf{n})\mathbf{B} - \frac{1}{2}(\mathbf{B}\cdot\mathbf{B})\mathbf{n}]dS . \tag{2.5.17}$$

If we now consider the fact that the supercurrent density is only distributed within a very thin layer (with the penetrating depth λ) near the boundary surface of the superconductor, we may introduce the concept of surface current by making a limiting process of letting $\lambda \to 0$ and denoting the surface current density by $\mathbf{K}_f = \lim \mathbf{J}\lambda$. Then eqn.(2.5.17) may be written as

$$\mathbf{F}^{em} = - \frac{1}{2\mu_0} \int_{\partial V} |\mathbf{B}^+|^2 \mathbf{n} \, dS = = - \frac{\mu_0}{2} \int_{\partial V} |\mathbf{K}_f|^2 \mathbf{n} \, dS \tag{2.5.18}$$

by noting that $\mathbf{B}^+$ is the magnetic induction field at the outer side of the boundary of the superconductor, and that the normal component of $\mathbf{B}^+$

vanishes at the boundary due to the perfect diamagnetism (**B**=0 in the superconductor). Here we have used the following magnetic interface condition

$$\mathbf{n} \cdot [\,\mathbf{B}\,] = 0 \qquad \text{and} \qquad \mathbf{n} \times [\,\mathbf{B}\,] = \mu_0 \mathbf{K}_f \qquad \text{on } \partial V \qquad (2.5.19)$$

which gives the expression $\mathbf{B}^+ = \mu_0 \mathbf{K}_f \times \mathbf{n}$ on ∂V.

Similarly, we can find the following interface relation at the interface S between a normal conductor and a superconductor with perfect diamagnetism, having the free surface current $\mathbf{K}_f$,

$$\mathbf{n} \cdot [\,\mathbf{t}^{em}\,] = -\frac{\mu_0}{2} |\mathbf{K}_f|^2 \mathbf{n} \qquad \text{on } S \qquad (2.5.20)$$

where **n** is a unit normal vector of the interface S, drawn from the superconductor to the normal conductor.

It is shown by eqn.(2.5.18) that the superconductor is subjected to an electromagnetic pressure with magnitude $\frac{\mu_0}{2} |\mathbf{K}_f|^2$ (or $\frac{1}{2\mu_0} |\mathbf{B}^+|^2$) on its boundary surface. To distinguish the electromagnetic pressure from normal magnetic pressure due to the jump in the magnetization of a magnetic body, we may call this electromagnetic pressure on the non-magnetic superconductor the superconducting pressure. The effect of superconducting pressure makes possible superconducting suspensions, bearings and motors by means of properly shaped magnetic fields (Newhouse (1964)). Typical examples of the effect of superconducting pressure are the experiment by Arkadiev (1945, 1947) which

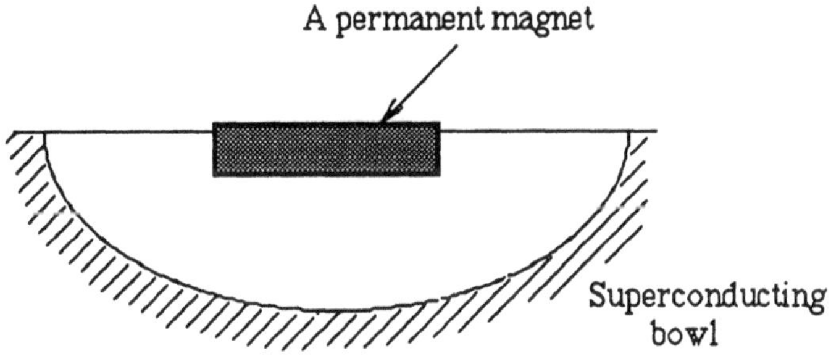

Figure 2.5.1 A bar magnet is floating above the bottom of a superconducting bowl.

showed that a permanent magnet can float in equilibrium above a superconducting cup (see **Figure** 2.5.1), and the experiment by Simon (1953) who showed the ability of suitably shaped magnetic fields to support a superconducting sphere.

2.6 Pippard coherence length and nonlocal relation

The London theory introduced before is a local theory in the sense that superconducting current density is related to the magnetic vector potential at the same point in space, shown by eqn.(2.2.10). It was later found by Pippard that the London theory had to be modified for certain superconductors in order to explain some experimental observations from his series of experiments on the measurement of the penetration depth of various types of superconductors, and on the dependence of the penetration depth on applied magnetic fields (Pippard (1950a)) as well as its anisotropy (Pippard (1953)). In particular, Pippard found that the penetration depth was also noticeably dependent upon the impurity content (Pippard (1950b)), which could not be explained by the local theory of London since the density of superelectron and its effective mass could only be weak functions of the impurity concentration. Thus, guided by the nonlocal expression relating current density and electric field in normal conductors (Reuter and Sondheimer (1948) and Chambers (1952)) for explaining anomalous skin effect

$$\mathbf{J}_n(\mathbf{x}, t) = \frac{3\sigma}{4\pi l_e}\int \frac{[\mathbf{r}\cdot\mathbf{E}(\mathbf{x}',t)]\mathbf{r}}{r^4}\exp(-r/l_e)d\mathbf{x}' \tag{2.6.1}$$

with $\mathbf{r} = \mathbf{x} - \mathbf{x}'$ and $r = |\mathbf{x} - \mathbf{x}'|$, which is reduced to Ohm's law $\mathbf{J}_n = \sigma\mathbf{E}$ for fields varying slowly over the electron mean free path l_e, Pippard proposed that the local $\mathbf{J}_s$ - $\mathbf{A}$ relation (2.2.10) should be replaced for certain superconductors by a nonlocal relation of the form

$$\mathbf{J}_s(\mathbf{x}, t) = \frac{-3}{4\pi\mu_o\lambda_L^2\xi_o}\int \frac{[\mathbf{r}\cdot\mathbf{A}(\mathbf{x}',t)]\mathbf{r}}{r^4}\exp(-r/\xi)d\mathbf{x}' \tag{2.6.2}$$

where the integral is to be taken over the whole volume of the superconductor.

In the nonlocal relation (2.6.2), Pippard introduced two new parameters ξ_0 and ξ. The parameter ξ_0 is called the intrinsic coherence length which is independent of impurities, and can be given empirically by

$$\xi_0 = \gamma \frac{\hbar v_F}{k_B T_c} \qquad (2.6.3)$$

with the empirical constant γ, which was found to be $\gamma = 0.18$ by the BCS microscopic theory. Here v_F is the electron velocity at the Fermi surface and k_B the Boltzmann constant. The numerical values of ξ_0 are found roughly of the order of 10^{-6} m for most pure metals.

The parameter ξ is called the effective coherence length which may be given by an empirical relation

$$\frac{1}{\xi} = \frac{1}{\xi_0} + \frac{1}{\alpha l_e} \qquad (2.6.4)$$

with l_e being the electron mean free path and α being a constant on the order of unity.

The introduction of the concept of the coherence length in the nonlocal relation (2.6.2) characterizes the fact that spatial variation of the density of superelectrons cannot occur over arbitrarily small distances and is only possible within a certain distance ξ. Furthermore, the nonlocal model shows that the superconducting current density at a certain point depends on an average of the magnetic vector potential **A** over a volume of radius about ξ around the point of interest.

To study quantitatively the variation of the penetration depth of superconductors with, for instance, the change of impurities in the superconductor, we may define, independent of any particular penetration law, an effective penetration depth by

$$\lambda = \frac{1}{B(0)} \int_0^\infty B(x)dx \qquad (2.6.5)$$

where B(0) is the magnetic induction at the surface, B(x) the magnetic field at a distance x inside the superconductor, where x is measured from the surface of the superconductor. It can be shown that eqn.(2.6.5) clearly yields the right value of λ when B(x) is exponential as shown, for instance, by the first example in section 2.3.

We now consider two extreme cases where Pippard's nonlocal relation (2.6.2) may be reduced to simpler local form. In the first extreme case of $\lambda >> \xi$, which implies the slow variation of the magnetic vector potential **A** over the region $r < \xi$, the nonlocal relation (2.6.2) can be reduced to the following local form

$$J_s = -\frac{A\xi}{\mu_0 \lambda_L^2 \xi_0}. \tag{2.6.6}$$

Comparison with the London relation (2.2.10) shows that the penetration depth λ for this case is

$$\lambda = \lambda_L \sqrt{\frac{\xi_0}{\xi}} \tag{2.6.7}$$

which implies that λ will be increased if ξ is reduced by impurities or geometry in accordance with experimental observation (Pippard (1953)). For very impure specimens or for thin film, ξ may be taken equal to the mean free path (limited by impurity and/or boundary scattering).

It is seen that the London theory is rigorous only in the case of $\lambda >> \xi$. In the other extreme case of $\lambda << \xi_0$, which holds for most pure bulk superconductors at temperatures not too close to T_c, Faber and Pippard have shown that

$$\lambda = [\frac{\sqrt{3}}{2\pi}\xi_0 \lambda_L^2]^{1/3}. \tag{2.6.8}$$

The intrinsic coherence length ξ_0 are, for instance, $\xi_0 = 2.1 \times 10^{-7}$ m for tin and 12.3×10^{-7} m for Al (Faber and Pippard (1955)). It is the contribution of Pippard to the physics of superconductivity that the concepts of his nonlocality and coherence play very important roles in determining the properties of the superconductors to be discussed in later sections.

2.7 Thermodynamics of phase transitions in superconductors

2.7.1 Thermodynamic functions for superconductors

In this section, some thermodynamic behaviors and properties of superconductors will be studied. The idea of applying thermodynamics to the transition between the superconducting and normal states was originally suggested and developed by Keesom (1924), Rutgers (1934) and Gorter and Casimir (1934), etc. To study the thermodynamics of the superconducting transition in various fields, the simplest way of obtaining relevant thermodynamic functions is to treat the superconductive material in the superconducting state as a magnetic substance described at the end of section 2.2 and in the normal state as a non-magnetic substance (here, we do not consider magnetic superconductors which will be studied in the next chapter).

For a magnetic body placed in a magnetic field whose (current) sources are fixed, the change of the Gibbs free energy of the body corresponding to an infinitesimal change in the field can be found as (Landau et al. (1984))

$$\delta G = - S\delta T - \int_V \mathbf{M}\cdot\delta\mathbf{B}^e \, dV \tag{2.7.1}$$

where V is the volume of the magnetic body. The last term on the r.h.s. of eqn.(2.7.1) represents the differential amount of work done on the body when the external field is increased by an amount of $\delta\mathbf{B}^e$ for fixed (current) sources, which are independent of the field that they produce. The magnetization vector **M** in eqn.(2.7.1) can be, in general, a nonlinear functional of the magnetic field. Here, we have ignored mechanical deformation effects by assuming a constant (rigid) material body under constant pressures. The mechanical effects on the superconductivity will, however, be studied in detail in the next chapter.

In the case of a constant temperature, one has then

$$\delta G = - \int_V \mathbf{M}.\delta\mathbf{B}^e \, dV = - \delta \int_V [\mu_0 \int_0^{H^e} \mathbf{M}\cdot d\mathbf{H}^e] dV \, . \tag{2.7.2}$$

Thus, one derives that any substance, which in an applied field $\mathbf{H}^e$ ($=\mathbf{B}^e/\mu_0$) acquires a magnetization $\mathbf{M}$, changes its Gibbs free energy per unit volume g at a given temperature by an amount

$$\delta g = -\mu_0 \delta \int_0^{H^e} \mathbf{M} \cdot d\mathbf{H}^e \,. \tag{2.7.3}$$

It is shown that in the case of the field producing a positive magnetization, i.e. the magnetization in the same direction as the magnetic field (for isotropic magnetic materials), the Gibbs free energy is lowered.

For a magnetic body of the shape of an ellipsoid of rotation with an axis of symmetry being parallel to the direction of the applied field $\mathbf{H}^e$, then the internal magnetization $\mathbf{M}$ and the field $\mathbf{H}$ in the body are everywhere uniform and can be expressed by

$$\mathbf{H} = \mathbf{H}^e - N\mathbf{M} \tag{2.7.4}$$

where N is the demagnetizing factor along the axis of the ellipsoid parallel to $\mathbf{H}^e$ and is a function of its ellipticity only. Some numerical values of the demagnetizing factor are, for instance, N=1/3 for a sphere, N=1/2 for a cylinder with its axis normal to the direction of the applied field, and N=0 for a cylinder with its axis parallel to the direction of the applied field or for a flat infinitely thin plate parallel to the direction of the applied field.

For a superconducting specimen, the application of a magnetic field produces a negative magnetization which, if penetration of the field is neglected, exactly cancels the flux due to the applied field, so that $\mathbf{M} = -\mathbf{H}$ in the superconducting body. In particular, for a superconducting body of the shape of an ellipsoid of rotation with an axis of symmetry being parallel to the direction of the applied field $\mathbf{H}^e$, we can find by eqn.(2.7.4)

$$\mathbf{M} = -\frac{\mathbf{H}^e}{1 - N} \tag{2.7.5}$$

and, therefore,

$$g_s(T, H^e) = g_s(T, 0) + \frac{\mu_o}{2}\frac{(H^e)^2}{(1 - N)} \; . \qquad (2.7.6)$$

where $g_s(T, 0)$ denotes the Gibbs free energy per unit volume of the superconducting body at the superconducting state and in the absence of the applied magnetic field. Eqn.(2.7.6) shows that if we apply a magnetic field to a superconductor, its Gibbs free energy increases due to the Meissner effect.

In the case of the superconducting specimen being a long cylinder with its axis parallel to the applied field, the Gibbs free energy per unit volume is then increased to the value

$$g_s(T, H^e) = g_s(T, 0) + \mu_o\frac{(H^e)^2}{2} \qquad (2.7.7)$$

with negligible demagnetization effects from the ends of the specimen. The specimen in normal state, however, is virtually non-magnetic and acquires negligible magnetization in an applied magnetic field. Consequently the application of a magnetic field does not change the Gibbs free energy of the normal state though it raises that of the superconducting state. If the field strength is increased enough, the Gibbs free energy of the superconducting state will be raised above that of normal state, and, in this case, the specimen will not remain superconducting but will become normal (see **Figure** 2.7.1). There is, therefore, a maximum magnetic field strength that can be applied to a superconductor if it is to remain in the superconducting state. This critical magnetic (intensity) field strength is given by

$$H_c(T) = \sqrt{\frac{2}{\mu_o}[g_n(T) - g_s(T, 0)]} \qquad (2.7.8)$$

where $g_n(T)$ is the Gibbs free energy density of the specimen in its normal state. This critical magnetic field strength can be measured simply by applying a magnetic field parallel to a wire or a long rod of superconductor and observing the strength at which resistance appears.

It is seen that this critical magnetic field is temperature-dependent and that it falls from some value, denoted by H_0, at low temperatures to

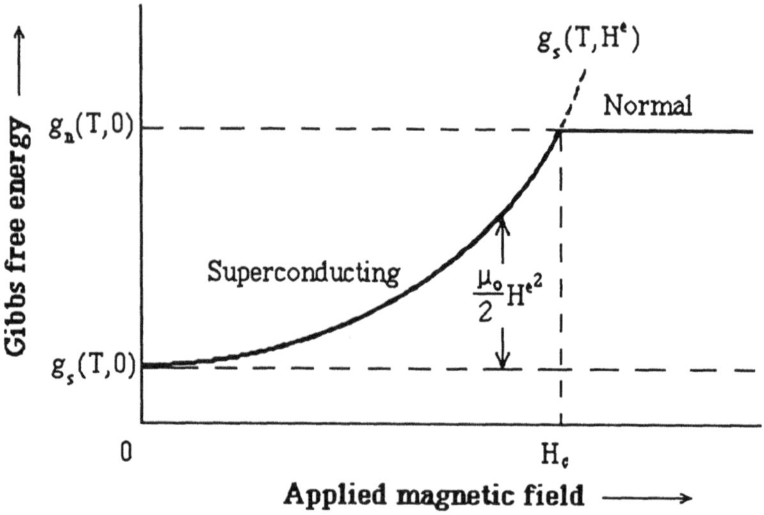

Figure 2.7.1 Effect of applied magnetic field on Gibbs free energy.

zero at the superconducting transition temperature T_C as shown by **Figure** 2.1.2. Experimentally, it has been found that the variation of the critical magnetic field can be conveniently described with good approximation by a parabolic curve given by eqn.(2.1.1).

The existence of a critical magnetic field has been made use of in a controlled switch called a cryotron, which is composed of a straight wire called the gate and a wire wound in a long single-layer coil called the control. The straight wire is inset in the wire coil and both wires are in superconducting state at low temperatures. The superconducting materials are chosen such that the critical magnetic field of the coil wire is higher than that of the straight wire. The current through the gate can then be controlled by a small current in the control using the fact that the gate may be driven to normal by the magnetic field generated by the control current in the coil while the control coil remains resistanceless. Such a device is analogous to a relay. Small cryotrons were first developed as fast acting switches for possible use in digital computers. Large cryotons can be used to control the currents in superconducting magnet circuits (Rose-Innes and Rhoderick (1969)).

2.7.2 First and second-order phase transitions in superconductors

From eqns.(2.7.7) and (2.7.8), one has, therefore, in an applied magnetic field of strength H^e a difference in the Gibbs free energy density between the normal and superconducting states,

$$g_n(T) - g_s(T, H^e) = \frac{1}{2}\mu_0[\, H_c^2 - H^{e2}] \,. \qquad (2.7.9)$$

If the applied magnetic field strength are kept constant but the temperature is varied by an amount δT there will also be a change of the Gibbs free energy by

$$\delta G = - S\, \delta T \qquad (2.7.10)$$

with

$$S = - (\frac{\partial G}{\partial T})_{H^e} \,. \qquad (2.7.11)$$

The entropy per unit volume can thus be written as

$$s = - (\frac{\partial g}{\partial T})_{H^e} \,. \qquad (2.7.12)$$

Substituting eqn.(2.7.9) into this expression, one gets for the superconductor (ignoring any flux penetration)

$$s_n - s_s = - \mu_0 H_c \frac{\partial H_c}{\partial T} \qquad (2.7.13)$$

which shows that the entropy of the superconducting state is less than that of the normal state, i.e. that the superconducting state has a higher degree of order than the normal state since the critical magnetic field always decreases with increase of temperature, so $\partial H_c/\partial T$ is always negative. The critical field H_c falls to zero as the temperature is raised towards T_c; therefore, according to eqn.(2.7.13), the entropy difference between the normal and superconducting states vanishes at this temperature. Furthermore, by the third law of thermodynamics, s_n must also equal s_s at T = 0, which means that $\partial H_c/\partial T$ must be zero at 0 °K since the critical field H_c is not zero. This is in accordance with the experimental observation that, for all superconductors, the slope of the H_c versus T curve appears to become zero as the temperature approaches 0 °K. Experimentally, it has been found that the critical magnetic field can be closely approximated by the parabolic relation (2.1.1).

In the presence of an applied magnetic field, there is a latent heat when a superconductive specimen undergoes the superconducting-normal transition. The latent heat L for the transition between two phase α and β is given by $L = vT(s_\alpha - s_\beta)$, so from eqn.(2.7.13) we have

$$L = -vT\mu_0 H_c \frac{\partial H_c}{\partial T} \tag{2.7.14}$$

where v is the volume per unit mass.

This latent heat arises because at temperatures between T_c and 0°K the entropy of the normal state is greater than that of the superconducting state, so heat must be supplied if the transition is to take place at constant temperature. Thus, in the presence of an applied magnetic field, the superconducting-normal transition is of the first-order, i.e. although g is continuous, $\partial g/\partial T$ is not. It can be seen that in the absence of any magnetic field the transition occurs at the transition temperature T_c and $H_c=0$, but if there is a magnetic field the transition occurs at some lower temperature T where $H_c > 0$.

In the absence of the applied magnetic field, since, at the transition temperature T_c, $s_n = s_s$, we have, for the superconducting-normal transition at T_c,

$$\left(\frac{\partial g}{\partial T}\right)_n = \left(\frac{\partial g}{\partial T}\right)_s \tag{2.7.15}$$

which shows that it is a phase transition of the second order (i.e. not only g is continuous but $\partial g/\partial T$ is also continuous at the transition).

2.7.3 **Discontinuity of specific heat at transition temperature**

The second-order phase transition has two important characteristics: at the transition there is no latent heat, and there is a jump in the specific heat. The first characteristic follows immediately from the fact that $đQ = T\,dS$ and we have seen that, at the transition temperature, there is no change in entropy and therefore no latent heat. The second condition follows from the fact that the specific heat of a material is given by

$$C_v = vT\frac{\partial s}{\partial T} \qquad (2.7.16)$$

and the difference in the specific heats of a superconductor in normal and superconducting states can be obtained by using eqn.(2.7.13):

$$C_{vs} - C_{vn} = vT\mu_0 H_c \frac{\partial^2 H_c}{\partial T^2} + vT\mu_0(\frac{\partial H_c}{\partial T})^2 . \qquad (2.7.17)$$

In particular, at the transition temperature T_c, we have H_c=0 and so have for the transition in the absence of an applied magnetic field

$$(C_{vs} - C_{vn})_{T_c} = vT_c\mu_0(\frac{\partial H_c}{\partial T})^2_{T_c} \qquad (2.7.18)$$

which is known as Rutgers' formula, and it predicts the value of the discontinuity in the specific heat of a superconductor at the transition temperature (see **Figure** 2.7.2). It should be noticed that here $\partial H_c/\partial T$ is a property of the material whose value does not depend on whether or not a field is actually present.

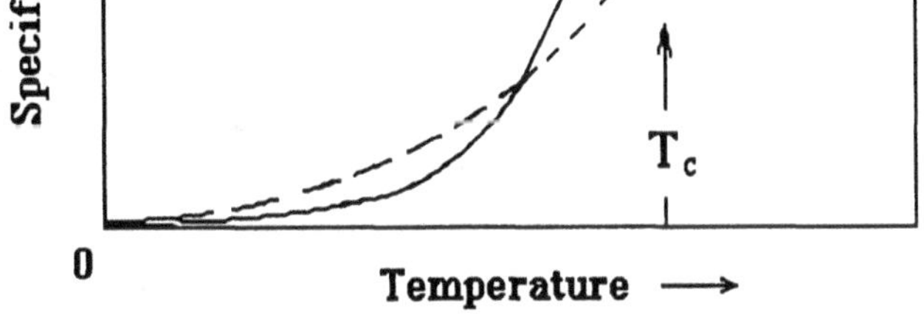

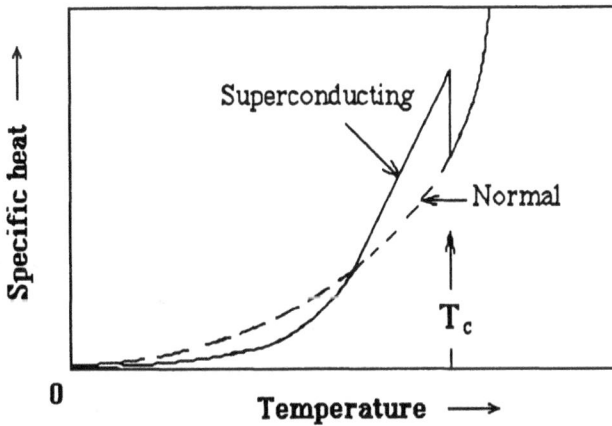

Figure 2.7.2 Specific heat of a superconductor in normal and superconducting states.

It is an interesting suggestion of Kok (1934) that if the specific heat of the metal in the superconducting state is assumed to vary as T^3 and the specific heat in the normal state is assumed to vary as $aT+bT^3$, then it is possible to predict the parabolic form (2.1.1) of describing the temperature-dependent behavior of the critical field $H_c(T)$, which is in fair agreement with experiment. In addition, the coefficient a in the linear term of the specific heat can be deduced from purely superconducting data.

2.8 Ginzburg-Landau theory of superconductors near T_c

It has been shown in section 2.2 that the fundamental assumptions used in London theory are the spatial invariance of the superconducting electron density n_s^* and the independence of n_s^* on the magnetic field. These assumptions are only justified for certain cases where the magnetic field is small and where no inhomogeneities are present. In many circumstances of physical or technological interest, however, these restrictions are not appropriate. In particular, one has seen from Pippard's work that the electromagnetic properties of superconductors in the superconducting state could only be understood if there existed a certain degree of long-range order, extending typically over the coherence distances. To account for the possible spatial variation of the superconducting electron density, Ginzburg and Landau (1950) proposed an extension of the London theory with the aid of Landau's general theory of second-order phase transitions (1937), in which Landau introduced an important concept of the order parameter.

2.8.1 Complex order parameter

It is assumed in the Ginzburg-Landau (G-L) theory that the behavior of the superconducting electrons may be described by a complex order parameter, being an "effective wave function" $\psi=|\psi|e^{i\theta}$ with the square of its amplitude $|\psi|$ equal to the superelectron density n_s^*, i.e. $|\psi|^2=n_s^*$. The order parameter ψ goes to zero at the superconducting-normal transition. Furthermore it is assumed that the free energy of the superconducting state differs from that of the normal state by an amount which can be expanded in powers of ψ with the coefficients of the expansion being regular functions of the temperature (where one ignores mechanical deformation effects) in the absence of the magnetic field. Thus at T close to T_c where ψ is small, they wrote

$$f_s - f_n = \alpha |\psi|^2 + \frac{1}{2}\beta |\psi|^4 + \gamma |\nabla\psi|^2 + \ldots \tag{2.8.1}$$

where f_s and f_n are respectively the free energies per unit volume in the superconducting and normal states. α, β and γ are phenomenological coefficients. The gradient term of ψ in eqn.(2.8.1) accounts for the non-uniformity of ψ in space. It is argued that if ψ varies slowly in space, it should be sufficient to keep only the leading term in $|\nabla\psi|^2$. To be in accordance with the Landau-Lifshitz general theory of second-order phase transitions (Landau and Lifshitz (1958)), only even powers may appear in the expansion of (2.8.1) due to the fact that the stability of the system at the transition point (at which $\psi = 0$) requires f_s to attain a minimum for $\psi = 0$. In addition, for the minimum in f_s to occur at finite values of $|\psi|^2$, we must have $\beta > 0$. Furthermore, since higher-order terms have been omitted, eqn.(2.8.1) is clearly only valid in the limit of small ψ, which implies temperatures close to the transition temperature T_c. To avoid inessential complications, we shall consider here only the superconducting crystal of cubic symmetry or the macroscopic isotropic superconducting solid. Superconductors of anisotropic behavior will, however, be considered in the next chapter.

In a homogeneous superconductor with no external field, the order parameter ψ is independent of the coordinates in space. In such a case, the gradient term $|\nabla\psi|^2$ disappears and the equilibrium value of $|\psi|^2$ is determined by the condition for the free energy density f_s to be a minimum. For $\alpha > 0$ the minimum occurs at $|\psi|^2 = 0$ corresponding to the normal state and to the case $T > T_c$. On the other hand, for $\alpha < 0$ the minimum occurs at

$$|\psi|^2 = |\psi_0|^2 = -\frac{\alpha}{\beta} \tag{2.8.2}$$

corresponding to the superconducting state where $T < T_c$. Thus, by eqns.(2.8.1) and (2.8.2), we get the free energy density difference

$$f_s - f_n = -\frac{\alpha^2}{2\beta} \tag{2.8.3}$$

which shows that the free energy of the superconducting state is lower than that of the normal state.

By noting that α must change its sign at $T = T_c$, one may write the expansion near the transition temperature T_c

$$\alpha(T) = \alpha_0(T - T_c)\,, \tag{2.8.4}$$

with the constant $\alpha_0 > 0$, and write

$$\beta(T) \approx \beta(T_c)\,. \tag{2.8.5}$$

By eqn.(2.8.2), we then have

$$|\psi_0|^2 = \frac{\alpha_0}{\beta(T_c)}(T_c - T)\,. \tag{2.8.6}$$

Thus, eqn.(2.8.3) may be written as

$$f_s - f_n = -\frac{\alpha_0^2}{2\beta(T_c)}(T_c - T)^2 \tag{2.8.7}$$

from which, we can find

$$\frac{\partial(f_s - f_n)}{\partial T} = \frac{\alpha_0^2}{\beta(T_c)}(T_c - T)\,. \tag{2.8.8}$$

which tends to zero for T -> T_c, indicating a phase transition at least of second order. The discontinuity in the specific heat at the transition point can then be found by

$$C_{vs} - C_{vn} = -vT_c\frac{\partial^2(f_s - f_n)}{\partial T^2} = vT_c\frac{\alpha_0^2}{\beta(T_c)}\,. \tag{2.8.9}$$

We may now define, in analogy to the definition of the critical magnetic field by (2.7.8), a thermodynamic critical field H_c for all types of superconductors by means of $f_n - f_s = \mu_o H_c^2/2$ at zero field. Thus, from eqn.(2.8.3), one may obtain

$$H_c(T) = \sqrt{\frac{\alpha^2}{\beta\mu_o}} \tag{2.8.10}$$

which can also be written, by eqns.(2.8.4) and (2.8.5) near the transition temperature, as

$$H_c(T) = \sqrt{\frac{\alpha_o^2}{\beta\mu_o}}\,|T_c - T|. \tag{2.8.11}$$

2.8.2 **Ginzburg-Landau equations**

In inhomogeneous superconductors, to take into account the non-local character (Pippard (1950, 1953)) of the superconducting state, i.e. of the fact that the values of $\psi(\mathbf{x})$ at any point are described by integral equations which involve the values of $\psi(\mathbf{x})$ at all neighboring points, the term of lowest order $|\nabla\psi(\mathbf{x})|^2$ accounting for the spatial variation in $\psi(\mathbf{x})$ has to be added to the free energy. If also magnetic field exists, Ginzburg and Landau proposed that a gauge invariant free energy density for the superconductor can be expanded in the following form

$$f_s(T, \mathbf{B}, \psi) = f_n + \alpha\,|\psi|^2 + \frac{1}{2}\beta\,|\psi|^4 + \frac{1}{2m^*}|(-i\hbar\nabla - e^*\mathbf{A})\psi|^2 + \frac{1}{2\mu_o}B^2 \tag{2.8.12}$$

where the fourth term on the r.h.s. represents presumably the kinetic energy of the superconducting electrons. This may be seen by identifying $|\psi|^2 = n_s^*$ with n_s^* being the number of the Cooper pairs of superconducting electrons per unit volume. The kinetic-energy density is then $m^*n_s^*v_s^2/2$ where m^* may be approximately identified as the mass of the Cooper pairs (the total mass of the two electrons) according to the experimental evidence by Zimmermann and Mercereau (1965)). The effect of a magnetic field is then introduced by resorting to a theorem in classical mechanics which states that the effect of the Lorentz force $(e^*\mathbf{v}_s \times \mathbf{B})$ on the motion of a charged particle in the magnetic field $\mathbf{B}$ may be completely accounted for by replacing the momentum **p**, wherever it occurs in the expression for the kinetic energy, by **p** - e***A**. Finally to make the transition to quantum-mechanics description, **p** is replaced by the operator $-i\hbar\nabla$ in the expression for the kinetic energy. It is shown that the phenomenological coefficient γ in eqn.(2.8.1) is identified here by $\gamma = \hbar^2/(2m^*)$.

The thermodynamic equilibrium of the system characterized by the free energy ($F = \int f dV$) as a functional of the three independent functions ψ, ψ^* and **A** (noting the relation $\mathbf{B} = \nabla \times \mathbf{A}$) requires the following set of differential equations to be satisfied, which are called the Ginzburg-Landau equations:

$$\alpha\psi + \beta|\psi|^2\psi + \frac{1}{2m^*}(-i\hbar\nabla - e^*\mathbf{A})^2\psi = 0 \qquad \text{in V} \tag{2.8.13}$$

$$\mathbf{J} = \frac{e^*\hbar}{i2m^*}(\psi^*\nabla\psi - \psi\nabla\psi^*) - \frac{e^{*2}}{m^*}|\psi|^2\mathbf{A} \qquad \text{in V} \tag{2.8.14}$$

where V denotes the volume of the superconductor. We have written **J** for $\mathbf{J}_s$ since in thermodynamic equilibrium there are no normal currents. Nonequilibrium cases where there are normal currents will, however, be studied in the next chapter.

By noting $\psi = |\psi|\exp(i\theta)$, eqn.(2.8.14) for the supercurrent density may also be wriiten as

$$\mathbf{J} = \frac{e^*|\psi|^2\hbar}{m^*}(\nabla\theta - \frac{e^*}{\hbar}\mathbf{A}) \tag{2.8.15}$$

which shows that the gradient of the phase of the wave function ψ determines the observable quantity, the supercurrent density. In the presence of an external magnetic field, eqn.(2.8.15) is shown to be of gauge-invariance as it shoud be.

To complete the theory, we have the following boundary conditions to the G-L equations

$$(\nabla\psi - \frac{ie^*}{\hbar}\mathbf{A}\psi)\cdot\mathbf{n} = \lambda_b\psi \qquad \text{on } \partial V \tag{2.8.16a}$$

where λ_b is a real constant, being zero for a superconductor-insulator (S-I) interface, i.e.,

$$(\nabla\psi - \frac{ie^*}{\hbar}\mathbf{A}\psi)\cdot\mathbf{n} = 0 \qquad \text{on } \partial V \tag{2.8.16b}$$

and non-zero for a superconductor-normal metal (S-N) interface, where the proximity effect can be of importance (de Gennes (1964) and Deutscher and de Gennes (1969)). It is easy to verify that, for real λ_b, the condition (2.8.16a) is consistent with the fact that the normal component of the superconducting current from eqn.(2.8.14) is zero across the S-N boundary since no nondissipative current can flow in a normal metal. However, the mutual effect of the electrons at the two sides of the interface between the superconductor and the normal metal can be extended much farther than the interatomic distance. The quantity $1/\lambda_b$ must be of the order of the coherence length ξ, but its exact value can only be determined from a microscopic theory of superconductivity. Some detail discussions based on the BCS microscopic theory for the coefficient λ_b may be found in the work of, for instance, Zaitsev (1965, 1966) and Deutscher and de Gennes (1969)). Due to the proximity effect, a sufficiently thin film of superconductor deposited on a bulk normal metal may cease to be superconductive, while a thin film of normal metal deposited on a bulk superconductor may become superconducting. These phenomena have been detected experimentally (Meissner (1958, 1959, 1960) and Werthamer (1963)), and have been used in the modelling of *in situ* composite superconductors (Carr (1983)). As to the boundary condition for the field, it is simply the continuity of the magnetic induction field **B** across the interface since we are considering non-magnetic superconductors.

2.8.3 Critical fluctuation and validity of Ginzburg-Landau theory

It is known that the physical quantities which describe a macroscopic body in equilibrium are, almost always, very nearly equal to their mean values. Nevertheless, deviations from the mean values, though probably small, do occur, which characterize the fluctuation of the physical quantities. In the discussion of a phase transition of the second kind, the fluctuation of the order parameter, which is the wave function ψ for the superconducting-normal transition, may increase rapidly when the transition point is approached due to the anomalous behavior of the thermodynamic functions of the body at the actual transition point

(Lifshitz and Pitaevskii (1980)). Thus, to have small fluctuation of the order parameter, one has to be restricted to situations outside of the fluctuation region near the transition point. On the other hand, the validity of the expansion (2.8.1) requires the order parameter to be small, which implies here the condition $|T - T_c| << T_c$ has to be satisfied. Therefore, for the G-L theory to be valid, the temperature of the superconductor to be considered should be within a certain range, away from the fluctuation region near the transition temperature so that both the order parameter and its fluctuation can be kept small. A quantitative estimate of the fluctuation region for the superconducting-normal transition was made by Ginzburg (1961), who showed that the fluctuation regions for common superconductors were extremely narrow so that the observed superconducting-normal transition was practically of a "normal" character (i.e. there are no anomalies of the specific heat and so on). From this it follows that the fluctuation is unimportant in the thermodynamics of bulk superconductors. Nevertheless, situations are possible where the role of fluctuation increases noticeably for some observed effects, say paraconductivity etc., in small objects such as thin films and filaments (Aslamazov and Larkin (1968) and Glover (1971)). Recently discovered high-T_c oxide superconductors may have relatively large fluctuation region due to their small coherence lengths (of the order of 10 Å), which implies some modifications of the G-L theory might be required (Inderhees et al. (1988)).

One may further ask what the range of the validity of G-L theory is on the low-temperature side. One condition is known to be $|1 - T/T_c| << 1$. The other is due to the limitation of G-L theory being a local approximation. To study the limitation of locality, we may introduce the following two characteristic lengths ξ and λ from the G-L equations (2.8.13) and (2.8.14). The first characteristic length ξ can be introduced by writing the G-L equation (2.8.13) in the following form

$$\xi^2 \nabla^2 \psi' - \psi' + \psi' |\psi'|^2 = 0 \tag{2.8.17}$$

where one has introduced a new non-dimensional function $\psi' = \psi / |\psi_0|$ with $|\psi_0|$ being the equilibrium solution given by eqn.(2.8.2). The quantity ξ is defined by

$$\xi(T) = \frac{\hbar}{\sqrt{2m^*|\alpha|}} \tag{2.8.18}$$

which clearly measures the range of variation of ψ' (i.e. which is the smallest distance over which the order parameter ψ can be of large fractional change). Therefore, it is also called the coherence length at temperature T (or the Ginzburg-Landau coherence length) in accordance with the concept of Pippard coherence length. However, it should be noticed that this coherence length $\xi(T)$ has a different significance from those of the (temperature-independent) Pippard coherence lengths, the intrinsic coherence length ξ_0 and the mean-free-path-dependent coherence length $\xi(l_e)$ given by eqn.(2.6.4).

The second characteristic length λ comes into play if we introduce electromagnetic effects. Consider a superconductor in a weak field: to the first order in **B**, $|\psi|^2$ can be replaced by $|\psi_0|^2$, and the G-L equation (2.8.14) may be written as

$$\mathbf{J} = \frac{e^*\hbar}{i2m^*}(\psi^*\nabla\psi - \psi\nabla\psi^*) - \frac{e^{*2}}{m^*}|\psi_0|^2 \mathbf{A} \ . \tag{2.8.19}$$

Taking the curl of **J**, one then obtains

$$\nabla \times \mathbf{J} = -\frac{e^{*2}}{m^*}|\psi_0|^2 \mathbf{B} \tag{2.8.20}$$

which is equivalent to the London equation (2.2.9) with the penetration depth defined by

$$\lambda(T) = \sqrt{\frac{m^*}{\mu_0 e^{*2}|\psi_0|^2}} = \sqrt{\frac{m^*\beta}{\mu_0 e^{*2}|\alpha|}} \ . \tag{2.8.21}$$

This temperature-dependent penetration depth characterizes the range of variation of electromagnetic induction field **B**.

It is shown that, by defining an effective density of the superelectron pairs $n^*_{s\,eff} = |\psi_0|^2$, the weak field solutions of the G-L theory are the same as those of the London theory when the gradient term of ψ is neglected. This is known to be able to give accurate results for local electrodynamics of the superconductors. For purposes of practical calculations, we may also assume it to be useful as an approximation in nonlocal cases by taking

$$|\psi_0(T)|^2 = \frac{m^*}{\mu_0 e^{*2} \lambda^2(T)} \tag{2.8.22}$$

with $\lambda(T)$ being the appropriate penetration depth, subject to direct measurement (see section 2.6).

In order for the G-L theory to be valid, both these lengths have to be large in comparison with the intrinsic coherence length ξ_0 so that all quantities vary sufficiently slowly in space. Such a condition can be, in general, satisfied near the transition point since, by noting eqns.(2.8.4) and (2.8.5), both lengths ξ and λ increase in proportion to $|T-T_c|^{-1/2}$ as the transition temperature is approached.

2.8.4 **Ginzburg-Landau parameter κ**

Due to the divergence of both two characteristic lengths $\xi(T)$ and $\lambda(T)$ as $T \rightarrow T_c$, it is thus of interest to consider their ratio

$$\kappa = \frac{\lambda(T)}{\xi(T)} = \frac{m^*}{e^*\hbar}\sqrt{\frac{2\beta}{\mu_0}} \tag{2.8.23}$$

which is called the Ginzburg-Landau parameter of the superconductive material. This parameter is independent of temperature within the framework of the G-L theory, and has the importance of characterizing superconducting materials. By eqns.(2.8.10) and (2.8.21), this parameter can also be related to observable quantities, the critical field H_c and the penetration depth λ, by

$$\kappa = \frac{\sqrt{2}\mu_0 e^*}{\hbar}\lambda^2 H_c \tag{2.8.24}$$

which shows that the G-L parameter κ may be determined by measuring the thermodynamic critical field H_c, and the penetration depth λ in low field. In practice, there are, however, other ways of obtaining the values of κ (Saint-James et al. (1969) and Kuper (1968)).

Further studies of the microscopic foundation of the G-L theory by Gor'kov (1959a,b) based on the BCS microscopic theory have helped to improve our understanding of the G-L equations and their limitations. It was found by Gor'kov that the G-L parameter κ may be expressed as

$$\kappa = \kappa_p = 0.96\,\frac{\lambda_L(0)}{\xi_o} \qquad \text{for pure superconductors } (\xi_o << l_e) \qquad (2.8.25)$$

and

$$\kappa = \kappa_d = 0.72\,\frac{\lambda_L(0)}{l_e} \qquad \text{for dirty superconductors } (\xi_o >> l_e) \qquad (2.8.26)$$

where $\lambda_L(0)$ is the London penetration depth at absolute zero, ξ_o Pippard's intrinsic coherence length introduced in (2.6.3), and λ_c the electron mean free path. Eqn.(2.8.26) shows that the κ of a superconductor is increased if the electron mean free path is shortened by a high impurity concentration, which also implies that the electric resistance of the superconductor in the normal state is increased. The general dependence of κ upon the normal state resistivity ρ_n has been given by Goodman (1962) as

$$\kappa = \kappa_p + 2.4 \times 10^6 \rho_n \sqrt{\gamma_e} \qquad (2.8.27)$$

where γ_e is the electronic specific heat constant in J m^{-3} deg^{-2} and ρ_n the normal state resistivity in Ωm. With the use of eqn.(2.8.27), good agreement with experiment results has also been reported by Livingston (1963). In addition, near T_c in the pure and dirty limit, the microscopic theory gives

$$\xi = \xi_p(T) = 0.74\,\xi_o \sqrt{\frac{T_c}{T_c - T}} \qquad \text{(pure)} \qquad (2.8.28)$$

and

$$\xi = \xi_d(T) = 0.85\sqrt{\xi_o l_e}\sqrt{\frac{T_c}{T_c - T}} \qquad \text{(dirty)} \qquad (2.8.29)$$

and

$$\lambda = \lambda_p(T) = \frac{\lambda_L(0)}{\sqrt{2}}\sqrt{\frac{T_c}{T_c - T}} \qquad \text{(pure)} \qquad (2.8.30)$$

and

$$\lambda = \lambda_d(T) = 0.615\, \lambda_L(0)\sqrt{\frac{\xi_0}{l_e}}\sqrt{\frac{T_c}{T_c - T}} \qquad \text{(dirty).} \tag{2.8.31}$$

Further generalization of the G-L theory to all temperatures has been made for dirty superconducting alloys in high magnetic fields based on the BCS microscopic theory (see, for instance, Maki (1964) and de Gennes (1966)), where one can find that the quantity κ is temperature-dependent. In the following section, we shall see that superconductors can be classified according to their particular values of the G-L parameter κ.

2.8.5 Upper critical field H_{c2}

It has been known that some pure superconductors undergo a first-order transition into the normal state at a critical field H_c, and the density of superelectrons has almost a constant value up to H_c and then drops abruptly to zero. In this section, with the use of G-L theory, we shall study superconductors which present a second-order phase transition at a certain critical field and that whose order parameter ψ approaches zero continuously when the critical field is reached. Let us now consider an infinite superconductive medium, and take the z-axis along the uniform applied magnetic field $\mathbf{B}^e$, i.e. $\mathbf{B}^e = (0, 0, B^e)$. When the applied field is sufficiently large the medium is in the complete normal state with $|\psi| = 0$. If the applied field is gradually reduced, one may ask what value this field must be reduced to in order that nucleation of superconducting regions starts to occur, i.e. that eqn.(2.8.13) may just have solutions other than ψ=0. Since during the first appearance of superconducting regions, ψ is small, we can linearize the G-L equation for ψ by dropping the nonlinear term $\beta|\psi|^2\psi$. In addition, the condition of small ψ implies that field penetration is virtually complete so that the magnetic field in the medium may be written approximately as $\mathbf{B} = \mathbf{B}^e$. Thus, the vector potential $\mathbf{A}$ appropriate to the field $\mathbf{B}$ may be chosen as

$$\mathbf{A} = (0, \mu_0 H^e x, 0). \tag{2.8.32}$$

The linearized form of G-L equation (2.8.13) now becomes

$$-\frac{\hbar^2}{2m^*}\left(\frac{\partial^2\psi}{\partial x^2}+\frac{\partial^2\psi}{\partial z^2}\right)+\frac{1}{2m^*}\left(-i\hbar\frac{\partial}{\partial y}-e^*\mu_0 H^e x\right)^2\psi=|\alpha|\psi \tag{2.8.33}$$

which has the form of the Schrödinger equation for a particle of mass m^*, charge e^* and zero spin moving in a uniform magnetic field directed along the z-axis (Landau and Lifshitz (1977)), where $|\alpha|$ plays the part of the eigenvalue. Thus, we may seek ψ in the following form

$$\psi(\mathbf{x}) = u(x)\exp(ik_y y)\exp(ik_z z) \tag{2.8.34}$$

where u(x) is yet to be determined.
By substituting eqn.(2.8.34) into eqn.(2.8.33), we find

$$-\frac{\hbar^2}{2m^*}\frac{\partial^2 u}{\partial x^2}+\frac{e^{*2}\mu_0^2(H^e)^2}{2m^*}(x-x_0)^2u=\left(|\alpha|-\frac{\hbar^2k_z^2}{2m^*}\right)u \tag{2.8.35}$$

which has the form of the Schrödinger equation for a particle moving in a harmonic potential well centered at

$$x_0=\frac{\hbar k_y}{e^*\mu_0 H^e}\ . \tag{2.8.36}$$

The eigenvalues of eqn.(2.8.35) are thus given by

$$|\alpha|=\left(n+\frac{1}{2}\right)\hbar\omega_c+\frac{\hbar^2k_z^2}{2m^*} \tag{2.8.37}$$

where ω_c is the cyclotron resonance frequency defined by

$$\omega_c = e^*\mu_0 H^e/m^*\ . \tag{2.8.38}$$

Eqn.(2.8.37) may also be expressed by

$$H^e=\frac{2m^*}{(2n+1)e^*\hbar\mu_0}\left(|\alpha|-\frac{\hbar^2k_z^2}{2m^*}\right). \tag{2.8.39}$$

Obviously the field of interest is the highest field for which superconductivity begins to occur. This corresponds to n=0 and k_z=0 in eqn.(2.8.39) and the field denoted by H_{c2} is

$$H_{c2} = \frac{2m^*|\alpha|}{e^*\hbar\mu_0} . \tag{2.8.40}$$

The corresponding eigenfunction is the ground-state wave function of the harmonic oscillator

$$u(x) = \exp[- \frac{(x - x_0)^2}{2\xi^2}] \tag{2.8.41}$$

where ξ is the G-L coherence length given by eqn.(2.8.18).
By noting eqns.(2.8.10) and (2.8.23), we can rewrite eqn.(2.8.40) as

$$H_{c2} = \kappa \sqrt{2}\, H_c \tag{2.8.42}$$

which is valid whatever the value of κ.

The result obtained above is for an infinite medium, where the influence of the specimen surfaces is neglected. In the case of a finite superconducting body, it has been shown (Saint-James and de Gennes (1963) and Saint-James et al. (1969)) that there exists a superconducting sheath with thickness of the order of ξ close to the surface of the body, in which $\psi \neq 0$ for magnetic fields up to $H_{c3}(\varphi)$ where φ is the angle between the direction of the field and the normal to the surface. For $\varphi = \pi/2$, i.e. for the surface parallel to the applied field, $H_{c3}(\varphi)$ has its maximum value of

$$H_{c3} = 1.695\, H_{c2} \tag{2.8.43}$$

which is larger than H_{c2}.

The presence of this superconducting sheath has been verified experimentally (Bon Mardion et al. (1964), Cardona and Rosenblum (1964) and Tomasch and Joseph (1964)). However, such a sheath does not exist if the surface of the superconductor is in contact with a normal metal rather than an insulator because of the different boundary condition on the normal derivative of ψ (see eqn.(2.8.16)).

2.8.6 **Classification of superconductors**

From eqn.(2.8.42), we can see that, if $\kappa > 1/\sqrt{2}$, there exist magnetic fields greater than H_c but less than H_{c2} in which the normal state is unstable with respect to the establishment of a certain degree of superconducting order ($\psi \neq 0$) (Ginzburg and Landau (1950)). We may, therefore, distinguish between two types of magnetic behavior of a superconductor in a decreasing field. If $\kappa < 1/\sqrt{2}$ (type I behavior) the normal state is the only stable state for $H^e > H_c$. When we decrease the field we first meet the value H_c, at which a complete Meissner effect takes place in the type I superconductors. If $\kappa > 1/\sqrt{2}$ (type II behavior) a new kind of superconducting state (characterized by the partial exclusion of a magnetic field) is stable for $H^e < H_{c2}$. The value $\kappa = 1/\sqrt{2}$ is thus the point that separates the two kinds of superconductors.

It may be shown that the surface energy associated with the interphase boundary between a normal and superconducting region is negative for type II superconductors, and is positive for type I superconductors (see, for instance, Saint-James et al. (1969)). A simple explanation of the origin of the surface energy may be given according to Rose-Innes and Rhoderick (1969). At the boundary of superconducting and normal phases, there is not a sudden change from fully normal behavior to full superconducting hehavior. The magnetic flux density penetrates a distance charaterized by λ into the superconducting region, and in the superconducting region the number of superelectrons per unit volume increases slowly over a distance characterized by ξ (see **Figure** 2.8.1). Thus, at the boundary, the degree of order (i.e. the number of superelectrons n_s^*) rises gradually over a distance determined by the coherence length ξ, so the decrease in free energy due to the increasing order of the electrons takes place over the same distance. On the other hand, the free energy rises over a distance of about the penetration depth λ due to the positive "magnetic" contribution. In general, ξ and λ are not the same so that the two contributions do not cancel near the boundary. If the coherence length is longer than the penetration depth, the total free energy density is increased close to the boundary; i.e., there is a positive surface energy. If the coherence is shorter than the penetration depth, the total free energy density is decreased close to the boundary; i.e. there is a negative surface energy.

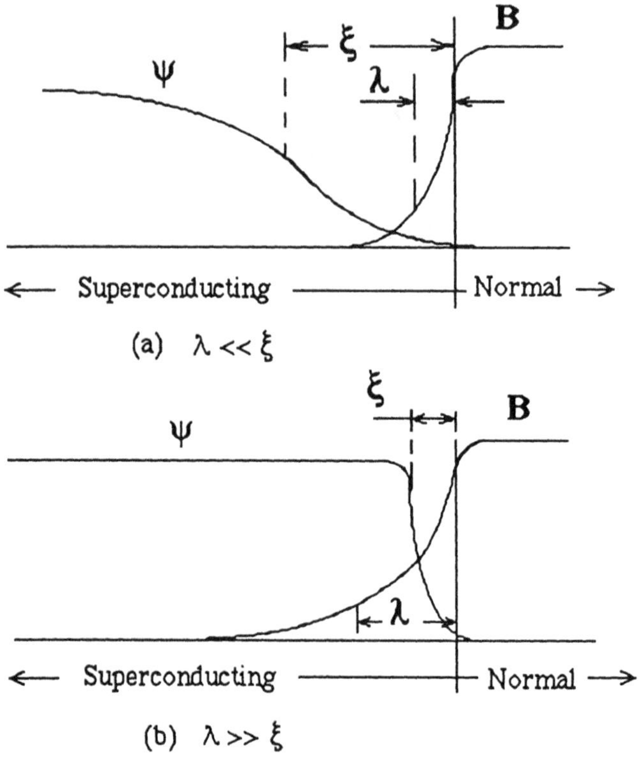

Figure 2.8.1 Schematic diagram of variation of B and ψ in a domain wall.
(a) $\lambda << \xi$ for positive surface energy (type I superconductor);
(b) $\lambda >> \xi$ for negative surface energy (type II superconductor).

This means that, for type I materials ($\kappa < 1/\sqrt{2}$), the existence of an interphase boundary leads to a raising of the Gibbs free energy over that of a uniformly superconducting sample. Thus it is energetically unfavorable for such interphase boundaries to exist in a sample of type I material with zero demagnetization factor. For samples of other shapes with non-zero demagnetization factor, such as a sphere, the energy involved in the distortion of the applied field makes a subdivision into superconducting and normal phases energetically favorable, and, thus, results in the presence of a so-called intermediate state in the superconducting sample. Such an intermediate state may be understood from the following consideration.

Consider a sphere of type I superconductor in a uniform applied magnetic intensity field $\mathbf{H}^e$. When the applied field $\mathbf{H}^e$ is raised gradually, the magnetic field strength at the equator of the sphere will first reach the critical value H_c for $H^e = 2H_c/3$ (see sect. 2.7). One might expect that the sphere would then be driven into a normal state. However, if it were so, the magnetic intensity field inside the sphere would be equal to $H^e = 2H_c/3$, which is less than H_c. One should then have the impossible situation of a completely normal body in a field smaller than H_c. Thus, the possible situation is that the magnetic flux may penetrate into the superconducting sphere such that the interior region of the sphere is divided into a certain structure of superconducting and normal domains. For the two phases to exist in equilibrium, the magnetic field at the interphase boundary must be tangential and equal to the critical field H_C. The positive free energy contribution of the interphase surface will determine the equilibrium configuration of superconducting and normal domains. Such a configuration is known as the intermediate state, which has a relatively coarse structure and has been confirmed experimentally (see Shoenberg (1952) and Bodmer et al. (1972) for more discussions).

For type II materials ($\kappa > 1/\sqrt{2}$), the presence of interface boundaries in the superconducting sample, even having a zero demagnetization factor, are, however, energetically favorable. Thus, when the type II superconductor is placed in a magnetic field above a certain value, the superconductor will be in a mixed state, i.e. the superconducting medium splits into some fine-scale mixture of superconducting and normal regions whose boundaries lie parallel to the applied field. It can be seen that the intermediate state in type I superconductors is fundamentally different from the mixed state in type II superconductors. In addition, the structure of the intermediate state is relatively coarse and the gross features may be made visible to the naked eye. The structure of the mixed state is, however, on a much finer scale with a periodicity generally less than 10^{-7} m, which was first theoretically predicted by Abrikosov (1957) and was later verified experimentally by Essmann and Träuble (1967) etc. The detail analysis of the mixed state will be given in the next section.

2.8.7 **Abrikosov mixed state in type II superconductors**

Here, we shall present the analysis of Abrikosov (1957) on the vortex structures in the mixed state of type II superconductors in a field near the upper critical field H_{c2}. In the analysis, only the regime with applied fields slightly less than H_{c2} is treated, where the solution ψ of the complete G-L equations must have strong similarity to a certain solution ψ_L of the linearized G-L equations. This may be seen by considering a sample of type II superconductive material with negligible demagnetization effect placed in an external magnetic field $\mathbf{B}^e$ ($=\mu_0\mathbf{H}^e$). If one let $\mathbf{B}^e$ decrease gradually, the nucleation of the superconducting phase in the interior of the sample will begin when the magnetic intensity field H^e in the sample becomes equal to the upper critical field H_{c2}. Thus, for the field H^e slightly less than H_{c2}, the order parameter $|\psi|$ will be small and the form of ψ may be obtained from the linearized G-L equations. However, if the field H^e decreases appreciably below H_{c2}, the order parameter $|\psi|$ will become larger and, then, the complete nonlinear G-L equations have to be used.

In the case of the applied field slightly less than H_{c2}, as the first-order approximation, we have the linearized G-L equation for the determination of ψ_L, which reads

$$\alpha\psi_L + \frac{1}{2m^*}(-i\hbar\nabla - e^*\mathbf{A}_o)^2\psi_L = 0 \qquad (2.8.44)$$

with $\mathbf{A}_o = (0, \mu_0 H_{c2}x, 0)$ so that

$$\nabla \times \mathbf{A}_o = (0, 0, \mu_0 H_{c2}) \qquad (2.8.45)$$

where we have assumed that the applied field is along the z-direction. From section 2.8.5, we have shown that the solutions of eqn.(2.8.44) are of the form

$$\psi_k = e^{iky}\exp[-\frac{(x-x_o)^2}{2\xi^2}] \qquad (2.8.46)$$

with

$$x_o = \frac{\hbar k}{e^*\mu_0 H_{c2}} = \frac{\Phi_o k}{2\pi\mu_0 H_{c2}} \qquad (2.8.47)$$

where k is an arbitrary parameter and Φ_0 is the basic quantum of magnetic flux defined by

$$\Phi_0 = \frac{2\pi\hbar}{e^*} = 2.07 \times 10^{-15} \quad \text{Wb}. \tag{2.8.48}$$

It can be seen from eqn.(2.8.46) that the solutions describe a Gaussian band of superconductivity of width $\xi(T)$ extending perpendicular to the x-axis at the location $x = x_0(k)$. A general solution ψ_L must be a linear combination of the ψ_k. Following the original calculation of Abrikosov, we shall consider the solution being periodic both in x- and y-direction. The periodicity in y-direction is achieved by setting

$$k = k_n = nq \tag{2.8.49}$$

yielding the period

$$\Delta y = \frac{2\pi}{q} \; . \tag{2.8.50}$$

Thus the general solution for ψ_L can be expressed as

$$\psi_L = \sum_n C_n e^{inqy} \exp[-\frac{(x - x_n)^2}{2\xi^2(T)}] \tag{2.8.51}$$

with

$$x_n = \frac{\Phi_0 nq}{2\pi\mu_0 H_{c2}} \; . \tag{2.8.52}$$

The periodicity in x-direction can be established if the coefficients C_n are periodic functions of n, such that $C_{n+v}=C_n$, where v is some integer. The particular choice of v determines the type of periodic lattice structure (v = 1: square lattice; v = 2: triangular lattice). From eqns. (2.8.50) and (2.8.52), we note that the periodicity in x-direction is

$$\Delta x = \frac{\Phi_0}{2\pi\mu_0 H_{c2}} \frac{2\pi}{\Delta y} \tag{2.8.53}$$

yielding

$$\Delta x \Delta y \, \mu_0 H_{c2} = \Phi_0 \tag{2.8.54}$$

which means that each unit cell of the periodic array contains one flux quantum.

From the form ψ_L in eqn.(2.8.51) we can draw some general conclusions independent of the choices of C_n and q. The first important conclusion concerns the current $\mathbf{J}_L$ associated with ψ_L. By inserting eqn.(2.8.51) into eqn.(2.8.14), we obtain for the current density

$$J_{Lx} = -\frac{e^*\hbar}{2m^*}\frac{\partial}{\partial y}|\psi_L|^2 \tag{2.8.55a}$$

$$J_{Ly} = \frac{e^*\hbar}{2m^*}\frac{\partial}{\partial x}|\psi_L|^2 \tag{2.8.55b}$$

which shows that the lines of current flow $\mathbf{J}_L$ coincide with the lines of constant $|\psi_L|$. Furthermore, the local magnetic field $\mathbf{B}_s$ associated with the supercurrents by $\nabla \times \mathbf{B}_s = \mu_0 \mathbf{J}_L$ may be found to be

$$\mathbf{B}_s = -\frac{\mu_0 e^*\hbar}{2m^*}|\psi_L|^2 \, \mathbf{z}^0 \tag{2.8.56}$$

where $\mathbf{z}^0$ is the unit vector along the direction of z-axis of coordinates. Eqn.(2.8.56) also indicates that the lines of constant $|\psi_L|$ coincide with the lines of constant local magnetic field $\mathbf{B}_s$.

We shall now study the correction to the first-order solution ψ_L from the perturbation of the field which is slightly less than the critical field H_{c2}. We consider the normalization of ψ_L, which is rather important due to the nonlinearity of the complete G-L equations. The normalization will determine the strength of the supercurrent and, therefore, the macroscopic magnetic flux density and the mean free energy density. We assume that the free energy F remains stationary if ψ_L is replaced by the function $(1+\varepsilon)\psi_L$, where ε is a small quantity independent of the spatial coordinate $\mathbf{x}$, To the first order in ε, the variation in the free energy is

$$\delta F = 2\varepsilon \int \{\alpha|\psi_L|^2 + \beta|\psi_L|^4 + \frac{1}{2m^*}|(-i\hbar\nabla - e^*\mathbf{A})\psi_L|^2\} dV \,. \tag{2.8.57}$$

By defining the following form of volume average integrals

$$<|\psi_L|^2> = \frac{1}{V}\int |\psi_L|^2 \, dV \tag{2.8.58}$$

with V being the macroscopic volume, the normalization condition $\delta F=0$ yields

$$\alpha <|\psi_L|^2> + \beta <|\psi_L|^4> + \frac{1}{2m^*}<|(-i\hbar\nabla - e^*\mathbf{A})\psi_L|^2> = 0 \,. \tag{2.8.59}$$

By setting

$$\mathbf{A} = \mathbf{A}_o + \mathbf{A}_1 \tag{2.8.60}$$

where $\mathbf{A}_o$ is the vector potential for $B=\mu_0 H_{c2}$. The correction $\mathbf{A}_1$ arises from the fact that the applied field is slightly less than $\mu_0 H_{c2}$ and that the supercurrents also contribute to the field. Noting that ψ_L must satisfy eqn.(2.8.44) and keeping only terms up to first order in $\mathbf{A}_1$, we obtain

$$\beta <|\psi_L|^4> - <\mathbf{A}_1 \cdot \mathbf{J}_L> = 0 \tag{2.8.61}$$

with

$$\mathbf{J}_L = \frac{e^*\hbar}{i2m^*}(\psi_L^*\nabla\psi_L - \psi_L\nabla\psi_L^*) - \frac{e^{*2}}{m^*}|\psi_L|^2 \mathbf{A}_o \,. \tag{2.8.62}$$

We note that $\mathbf{J}_L$ is the current associated with the unperturbed solution. Integrating the second term in (2.8.61) in parts and setting $\nabla \times \mathbf{A}_1 = \mathbf{B}_1$ and $\nabla \times \mathbf{B}_s = \mu_0 \mathbf{J}_L$, we find

$$\beta <|\psi_L|^4> - \frac{1}{\mu_o}<\mathbf{B}_1 \cdot \mathbf{B}_s> = 0 \,. \tag{2.8.63}$$

Noting that $\mathbf{B}_1$ and $\mathbf{B}_s$ are everywhere parallel to the z-direction, we can write

$$B_1(x) = B^e - \mu_o H_{c2} + B_s(x) . \qquad (2.8.64)$$

Inserting (2.8.56) and (2.8.64) into (2.8.63), we have

$$\beta <|\psi_L|^4> + \frac{e^*\hbar}{2m^*} <|\psi_L|^2(B^e - \mu_o H_{c2} - \frac{\mu_o e^*\hbar}{2m^*}|\psi_L|^2)> = 0 \qquad (2.8.65a)$$

which can be further written, by noting eqn.(2.8.23) for κ, as

$$(2\kappa^2 - 1)\frac{\mu_o e^*\hbar}{2m^*} <|\psi_L|^4> - (\mu_o H_{c2} - B^e)<|\psi_L|^2> = 0 . \qquad (2.8.65b)$$

Setting again $|\psi_L| = |\psi_o|\psi_L'$ and using eqns.(2.8.21), (2.8.24) and (2.8.42), we can get

$$<\psi_L'^4> (1 - \frac{1}{2\kappa^2}) - <\psi_L'^2> (1 - \frac{B^e}{\mu_o H_{c2}}) = 0 . \qquad (2.8.66)$$

This equation represents a rather general result, which is independent of the detailed behavior of the function ψ_L, i.e., independent of the type of the periodic lattice configuration of ψ_L.

For a particular lattice type, as determined by selecting the wave number q and the periodicity of the coefficients C_n, we can calculate the quantity

$$\beta_A = \frac{<\psi_L'^4>}{<\psi_L'^2>^2} \qquad (2.8.67)$$

which is a function of the geometry of the vortex arrangement only, and is larger than 1 (Schwartz inequality). The ratio β_A takes the value unity if ψ is spatially constant and becomes increasingly large for functions which are more and more peaked locally. It is only the quantity β_A which must be determined numerically for obtaining $<\psi_L'^2>$ and $<\psi_L'^4>$ from eqns.(2.8.66) and (2.8.67). The macroscopic magnetic flux density <B> and the free energy density can then be calculated immediately. By using (2.8.56), the macroscopic magnetic flux density <B> may be expressed as

$$\langle B\rangle = B^e + \langle B_s\rangle = B^e - \frac{\mu_o e^* \hbar}{2m^*}\langle|\psi_L|^2\rangle . \tag{2.8.68}$$

Using eqns.(2.8.22) and (2.8.24), we find

$$\langle B\rangle = B^e - \frac{\mu_o H_c}{\sqrt{2}\kappa}\langle\psi_L'^2\rangle) . \tag{2.8.69}$$

Eliminating $\langle\psi_L'^2\rangle$ by using eqns.(2.8.66) and (2.8.67), we finally find

$$\langle B\rangle = B^e - \frac{\mu_o(H_{c2} - H^e)}{(2\kappa^2 - 1)\beta_A} \tag{2.8.70}$$

and equivalently the macroscopic magnetization M

$$M = \frac{1}{\mu_o}\langle B\rangle - H^e = -\frac{H_{c2} - H^e}{(2\kappa^2 - 1)\beta_A} . \tag{2.8.71}$$

The magnetization is shown to vanish at $H^e = H_{c2}$ and the transition is of the second order. The slope of the magnetization is finite and is given by

$$\frac{dM}{dH^e} = \frac{1}{(2\kappa^2 - 1)\beta_A} \tag{2.8.72}$$

which becomes very large when κ approaches the value $1/\sqrt{2}$ from above, and diverges at this value. It is such a discontinuous rise in M which characterizes type I superconductors as shown in **Figure** 2.8.2. Using the identity for the mean Gibbs free energy density g

$$\left(\frac{\partial g}{\partial H}\right)_T = -\mu_o M \tag{2.8.73}$$

we may calculate g by integrating down from the normal state at H_{c2}, where $g_s(T, H_{c2}) = g_n(T, H_{c2})$, and obtain

$$g_s(T, H^e) - g_n(T, H_{c2}) = -\frac{\mu_o}{2}\frac{(H_{c2} - H^e)^2}{(2\kappa^2 - 1)\beta_A} \tag{2.8.74}$$

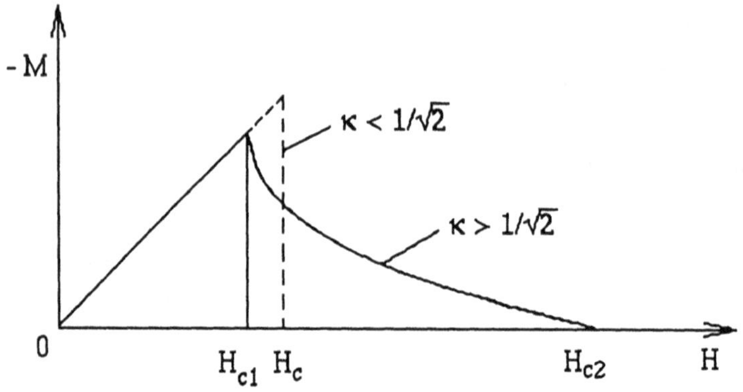

Figure 2.8.2 Variation of the magnetization versus the field H in a superconductor.

which applies to the regime $H^e < H_{c2}$ and $\kappa > 1/\sqrt{2}$. Eqn.(2.8.74) indicates that the configuration with the smallest value of β_A is thermodynamically most stable. Numerical calculations show that the square lattice and the triangular lattice yield the values $\beta_A = 1.18$ and $\beta_A = 1.16$, respectively (Kleiner et al. (1964)). This means that a triangular flux arrangement will be lowest in energy and, therefore, will be the most stable form, at least in isotropic superconductors. The theoretical result has been well supported experimentally. **Figure** 2.8.3 shows the triangular pattern of fluxons in a type II superconductor.

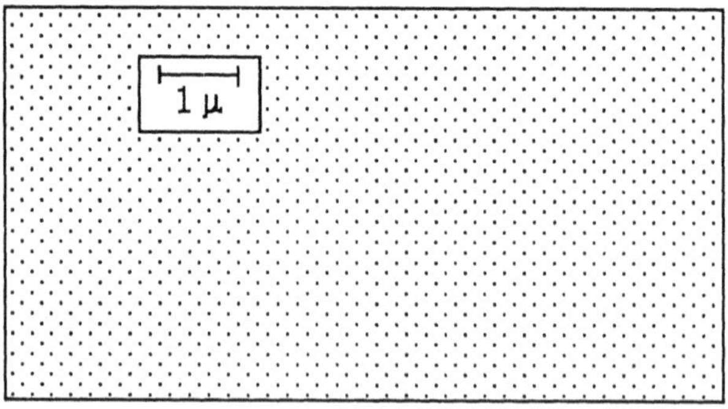

Figure 2.8.3 The triangular pattern of vortex lattice in a type II superconductor.

The experimental observation of such triangular vortex lattice in a Nb disk in a perpendicular magnetic field was made by using a high-resolution Bitter method (see Essmann and Träuble (1967)).

2.9 London model of vortex and lower critical field H_{c1}

This section is concerned with the study of vortex lines in type II superconductors in fields slightly higher than a critical field H_{c1}, called the lower critical field, defined by the lowest field at which the first vortex line appears in the superconductor. In such a case, only a few vortices occur and, thus, the separation of these vortices may be much greater than the penetration depth λ so that it is possible to treat them individually as isolated vortex by neglecting the interaction among these vortices as the first approximation. In what follows we shall introduce the London model for the study of these vortices. It is assumed in this model that the diameter ($\approx 2\xi$) of the normal vortex core is very small compared with the penetration depth λ (i.e. $\kappa >> 1$). With this restriction, the London model represents a good approximation at all temperatures (not too close to T_c) and electron mean free paths, and for magnetic fields in the range $H_{c1} < H << H_{c2}$, where the interaction between vortices is not too strong.

First, let us study an isolated vortex line. For $\kappa >> 1$, the local magnetic field **b** may be described by the following modified London equation

$$\mu_0\lambda^2 \nabla \times \mathbf{J} + \mathbf{b} = \Phi_0\delta(\mathbf{r} - \mathbf{r}_0)\mathbf{z}^o \tag{2.9.1}$$

where Φ_0 is the flux quantum of the vortex line defined in eqn.(2.8.48), $\mathbf{z}^o$ is the unit vector along the axis of the vortex line, and $\delta(\mathbf{r} - \mathbf{r}_0)$ is the two-dimensional delta function with $\mathbf{r} - \mathbf{r}_0$ being the radial distance vector from the center of the vortex line. With the use of Maxwell's equations $\nabla \times \mathbf{b} = \mu_0\mathbf{J}$ and $\nabla\cdot\mathbf{b} = 0$, eqn.(2.9.1) becomes

$$\lambda^2 \nabla^2\mathbf{b} - \mathbf{b} = -\Phi_0\delta(\mathbf{r} - \mathbf{r}_0)\mathbf{z}^o . \tag{2.9.2}$$

The solution for this equation in the circular cylindrical coordinates (r,φ,z) with its origin on the vortex line may be found as

$$\mathbf{b}(r) = \frac{\Phi_0}{2\pi\lambda^2} K_0(r/\lambda)\, \mathbf{z}^0 \tag{2.9.3}$$

and the circulating supercurrent reads

$$\mathbf{J}(r) = \frac{\Phi_0}{2\pi\mu_0\lambda^3} K_1(r/\lambda)\, \boldsymbol{\varphi}^0 \tag{2.9.4}$$

where K_0 and K_1 are the Hankel functions of imaginary argument of zero order and of first order respectively (Morse and Feshbach (1953)). With the use of the asymptotic approximations for K_0, we have

$$b(r) \approx \frac{\Phi_0}{2\pi\lambda^2} \ln(\lambda/r), \qquad \text{for } \xi \leq r << \lambda \tag{2.9.5}$$

and

$$b(r) \approx \frac{\Phi_0}{2\pi\lambda^2} \sqrt{\frac{\pi\lambda}{2r}} \exp(-r/\lambda), \qquad \text{for } r >> \lambda\,. \tag{2.9.6}$$

In the London model of vortex, it is assumed that $|\psi|$ is constant everywhere except at $r = 0$, where there is a singularity. To eliminate the artificial singularity, one has used eqn.(2.9.5) by cutting off the solution at $r = \xi$.

The energy E_v per unit length of the vortex line can be calculated from

$$E_v = \int_{S-S_0} \frac{1}{2\mu_0} [\mathbf{b}^2 + \lambda^2(\nabla \times \mathbf{b})^2]\, dS \tag{2.9.7}$$

where the plane surface S is perpendicular to the vortex line, and S_0 denotes the small cross surface of the core of the vortex $(0 \leq r < \xi)$, which is excluded from the integration. After some manipulation, we can find

$$E_v = \frac{\Phi_0^2}{4\pi\mu_0\lambda^2} \ln(\kappa) \tag{2.9.8}$$

which shows that the energy of a vortex line depends logarithmically upon the G-L parameter κ, and quadratically upon the flux quantum Φ_0 per vortex line.

The neglected contribution of the normal core to the energy may be estimated approximately to be $(\mu_0 H_c^2/2)\pi\xi^2$ per unit length. Eqn.(2.9.8) may be transformed, with the aid of eqns.(2.8.24) and (2.8.48), to

$$E_v = \frac{\mu_0}{2} H_c^2 \pi \xi^2 [4\ln(\kappa)] \tag{2.9.9}$$

which shows that the contribution of the normal vortex core is smaller than this value by a factor of about $4\ln(\kappa)$.

Now, from the energy E_v of one vortex line, we can calculate the lower critical field H_{c1} by noting the fact that the Gibbs free energies must be equal at the common boundary between the Meissner state (or phase) with no vortices and the mixed state with a few vortices at H_{c1}, i.e.

$$nE_v = H_{c1} \int_S b\, dS = H_{c1} n \Phi_0 \tag{2.9.10}$$

where n is the number of vortices in the superconductor, which is assumed to be a long and thin sample parallel to the magnetic field so that the internal field is uniform and equal to the applied field. From eqn.(2.9.10), we can find

$$H_{c1} = E_v/\Phi_0 \approx \frac{\Phi_0}{4\pi\mu_0\lambda^2} \ln(\kappa) \qquad \text{for } \kappa >> 1 \tag{2.9.11}$$

which may also be expressed in terms of H_c

$$H_{c1} \approx \frac{H_c}{\sqrt{2}\,\kappa} \ln(\kappa)\,. \tag{2.9.12}$$

This equation corresponds to the relation (2.8.42) between H_{c2} and H_c.

2.10 Critical field and critical current in thin films

Superconducting films with the advantage of high switching speed, compared to that of bulk superconductors limited by eddy currents, have led to their use in superconductive computer devices. Because of their possible device applications and theoretical interest they have been studied intensively. In this section, we shall present some basic properties of thin films by using the Ginzburg-Landau theory.

2.10.1 Solution of Ginzburg-Landau equations for thin film

To begin with, we introduce the following non-dimensional quantities

$$x' = \frac{x}{\lambda}, \qquad \psi' = \frac{\psi}{|\psi_0|} \qquad \text{and} \qquad \mathbf{A}' = \frac{\mathbf{A}}{\sqrt{2}\mu_0 H_c \lambda} \tag{2.10.1}$$

where λ, $|\psi_0|^2$ and H_c are respectively given by eqn.(2.8.21), (2.8.2) and (2.8.10). Thus, the G-L equation (2.8.13) may be written in the following non-dimensional form

$$\psi'|\psi'|^2 - \psi' + (-\frac{i}{\kappa}\nabla' - \mathbf{A}')^2\psi' = 0 \qquad \text{in V} \tag{2.10.2}$$

and the insulator-superconducting boundary condition (2.8.16b) is

$$\mathbf{n} \cdot (-\frac{i}{\kappa}\nabla' - \mathbf{A}')\psi' = 0 \qquad \text{on } \partial V \tag{2.10.3}$$

where the parameter κ is defined by eqn.(2.8.23).

Suppose now the film with the thickness d occupies the region $-d/2 < x < d/2$ and the applied field is along the z-axis, i.e. $\mathbf{B}^e=(0, 0, B^e)$, we may choose $\mathbf{A}=(0, A(x), 0)$ so that the magnetic field in the film may be expressed as $\mathbf{B}=(0, 0, dA/dx)$. Thus, eqns.(2.10.2) and (2.10.3) become

$$\frac{d^2\psi'}{dx'^2} = \kappa^2\psi'(A'^2 - 1 + |\psi'|^2) \qquad \text{for } -\frac{d}{2\lambda} < x' < \frac{d}{2\lambda} \tag{2.10.4}$$

and

$$\frac{d\psi'}{dx'} = 0 \qquad \text{for} \;\; x' = \pm\frac{d}{2\lambda}\;. \tag{2.10.5}$$

Since κ << 1 for many pure (type I) superconductors, we shall solve the problem in this case. At the zeroth approximation, we may neglect the right-hand side of eqn.(2.10.4), and find

$$\psi' = \psi'_0 = \text{const.} \tag{2.10.6}$$

Substituting this, as yet unknown, constant into eqn.(2.8.14) and using Maxwell's eqnation of Ampere's law, which now reads

$$\frac{d^2A'}{dx'^2} = |\psi'_0|^2 A' \qquad \text{for} \;\; -\frac{d}{2\lambda} < x' < \frac{d}{2\lambda} \tag{2.10.7}$$

together with the boundary condition $B(\pm\frac{d}{2}) = B^e$, we can obtain

$$B(x) = B^e \frac{\cosh(|\psi'_0|\frac{x}{\lambda})}{\cosh(|\psi'_0|\frac{d}{2\lambda})} \tag{2.10.8}$$

$$A(x) = \frac{B^e\lambda}{|\psi'_0|}\frac{\sinh(|\psi'_0|\frac{x}{\lambda})}{\cosh(|\psi'_0|\frac{d}{2\lambda})}\;. \tag{2.10.9}$$

Consider the next approximation by substituting $\psi' = \psi'_0 + \psi'_1$ into eqn.(2.10.4), we get

$$\frac{d^2\psi'_1}{dx'^2} = \kappa^2\psi'_0(A'^2 - 1 + |\psi'_0|^2). \tag{2.10.10}$$

Noting the boundary condition (2.10.5), we find

$$\int_{-\frac{d}{2\lambda}}^{\frac{d}{2\lambda}} (A'^2 - 1 + |\psi'_0|^2)dx' = 0 \tag{2.10.11}$$

which then gives

$$\left(\frac{B^e}{\sqrt{2}\mu_o H_c}\right)^2 = \frac{2|\psi_o'|^2(1-|\psi_o'|^2)\cosh^2(|\psi_o'|\frac{d}{2\lambda})}{\frac{\lambda}{|\psi_o'|d}\sinh(|\psi_o'|\frac{d}{\lambda}) - 1} . \tag{2.10.12}$$

This relation is valid at any B^e and gives an implicit relation for $|\psi_o'|$ in terms of B^e. It is shown that, though the order parameter $|\psi_o'|$ is treated as independent of space coordinate, it depends on the applied magnetic field.

From eqn.(2.10.8), we can calculate the mean magnetization of the film by

$$M = \frac{1}{\mu_o}\left(\frac{1}{d}\int_{-\frac{d}{2}}^{\frac{d}{2}} B dx - B^e\right) = -\frac{B^e}{\mu_o}\left[1 - \frac{2\lambda}{|\psi_o'|d}\tanh(|\psi_o'|\frac{d}{2\lambda})\right] \tag{2.10.13}$$

which shows that, for $|\psi_o'|d/\lambda >> 1$, we have $M \approx -H^e = -B^e/\mu_o$, corresponding to the Meissner effect. If, however, $|\psi_o'|d/\lambda << 1$, then, from eqns.(2.10.12) and (2.10.13), we find

$$\left(\frac{B^e}{\sqrt{2}\mu_o H_c}\right)^2 \approx \frac{12\lambda^2(1-|\psi_o'|^2)}{d^2} \quad \text{or} \quad |\psi_o'|^2 \approx 1 - \frac{d^2}{12\lambda^2}\left(\frac{B^e}{\sqrt{2}\mu_o H_c}\right)^2 \tag{2.10.14}$$

and

$$M \approx -\frac{B^e}{\mu_o}\frac{|\psi_o'|^2 d^2}{12\lambda^2} \approx -\frac{B^e}{\mu_o}\frac{d^2}{12\lambda^2}\left[1 - \frac{d^2}{12\lambda^2}\left(\frac{B^e}{\sqrt{2}\mu_o H_c}\right)^2\right] . \tag{2.10.15}$$

1.10.2 **Critical field for thin film**

The results show that, for sufficiently thin film, the order parameter $|\psi_o'|$, and therefore the magnetization, goes smoothly to zero with the increase of the magnetic field. Thus, the corresponding superconducting-normal phase transition is of the second-order, even for the film of type I superconductors. The corresponding critical field denoted by

H_{cf} for the thin film may be found from eqn.(2.10.14) by setting $|\psi_0'|^2=0$, i.e.,

$$H_{cf} = \frac{\lambda\sqrt{24}}{d} H_c \qquad (2.10.16)$$

which shows that, for very thin films, there is a considerable increase in the critical field H_{cf} over the bulk value H_c for the same material.

We may now calculate the Gibbs free energy, got by adding $-\mathbf{B}^e \cdot \mathbf{B}/\mu_0 + (B^e)^2/(2\mu_0)$ to eqn.(2.8.12). With ψ being constant, we have

$$\frac{G_s - G_n}{V} = \frac{1}{d}\int_{-\frac{d}{2}}^{\frac{d}{2}} \{\alpha|\psi|^2 + \frac{\beta}{2}|\psi|^4 + \frac{e^{*2}}{2m^*}A^2|\psi|^2 + \frac{1}{2\mu_0}(B - B^e)^2\}dx \qquad (2.10.17)$$

where V is the volume of the thin film. Substituting eqns.(2.10.8) and (2.10.9) into eqn.(2.10.17) and using eqn.(2.10.1), we can find

$$\Delta g' = \frac{G_s - G_n}{V\mu_0 H_c^2/2} = |\psi_0'|^4 - 2|\psi_0'|^2 + (\frac{B^e}{\mu_0 H_c})^2[1 - \frac{2\lambda}{|\psi_0'|d}\tanh(|\psi_0'|\frac{d}{2\lambda})] \,. \qquad (2.10.18)$$

In a field equal to the critical field H_{cf}, we must have $\Delta g'=0$, which means

$$(\frac{H_{cf}}{H_c})^2 = \frac{|\psi_0'|^2(2 - |\psi_0'|^2)}{1 - \frac{2\lambda}{|\psi_0'|d}\tanh(|\psi_0'|\frac{d}{2\lambda})} \,. \qquad (2.10.19)$$

Now consider the limiting case where the film thickness is large ($d \gg \lambda$). We can expect a first-order phase transition for the film of type I superconductor, which implies that $|\psi_0'|$ is close to unity at the transition point. Thus, we may let $|\psi_0'| = 1 - \varepsilon'$, where $\varepsilon' \ll 1$. We can find ε' approximately from eqn.(2.10.12), and then, from eqn.(2.10.19), find the critical field for the film with $d \gg \lambda$ by

$$H_{cf} = (1 + \frac{\lambda}{d})H_c \,. \qquad (2.10.20)$$

It can be expected that, at a certain thickness d_c of the film, the first-order phase transition changes to a second-order phase transition.

Indeed, a representative form of the variation of $|\psi_c/\psi_0|$ with respective to d/λ may be shown in **Figure** 2.10.1, where the value of d_c is found to be $\sqrt{5}\lambda$ (Douglass (1962) and Abrikosov (1988)). Since λ is temperature-dependent, even relatively large samples may undergo a second-order transition at temperatures sufficiently close to the transition temperature T_c.

In the above analysis, we have calculated the critical field of thin films of type I superconductors with $\kappa \ll 1$, which enabled us to assume $\psi \approx$ const. throughout the film. The result obtained may also be generalized to thin films of type II superconductors with their thickness d less than the coherence length ξ since for these the order parameter ψ must not vary throughout the film thickness. Practically, experiments with films of thicknesses down to about 10^{-8} m (in comparison with ξ being of the order of 10^{-6} m in pure crystals) are entirely feasible. Thus, for such thin films of type II superconductors with $\xi \ll \lambda$, one can use the formulas for $d \ll \lambda$. In particular, the critical field of the transition to the normal state is expressed by eqn.(2.10.16).

We have, so far, been discussing the case of a field parallel to the surface of the film. The actual behavior of a thin film in a magnetic field depends markedly upon the direction of that field. For fields normal to

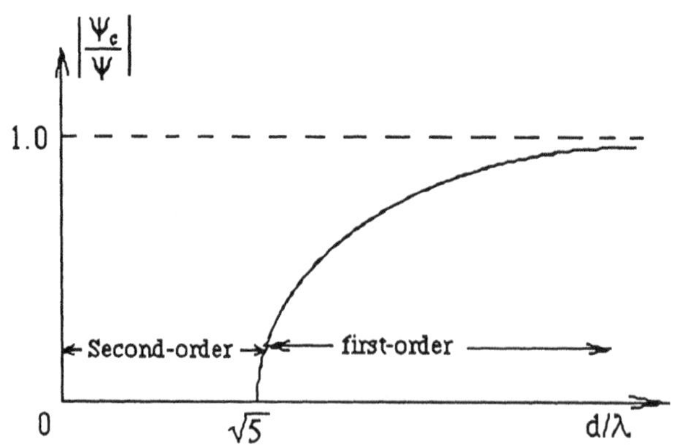

Figure 2.10.1 Critical value of order parameter at the superconducting-normal transition as a function of film thickness.

the film, the flux line picture of Abrikosov is likely to be of great importance even for films of type I superconductors (Tinkham (1963)). The field penetrates the film by way of individual flux lines, spaced some distance apart. Eventually, an "upper critical field" is reached at which the flux lines touch and the film makes the transition to the normal state. According to Tinkham (1963), the perpendicular critical field $H_{c\perp}$ for sufficiently thin films ($d < d_c$) of type I superconductors may be written as

$$H_{c\perp} = \sqrt{2}\,\kappa(d)H_c \tag{2.10.21}$$

where the G-L paramcter $\kappa(d)$ depends on the film thickness d and can be expressed approximately in the following form

$$\kappa(d) = 0.715\lambda_L(0)[\frac{1}{l_{eb}} + \frac{C}{d}]H_c \tag{2.10.22}$$

Here, l_{eb} is the electron mean free path of the bulk material, $\lambda_L(0)$ is the London penetration depth at absolute zero, and C is a constant. It can be seen that, for very thin films, $H_{c\perp}$ is proportional to $1/d$.

In the case of thick films ($d > d_c$) which behave like bulk slabs, one may write

$$H_{c\perp} = (1 - \sqrt{\frac{C'\delta}{d}})H_c \tag{2.10.23}$$

where δ is an interphase surface-energy parameter and C' is a constant of order 1 (Cody and Miller (1968)). The critical thickness d_c may be estimated by equating the two transition fields as

$$d_c \approx \frac{C'\delta}{(1 - 2\kappa^2)^2} \tag{2.10.24}$$

which is usually of the order of $10^{-6} \sim 10^{-7}$ m. Experimentally, the transition from type I to type II behavior in superconducting films in a perpendicular magnetic field has been studied by, for instance, Guyon et al. (1964), Miller and Cody (1968), Kunze et al. (1974), Dolan

(1974) and Gray (1974)).

1.10.3 Critical current in thin film

We shall now apply the G-L equations to calculate the critical current in a superconducting thin film at which superconductivity breaks down. First, we consider thin films $d << \xi$ and $d << \lambda$ so that $|\psi|$ and $\mathbf{J}_s$ may be supposed to be constant and uniform over the sample cross section of the thin films. We can set $\psi = |\psi| e^{i\theta(\mathbf{x})}$ with $|\psi|$ being independent of $\mathbf{x}$. Eqn.(2.8.15) for the supercurrent yields

$$\mathbf{J} = \frac{e^*}{m^*}(\hbar\nabla\theta - e^*\mathbf{A})\,|\psi|^2 = e^*|\psi|^2\,\mathbf{v}_s \tag{2.10.25}$$

where $\mathbf{v}_s$ denotes the mean velocity of the superconducting pair of electrons. The free energy density in eqn.(2.8.12) may thus be expressed as

$$f_s = f_n + (\alpha + \frac{1}{2}\beta|\psi|^2 + \frac{1}{2}m^*v_s^2)|\psi|^2 + \frac{1}{2\mu_0}B^2 \ . \tag{2.10.26}$$

Minimizing f_s with respect to $|\psi|$, we have

$$\alpha + \beta|\psi|^2 + \frac{1}{2}m^*v_s^2 = 0 \tag{2.10.27}$$

and, therefore,

$$v_s^2 = \frac{2\alpha}{m^*}(|\psi'|^2 - 1) \tag{2.10.28}$$

and

$$J_s = e^*|\psi|^2 v_s = e^*|\psi_0|^2\sqrt{\frac{2|\alpha|}{m^*}}\,|\psi'|^2\sqrt{1 - |\psi'|^2}\ . \tag{2.10.29}$$

It is shown that J_s equals zero for $|\psi'|^2 = 1$, and J_s reaches a maximum for $|\psi'|^2 = 2/3$ corresponding to the critical current density

$$J_c = \frac{2e^*|\psi_0|^2}{3\sqrt{3}}\sqrt{\frac{2|\alpha|}{m^*}} \tag{2.10.30}$$

which may be transformed, with the aid of eqns.(2.8.10), (2.8.18) and (2.8.21), into the following expression containing more readily available quantities as

$$J_c = \frac{2\sqrt{2}}{3\sqrt{3}} \frac{H_c(T)}{\lambda(T)} . \tag{2.10.31}$$

It is shown that, near the transition temperature T_c, J_c varies as $(T_c - T)^{3/2}$ since H_c varies as $(T_c - T)$ and λ varies as $(T_c - T)^{-1/2}$. The $(T_c - T)^{3/2}$ behavior of the critical current density near T_c has often been observed experimentally (Huebener (1979)). For current densities larger than J_c there do not exist solutions with $|\psi'| \neq 0$. As a consequence, for $J > J_c$ the superconductor becomes normal, and $|\psi'|$ changes abruptly from $\sqrt{2/3}$ to zero. The value of the critical current density given by eqn.(2.10.31) can be quite large. For typical values of $B_c = \mu_0 H_c = 0.05$ Tesla and $\lambda = 10^{-7}$ m, we have $J_c \approx 2 \times 10^{11}$ A/m^2.

Next, we consider thick films of type I superconductors ($d >> \lambda$). The critical current density J_c for such thick films may be found simply by using Silsbee's criterion for bulk superconductors, which states that a superconductor loses its zero resistance when, at any point on the surface, the total magnetic field strength due to transport current and applied magnetic field exceeds the critical field strength of the superconductor. Thus, by using eqn.(2.10.8) and Maxwell's equation $\nabla \times \mathbf{B} = \mu_0 \mathbf{J}$, we can find

$$J_c = \frac{H_c|\psi_0'|}{\lambda} \tanh(|\psi_0'|\frac{d}{2\lambda}) \approx \frac{H_c(T)}{\lambda(T)} \tag{2.10.32}$$

for sufficiently large d ($>>\lambda$).

The comparison of eqn.(2.10.31) and eqn.(2.10.32) shows that the critical current density for the very thin film is reduced, in contrast with the fact that its critical magnetic field is increased. It is worth a mention here that Silsbee's criterion is not valid for type II superconductors in high magnetic fields. The critical currents of type II superconductors in high magnetic fields are, in fact, almost completely controlled by the imperfections in the materials, which we shall discuss in the next section.

2.11 Flux flow, pinning force and critical current density

It has been shown that, in a high magnetic field greater than the lower critical field H_{c1}, the type II superconductor is in a mixed state with a distribution of flux vortices. For an ideal (soft) type II superconductor carrying no bulk transport currents, the vortices will interact and arrange themselves in equilibrium with no net force exerted on any vortex. If, however, there exists a bulk transport current $\mathbf{J}^{ext}$ flowing in the superconductor, the vortices will move under the action of the Lorentz body force acting on the vortex lines (see **Figure** 2.11.1). The force that acts on a unit length of the vortex is

$$\mathbf{f}_L = \mathbf{J}^{ext} \times \Phi_0 \mathbf{b} \tag{2.11.1}$$

where **b** is a unit vector along the vortex axis.

Multiplying by the number n of vortices per unit cross-sectional area, we find the Lorentz body force

$$\mathbf{F}_L = \mathbf{J}^{ext} \times \mathbf{B} \tag{2.11.2}$$

where $\mathbf{B}=n\Phi_0\mathbf{b}$ denotes the mean magnetic induction field.

Suppose that the motion of the vortex is opposed by a viscous friction force defined by

$$\mathbf{f}_v = -\eta \mathbf{v}_L \tag{2.11.3}$$

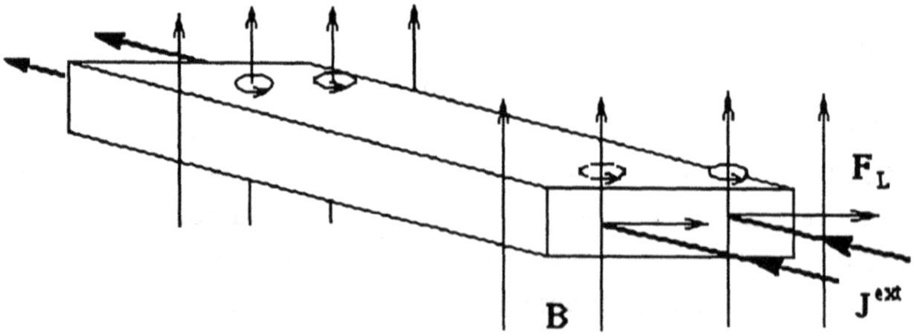

Figure 2.11.1 A type II superconductor carrying current in the mixed state.

where $\mathbf{v}_L$ denotes the mean velocity of vortex motion, and η is the viscosity coefficient, which may, in general, be a coefficient tensor for anisotropic superconductors.

By the force balance equation $\mathbf{f}_v + \mathbf{f}_L = 0$, $\mathbf{v}_L$ may be determined as

$$\mathbf{v}_L = \frac{\Phi_0}{\eta} \mathbf{J}^{ext} \times \mathbf{b} \tag{2.11.4}$$

which shows that the vortices move perpendicular to **B** and $\mathbf{J}^{ext}$.

The motion of the vortex lattice in the mean magnetic induction field **B** leads to the generation of an electric field by

$$\mathbf{E} = - \mathbf{v}_L \times \mathbf{B} \ . \tag{2.11.5}$$

In the case of **B** being perpendicular to $\mathbf{J}^{ext}$, we find

$$\mathbf{E} = \frac{\Phi_0 B}{\eta} \mathbf{J}^{ext} \ . \tag{2.11.6}$$

The generation of the electric field along $\mathbf{J}^{ext}$ implies the appearance of electric resistance defined by

$$\rho_f = \frac{\partial E}{\partial J^{ext}} = \frac{\Phi_0 B}{\eta} \ . \tag{2.11.7}$$

For a homogeneous superconductor, when it becomes normal at the critical field H_{c2}, the flow resistivity ρ_f is equal to ρ_n, the normal-state resistivity of the material. Thus, we may find

$$\eta = \frac{\Phi_0 \mu_0 H_{c2}}{\rho_n} \tag{2.11.8}$$

and

$$\rho_f = \rho_n \frac{B}{\mu_0 H_{c2}} \tag{2.11.9}$$

which shows that, for a given strength of applied magnetic field, the flow resistivity ρ_f is proportional to the normal-state resistivity ρ_n of the material. Furthermore, the flow resistivity increases with increasing strength of the applied magnetic field, approaching the normal-state resistivity as the applied field strength approaches $\mu_0 H_{c2}$. The simple expressions (2.11.8) and (2.11.9) are valid by order of magnitude at low temperature $T << T_c$ and small magnetic field $H << H_{c2}$ since the true viscosity coefficient η depends on many mechanisms and not only on the energy dissipation in the normal cores of the vortices. Some theoretical efforts to evaluate η in terms of material parameters may be referred to the work of, for instance, Bardeen and Stephen (1965).

It is shown that an ideal type II superconductor, which is free of lattice defects, in the mixed state is not able to carry transport currents without losses since the Lorentz force acting on the vortex structure will move the vortex lines, which generates electric fields in the superconductor resulting in dissipation. In order to transport larger resistanceless currents in type II superconductors, the motion of vortex lines has to be impeded since, when the vortices are pinned, there is no electromotive force induced and no current flows in the normal cores of the vortices, and the transport current flows around the cores without energy loss. Thus, for engineering applications, most type II superconductors contain imperfections, such as dislocations, grain boundaries, precipitates, inhomogeneities, etc., which are known to be able to create various pinning centers to impede the motion of flux vortices (Campbell and Evetts (1972) and Ullmaier (1975)). The local changes of the superconductor from these defects result in position-dependence of the free energy of the flux lines. The difference between the free energy of a flux line in the pinning center and in the surrounding medium characterizes the strength of the pinning center. In such (hard) superconductors, the vortex line driving forces from transport currents or temperature gradients are thus balanced by the pinning forces at the pinning centers. Microscopically not every individual vortex core is directly pinned to the material, but the interaction between the vortices is sufficient to give the vortex lattice a certain rigidity, so that if only a few vortex core are pinned the whole pattern may still be immobilized. What matters, therefore, is the average pinning force per vortex core. When the driving force density exceeds a critical value of the mean pinning force density, the flux lines in the superconductors will move and result in dissipation in the superconductor.

To provided a phenomenological description of the behavior of hard superconductors, the concept of the "critical state" was introduced by London (1962), Bean (1962) and Kim et al (1962), which reduced the variables to a single material-sensitive property, the mean pinning force density $\mathbf{P}_v$. In the equilibrium state, we can write the local force balance equation in the mixed state as

$$\mathbf{J}^{ext} \times \mathbf{B} + \mathbf{P}_v = 0 \qquad (2.11.10)$$

where $\mathbf{J}^{ext}$ is the transport current density, and $\mathbf{B}$ the mean flux density in the superconductor. The mean pinning force density $\mathbf{P}_v$ can take any value up to a certain maximum, leading to a maximum current density, or flux density gradient, given by the following critical state equation

$$\mathbf{B} \times \mathbf{J}_c = \mathbf{P}_{vmax} \qquad (2.11.11)$$

which may be further written, if $\mathbf{B}$ is perpendicular to $\mathbf{J}^{ext}$, as

$$J_c = \frac{P_{vmax}}{B} \qquad (2.11.12)$$

where J_c is called the critical current density for the bulk superconductor since it determines the maximum transport current density that a hard superconductor can carry without losses.

The theoretical calculation of the mean pinning force may be divided into two steps: calculating the force of interaction of the vortex lattice with a separate defect and, then, averaging the force over different randomly distributed pinning centers. However, it is often convenient and sufficient to assume that the maximum pinning force density $\mathbf{P}_{vmax}$ is a phenomenological function of the mean magnetic field at a given temperature. For instance, in the Bean model (Bean (1962)), P_{vmax} is supposed to be proportional to B. On the other hand in the Kim model (Kim et al. (1962)), P_{vmax} is assumed to be independent of B.

In general, to obtain high critical currents in practical type II superconductors, one may produce the maximum pinning forces in the material by introducing proper metallurgical defects, such as: point defects as voids and small second-phase particles, line defects as dislocations, surface defects as grain boundaries and volume defects as large precipitates, etc. Optimum pinning occurs when the defect size is comparable to the vortex core size, that is about ξ in size. If the defect size is smaller than the vortex core size, the core can span several defects at once, averaging out their effect and drastically reducing the pinning force. Therefore, the maximum pinning force is reduced rapidly when the magnetic field applied approaches the critical value H_{c2} of the material since, in such a case, the vortex core size ξ becomes very large.

If the transport current density in the superconducting material is slightly larger than its critical value J_c, the flux vortices begin to move in the hard type II superconductors. The mean flux flow velocity $\mathbf{v}_L$ may be obtained from the balance of three forces: the driving force (Lorentz force), the maximum pinning force and the viscous friction force, as

$$\mathbf{v}_L = \frac{\Phi_0}{\eta}(\mathbf{J}^{ext} - \mathbf{J}_c) \times \mathbf{b} \; . \tag{2.11.13}$$

The electric field is then given from eqn.(2.11.5) as

$$\mathbf{E} = \frac{\Phi_0 B}{\eta}(\mathbf{J}^{ext} - \mathbf{J}_c) \tag{2.11.14}$$

in the case of $\mathbf{B}$ perpendicular to $\mathbf{J}^{ext}$.

We may find that the flow resistivity ρ_f of the hard type II superconductor is still given by eqn.(2.11.7). This means that the (differential) flow resistivity ρ_f, shown in the simple analysis, depends only on the parameters of ideal type II superconductors and is not influenced by pinning effects, which is qualitatively in accordance with experimental observations (Kim et al. (1965) and Kim and Stephen (1969)).

It has been shown that, as long as the driving force (the Lorentz body force) is smaller or just equal to the maximum pinning force density, the vortex arrangement remains stationary and no heat is generated in the superconductor due to the dissipation of energy from the motion of flux vortices. This critical state model does provide a sufficiently accurate description of the behavior of a hard superconductor for most practical purposes. However, very sensitive flux measurements reveal a very slow motion of flux at current densities even slightly lower than the critical current density. Such a phenomenon called the flux creep, attributed to thermally activated jumps of so-called flux bundles, each containing many vortices, was first observed by Kim et al. (1962) and Anderson (1962). The amount of heat generated by flux creep is usually sufficiently small not to be of practical importance, but because of the very small specific heat and thermal conductivity of the superconductors, the heat may produce relatively large local changes in temperature. Since the mean pinning force density decreases rapidly as the temperature rises, the driving force may become larger than the maximum mean pinning force, which results in the flow of the vortices and, then, causes the power dissipation in the superconductor. When the power dissipation exceeds a certain limit, thermomagnetic instability appears and may result in catastrophic consequences if the magnetic energy stored in, for instance, a superconducting magnet is suddenly converted into thermal energy. Thus, practically used superconducting materials have to be properly stabilized so that they can tolerate some individual flux jumps, which are a common feature of mixed-state behavior for type II superconductors with currents near critical levels where the flux creep rate is enhanced, without incurring permanent breakdown of superconductivity. This may be achieved by using composite superconducting materials, such as composite superconducting cables with many thin superconducting filaments imbedded in a copper matrix which has high thermal and electrical conductivity. The high conductivity matrix assists in removing heat from the superconductor and slows down the propagation of magnetic disturbances. In addition, to reduce the eddy current losses due to the coupling between superconducting filaments via the normal conducting matrix, most modern superconducting materials have twisted filaments.

2.12 AC losses and London-Bean model

It has been shown in the above sections that a type II superconductor exhibits the Meissner effect in small fields. With the increase of the applied field, the Meissner effect breaks down and magnetic flux begins to penetrate into the interior of the superconductor, but the superconductor continues to show essentially no dc resistance up to a much higher "upper" critical field H_{c2} where bulk superconductivity finally disappears. The macroscopic average magnetization density $\mathbf{M}_{vor}$ from the vortex structure formed in the superconductor becomes a small fraction of the magnetic field **B** for applied field well above H_{c1} (see **Figure** 2.8.2) and may, therefore, be neglected compared with the **H** contribution for most cases of interest.

In time-varying fields, ac losses may, however, appear at all frequencies for practical superconducting materials. Though these losses are often small compared with the loss in a copper wire at room temperature, they may generate heat that occurs at low temperature and requires a large amount of refrigerator power for its removal in order to avoid the increase of temperature of the superconductor, which may lead to the catastrophic destruction of superconductivity.

To analyze the ac losses in superconductors, we shall present a simple model (often called the London-Bean model) for the study of ac losses in hard type II superconductors in high fields. In this model, it is assumed that the constitutive relations for the description of the electromagnetic behavior of hard superconductors in large magnetic fields close to the upper critical field H_{c2} may be expressed by

$$\mathbf{B} = \mu_0 \mathbf{H} \qquad (2.12.1)$$

neglecting magnetization from vortices, and

$$\mathbf{J} = J_c \frac{\mathbf{E}}{E} \qquad \text{for } \mathbf{E} \neq 0 \qquad (2.12.2)$$

with $\mathbf{J} = 0$ or $|\mathbf{J}| = J_c$ for $\mathbf{E} = 0$ depending upon the magnetic history. If **E** is zero due to decay from some previous nonzero value, $|\mathbf{J}| = J_c$ and its direction is given by $\mathbf{E}/E$ as E approaches zero. It can been seen that, in this model, the magnitude of **J** can take only its critical value or zero, and the critical value exists everywhere in a material that has been fully penetrated. The critical current density J_c is supposed to be independent of magnetic field in the London-Bean model. In addition, the vortex structure and the Meissner effect have been ignored in the model so that the magnetic field in the superconductor is only the sum of the applied field and the field from bulk transport and shielding currents. Obviously the model will be particularly good for treating the case of a small ac field superimposed upon a dc bias such that the total magnetic field is close to the upper critical field H_{c2}.

The macroscopic Maxwell's equations with the magneto-quasistatic approximation are

$$\nabla \times \mathbf{E} = -\frac{\partial \mathbf{B}}{\partial t} \qquad \text{and} \qquad \nabla \cdot \mathbf{E} = 0 \qquad (2.12.3)$$

$$\nabla \times \mathbf{H} = \mathbf{J} \qquad \text{and} \qquad \nabla \cdot \mathbf{B} = 0\,. \qquad (2.12.4)$$

As an example, we shall consider the calculation of the ac loss in a thin plate in the applied field B^e in the plane of the plate along the positive direction of the y-axis, as shown in **Figure** 2.12.1.

In addition, the plate may have a net transport current per unit width of the plate along the z-axis, given by

$$I = \int_{-x_0}^{x_0} J_z \, dx \;. \qquad (2.12.5)$$

In such a case, the magnetic fields on the plate surfaces are

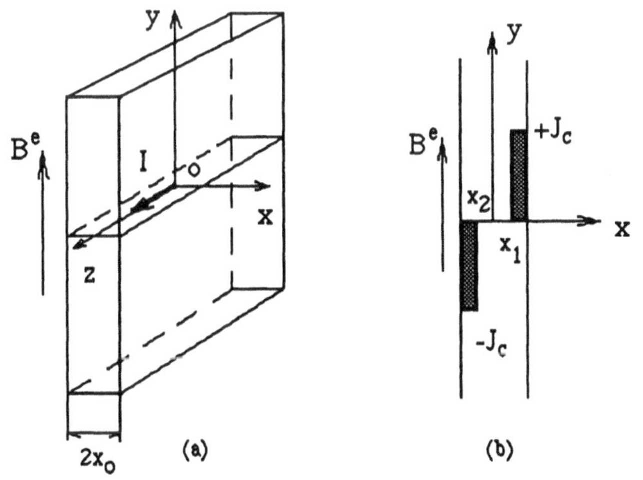

Figure 2.12.1 Configuration of a superconducting plate with the magnetic field applied along y-axis and transport current flowing along z-axis.

$$B_y(x_o) = B^e + \frac{1}{2}\mu_0 I \tag{2.12.6}$$

$$B_y(-x_o) = B^e - \frac{1}{2}\mu_0 I\,. \tag{2.12.7}$$

Consider that, for a small magnetic field and small transport currents, the flux penetration and also the shielding current penetration are only appreciable within the penetration depth λ. However, with the increase of the magnetic field and the transport current, the total current density from the transport current and the induced shielding current may reach the value of the critical current density J_c. If one further increases the field, the London-Bean model assumes that the total current density remains constant and the flux penetrates deeper into the material, where more shielding currents are induced. In such a case (as shown in **Figure** 2.12.1(b) in which one has assumed that the applied magnetic field is dominant), the magnetic field in the region of $-J_c$ may be found as

$$B_y(x) = B^e - \frac{1}{2}\mu_0 I - \mu_0 J_c(x + x_o) \qquad (-x_o < x < x_2) \tag{2.12.8}$$

and, in the region of $+J_c$, it is

$$B_y(x) = B^e + \frac{1}{2}\mu_0 I + \mu_0 J_c(x - x_o) \qquad (x_1 < x < x_o) \tag{2.12.9}$$

which shows that $B_y(x)$ is a linear function of the coordinate x in the London-Bean model.

The quantities x_1 and x_2, characterizing the field penetration, are determined by the conditions that the magnetic field is zero at $x=x_1$ and at $x=x_2$ in the case of virgin state, i.e.

$$0 = B^e - \frac{1}{2}\mu_0 I - \mu_0 J_c(x_2 + x_o) \tag{2.12.10}$$

and

$$0 = B^e + \frac{1}{2}\mu_0 I + \mu_0 J_c(x_1 - x_o) \tag{2.12.11}$$

which give

$$x_1 = x_o - \frac{1}{\mu_o J_c}(B^e + \frac{1}{2}\mu_o I) \tag{2.12.12}$$

and

$$x_2 = - x_o + \frac{1}{\mu_o J_c}(B^e - \frac{1}{2}\mu_o I) . \tag{2.12.13}$$

In particular, the penetration field B^p at which full penetration occurs may be determined by the condition $x_1 = x_2$

$$B^p = x_o \mu_o J_c \tag{2.12.14}$$

for the case described by eqns.(2.12.12) and (2.12.13), where the applied magnetic field is dominant. In general, the penetration field B^p will be dependent on the previous history of the superconductor.

For time-varying fields, the electric field in the superconductor may be determined from Maxwell's equation (2.12.3) as

$$E_z(x) = (\frac{\partial B^e}{\partial t} - \frac{1}{2}\mu_o \frac{\partial I}{\partial t})(x - x_2) \qquad (-x_o < x < x_2) \tag{2.12.15}$$

and

$$E_z(x) = (\frac{\partial B^e}{\partial t} + \frac{1}{2}\mu_o \frac{\partial I}{\partial t})(x - x_1) \qquad (x_1 < x < x_o) \tag{2.12.16}$$

where we have used the conditions that E_z must vanish at the inner moving boundaries x_1 and x_2. Eqns.(2.12.15) and (2.12.16) determine, in general, the sign of the current density.

The instantaneous power loss may now be calculated by

$$P(t) = \int \mathbf{J} \cdot \mathbf{E} \, dV \tag{2.12.17}$$

where the integral is over the volume of superconductor.

With the use of eqn.(2.12.2), the loss per unit total volume for the case considered may be given by

$$\frac{P}{V} = \frac{1}{2x_0}\int_{-x_0}^{x_0} \mathbf{E}\cdot\mathbf{J}\,dx = \frac{J_c}{2x_0}\int_{-x_0}^{x_0} |E_z|\,dx \tag{2.12.18}$$

which holds quite generally for the London-Bean model.
In the case of E_z given by eqns.(2.12.15) and (2.12.16), we have

$$\frac{P}{V} = \frac{J_c}{2x_0}\{|\frac{\partial B^e}{\partial t} - \frac{1}{2}\mu_0\frac{\partial I}{\partial t}|\int_{-x_0}^{x_2}|x - x_2|dx + |\frac{\partial B^e}{\partial t} + \frac{1}{2}\mu_0\frac{\partial I}{\partial t}|\int_{x_1}^{x_0}|x - x_1|dx\}$$

$$= \frac{J_c}{4x_0}\{|\frac{\partial B^e}{\partial t} - \frac{1}{2}\mu_0\frac{\partial I}{\partial t}|(x_0 + x_2)^2 + |\frac{\partial B^e}{\partial t} + \frac{1}{2}\mu_0\frac{\partial I}{\partial t}|(x_0 - x_1)^2\}\,. \tag{2.12.19}$$

For the particular simple case of the virgin state, where the fields are initially equal to zero, then the use of eqns.(2.12.12) and (2.12.13) leads to

$$\frac{P}{V} = \frac{1}{4x_0\mu_0^2J_c}\{|\frac{\partial B^e}{\partial t} - \frac{1}{2}\mu_0\frac{\partial I}{\partial t}|(B^e - \frac{1}{2}\mu_0I)^2 + |\frac{\partial B^e}{\partial t} + \frac{1}{2}\mu_0\frac{\partial I}{\partial t}|(B^e + \frac{1}{2}\mu_0I)^2\}$$

$$= \frac{1}{6\mu_0^2I_c}\{|\frac{\partial}{\partial t}(B^e - \frac{1}{2}\mu_0I)^3| + |\frac{\partial}{\partial t}(B^e + \frac{1}{2}\mu_0I)^3|\} \tag{2.12.20}$$

where $I_c = 2x_0J_c$ denotes the critical current per unit width.

For the case of a cyclic state with partial penetration, the positions of the moving boundaries x_1 and x_2 may be found as

$$x_1 = x_0 - \frac{1}{2\mu_0J_c}|B^e(t) + \frac{1}{2}\mu_0I(t) - B^e(t_1) - \frac{1}{2}\mu_0I(t_1)| \tag{2.12.21}$$

and

$$x_2 = -x_0 + \frac{1}{2\mu_0J_c}|B^e(t) - \frac{1}{2}\mu_0I(t) - B^e(t_2) + \frac{1}{2}\mu_0I(t_2)| \tag{2.12.22}$$

where t_1 is the time when $x_1 = x_0$ and t_2 the time when $x_2 = -x_0$.
It is shown that the speed of the moving boundary for the cyclic state is one-half that for the virgin state, assuming the same rate of change of the field at the surface.

Substitution of eqns.(2.12.21) and (2.12.22) into eqn.(2.12.19) gives

$$\frac{P}{V} = \frac{1}{8\mu_0^2 I_c}\{|\frac{\partial B^e}{\partial t} - \frac{1}{2}\mu_0\frac{\partial I}{\partial t}|[B^e - \frac{1}{2}\mu_0 I - B^e(t_2) + \frac{1}{2}\mu_0 I(t_2)]^2$$

$$+ |\frac{\partial B^e}{\partial t} + \frac{1}{2}\mu_0\frac{\partial I}{\partial t}|[B^e + \frac{1}{2}\mu_0 I - B^e(t_1) - \frac{1}{2}\mu_0 I(t_1)]^2\} \tag{2.12.23}$$

or

$$\frac{P}{V} = \frac{1}{24\mu_0^2 I_c}\{|\frac{\partial}{\partial t}[B^e - \frac{1}{2}\mu_0 I - B^e(t_2) + \frac{1}{2}\mu_0 I(t_2)]^3| + |\frac{\partial}{\partial t}[B^e + \frac{1}{2}\mu_0 I - B^e(t_1) - \frac{1}{2}\mu_0 I(t_1)]^3|\} \tag{2.12.24}$$

In particular, consider a sample in a cyclic state where the applied field, in the plane of the plate, is cycled from $B^e = - B^m$ to $B^e = B^m$ and back to $- B^m$ without reentrant loops, i.e. the sign of $\partial B^e/\partial t$ changes only at the ends of the half-cycles. If B^m is insufficient to cause full penetration, the instantaneous power loss may be calculated by eqn.(2.12.24) with $I = 0$ and $t_1 = t_2$ in the London-Bean approximation

$$\frac{P}{V} = \frac{1}{12\mu_0^2 I_c}|\frac{\partial}{\partial t}[B^e(t) - B^e(t_1)]^3| \tag{2.12.25}$$

where in the first half-cycle $B^e(t_1) = - B^m$ and for the second half-cycle $B^e(t_1) = B^m$. Thus, the loss per cycle can be found by

$$\int \frac{P}{V}dt = 2\int_{-B^m}^{B^m} \frac{P}{V}dt = \frac{1}{6\mu_0^2 I_c}\int_{-B^m}^{B^m} \frac{\partial}{\partial t}[B^e(t) + B^m]^3 dt = \frac{4(B^m)^3}{3\mu_0^2 I_c} \tag{2.12.26}$$

which shows that the loss is characterized by the appearance of the cubic terms in the field. For further details about the calculation of some other types of ac loss problems for superconductors in time-varying fields as well as some improvements for the ac loss calculation, readers are referred to the work of Carr (1983) and Wilson (1983).

2.13 Electrodynamics of Josephson junctions

2.13.1 Macroscopic quantum phenomena of superconductivity

Superconductivity is a macroscopic quantum phenomenon. This concept was first suggested by London (1948). The central idea of the macroscopic quantum state is represented by assigning a macroscopic number of electrons to a single "wave" function $\psi = |\psi| e^{i\theta}$ (see also section 2.2.2). These electrons are assumed somehow to have condensed into a single state due to the fact that, in the BCS theory, the superelectrons are considered to be formed in Cooper pairs which behave like Bose particles. This condensation results in a macroscopic density of Cooper-pair particles $|\psi|^2$ sharing the same quantum phase θ for all the pairs. Both $|\psi|^2$ and θ may be functions of space and time. In the phenomenological Ginzburg-Landau theory for superconductivity, the current-phase relation is expressed by eqn.(2.8.15) as

$$\mathbf{J}_s = \frac{e^*\hbar}{m^*}|\psi|^2\nabla\theta - \frac{e^{*2}}{m^*}|\psi|^2 \mathbf{A} \tag{2.13.1}$$

where $\mathbf{J}_s$ is the supercurrent density, and $\mathbf{A}$ the magnetic vector potential.

Consider now a multiply connected superconductor in weak magnetic fields, and let L be a closed curve located entirely within the superconductor and embracing one of these holes. Any surface S bounded by L will be partially in a superconducting region and partially in a non-superconducting region. Integrating eqn.(2.13.1) over the surface S for a uniform $|\psi|$ $(=|\psi_0|)$ and by noting $\mathbf{B} = \nabla \times \mathbf{A}$ and applying Stokes' theorem, one can get

$$\Phi_s + \int_L \mu_o\lambda_L^2\, \mathbf{J}_s\cdot d\mathbf{L} = \frac{\hbar}{e^*}\int_L \nabla\theta\cdot d\mathbf{L} \tag{2.13.2}$$

where λ_L is the penetration depth defined by eqn.(2.2.7) with $n_s^* = |\psi_0|^2$ and Φ_s is the magnetic flux defined by

$$\Phi_s = \int_S \mathbf{B}\cdot d\mathbf{S}\ . \tag{2.13.3}$$

The physical requirement that the wave function ψ is a single-valued function at each point indicates that the term on the r.h.s. of eqn.(2.13.2) can only be of the form

$$\frac{\hbar}{e^*}\int_L \nabla\theta\cdot d\mathbf{L} = n\Phi_0 \qquad (n = 0, 1, 2, ...) \tag{2.13.4}$$

where Φ_0 is the flux quantum (or fluxon) defined in eqn.(2.8.48). Here, n = 0 corresponds to a simply connected superconductor. In many cases, for instance, when the cross-section of the superconductor is large in comparison with the penetration depth, one may choose a loop with $\mathbf{J}_s$ = 0 on L so that the second term on the l.h.s. of eqn.(2.13.2) vanishes, and one thus has

$$\Phi_s = n\Phi_0 \qquad (n = 0, 1, 2, ...) \tag{2.13.5}$$

which shows that the magnetic flux enclosed by the loop is quantized, i.e. any magnetic flux contained within a superconductor should only exist as multiples of a quantum, the fluxon, Φ_0.

In general, $\mathbf{J}_s$ is generated to just produce the nearest integral number of fluxons (see Byers and Yang (1961)). It can be seen that this predicted value of the fluxon is extremely small. The value of the fluxon has been measured experimentally by Deaver and Fairbank (1961) and Doll and Nabauer (1961). The fact that the value of the fluxon is found to be, as predicted, Planck's constant ($h=2\pi\hbar$) divided by twice the electronic charge is strong evidence that the supercurrent is correctly considered to be carried by pairs of electrons. The measurements also show the existence of the predicted quantum-mechanical effect on a large scale.

In the approximation of London theory, one further has

$$\frac{\partial}{\partial t}\left[\int_S \mathbf{B}\cdot d\mathbf{S} + \mu_0\lambda_L^2\int_L \mathbf{J}_s\cdot d\mathbf{L}\right] = \frac{\partial}{\partial t}\int_S \mathbf{B}\cdot d\mathbf{S} + \int_L \mathbf{E}\cdot d\mathbf{L}$$

$$= \int_S \left(\frac{\partial \mathbf{B}}{\partial t} + \nabla\times\mathbf{E}\right)\cdot d\mathbf{S} = 0 \tag{2.13.6}$$

by using Maxwell's equation $\partial \mathbf{B}/\partial t = -\nabla \times \mathbf{E}$, which is valid on the whole surface S. Introducing a conservative quantity "fluxoid" (London (1950)) by

$$\Phi_L = \int_S \mathbf{B}\cdot d\mathbf{S} + \mu_0 \lambda_L^2 \int_L \mathbf{J}_s \cdot d\mathbf{L} \tag{2.13.7}$$

eqn.(2.13.6) states

$$\frac{\partial \Phi_L}{\partial t} = 0 \tag{2.13.8}$$

which means that the fluxoid is a constant in time. It can be shown that the fluxoid does not depend on the shape of the curve as long as they embrace the same hole but once, which comes from the fact that the fluxoid within any closed curve not surrounding a non-superconducting region is zero. The fluxoid is consequently not a property of the curve L but rather one of the hole in question.

2.13.2 Macroscopic quantum state and DC Josephson effect

In what follows, we shall present the theoretical basis of electrodynamics for weak links between superconductors. To study Josephson junctions where two superconductors are separated from each other by an insulator (a semiconductor or a normal metal), we may use the concept of macroscopic quantum state (Feynman (1965) and Mercereau (1969)) for a superconductor in which all electron pairs may be considered to have condensed into the same quantum state described by $\psi = |\psi| e^{i\theta}$. All superelectron pairs also have the same energy U. Since all the electrons must do exactly the same thing, the time rate for the macroscopic state ψ must be the same as for a single pair, i.e. $i\hbar\partial\psi/\partial t = U\psi$. To analyze a junction between two superconductors, one writes ψ_1 being the common wave function of all the electrons on one side, and ψ_2 being the corresponding function on the other side. Thus, in the absence of applied magnetic field, one may have the following relations

$$i\hbar \frac{\partial \psi_1}{\partial t} = U_1 \psi_1 + K\psi_2 \tag{2.13.9a}$$

$$i\hbar \frac{\partial \psi_2}{\partial t} = U_2\psi_2 + K\psi_1 \tag{2.13.9b}$$

where the constant K is a characteristic of the junction. If K were zero, these two equations would just describe the lowest energy state with energy U of each superconductor. In the case of $K \neq 0$, by substituting $\psi = |\psi| e^{i\theta}$ into eqn.(2.13.9) and equating real and imaginary parts, we can derive

$$\frac{\partial |\psi_1|^2}{\partial t} = \frac{2K}{\hbar} |\psi_1||\psi_2| \sin\varphi' \tag{2.13.10a}$$

$$\frac{\partial |\psi_2|^2}{\partial t} = -\frac{2K}{\hbar} |\psi_1||\psi_2| \sin\varphi' \tag{2.13.10b}$$

and

$$\frac{\partial \theta_1}{\partial t} = -\frac{K}{\hbar}\frac{|\psi_2|}{|\psi_1|}\cos\varphi' - \frac{U_1}{\hbar} \tag{2.13.11a}$$

$$\frac{\partial \theta_2}{\partial t} = -\frac{K}{\hbar}\frac{|\psi_1|}{|\psi_2|}\cos\varphi' - \frac{U_2}{\hbar} \tag{2.13.11b}$$

where $\varphi' = \theta_2 - \theta_1$. Eqn.(2.13.10) shows that $\partial|\psi_1|^2/\partial t = -\partial|\psi_2|^2/\partial t$ which means that one side loses charge at the same rate as the other side accumulates it. But whatever charge is lost it will be immediately replenished by the active element (voltage or current source) in the circuit. Thus one may write, for the current density being normal to the barrier flowing from side 1 to side 2

$$J = J_0 \sin\varphi' \tag{2.13.12}$$

where J_0 denotes the maximum current density which can flow across the junction and is defined by $J_0 = 2K|\psi_0|^2/\hbar$ with $|\psi_0|^2 = |\psi_1|^2 = |\psi_2|^2$. This equation tells us that if $\varphi' \neq 0$ (there is no reason why φ' should be equal to zero) then a finite current flows across the insulator without causing any voltage drop. So, in fact, the insulator behaves as a superconductor.

Based on a microscopic theory under BCS approximation, the magnitude of J_0 for an insulating barrier was derived by Ambegaokar and Baratoff (1963) as

$$J_0(T) = \frac{\pi\Delta(T)}{e^*R_n} \tanh \frac{\Delta(T)}{2k_BT} \tag{2.13.13a}$$

for identical superconductors on the two sides of the junction. Here, R_n is the resistance per unit area of the junction in the normal state, T the absolute temperature, k_B the Boltzmann constant, and Δ the energy gap. In particular, at absolute zero, one has

$$J_0(0) = \frac{\pi\Delta(0)}{e^*R_n} \tag{2.13.13b}$$

where $\Delta(0)$ is the energy gap at zero temperature, and, near T_c, one has

$$J_0(T) = \frac{\pi\Delta^2(T)}{2e^*R_nT_c} \, . \tag{2.13.13c}$$

The other pair of equations (2.13.11) tells us

$$\frac{\partial\varphi'}{\partial t} = \frac{e^*V}{\hbar} \tag{2.13.14}$$

which can be also integrated at constant temperature as

$$\varphi'(t) = \varphi'_0 + \frac{e^*}{\hbar} \int V(t)\, dt \tag{2.13.15}$$

where $V = (U_1 - U_2)/e^*$ denotes the voltage across the junction.

It can been seen from eqns.(2.13.12) and (2.13.15) that if one has zero voltage across the junction, one can get a current! With no voltage the current density can be any amount between $+J_0$ and $-J_0$, depending on the value of φ'_0. This is the essence of the dc Josephson effect (Josephson (1962)). **Figure** 2.13.1 shows schematically the current-voltage behavior of a Josephson junction. It is seen that, as the current

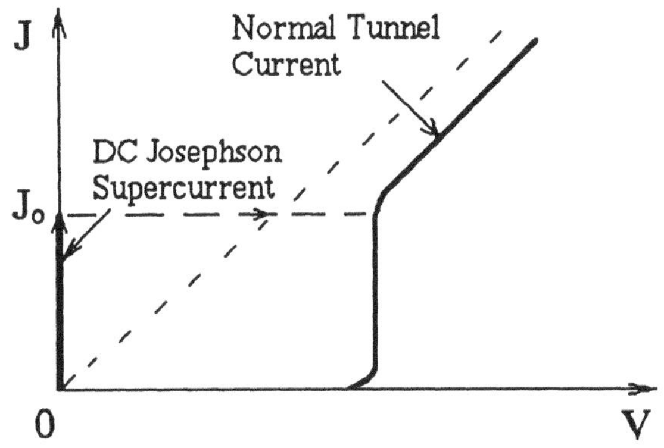

Figure 2.13.1 J-V characteristic of a Josephson Junction.

is increased from zero, no voltage appears across the junction for currents less than the critical current J_0. When the critical current is exceeded, there is a jump from zero voltage to a finite value of voltage as shown by the normal tunnel current curve. The first observation of such a strange behavior was made by Anderson and Rowell (1963). An experimental evaluation of the voltage across the junction in the dc Josephson regime was given by Smith (1965).

2.13.3 **Electrodynamic equations for Josephson junctions**

We shall now derive a set of electrodynamic equations for the study of Josephson junctions in the presence of applied magnetic fields. In such a case, a gauge invariance condition has to be imposed on eqns. (2.13.12) and (2.13.14). The gauge invariance means that the physics of any situation is unchanged if the magnetic vector potential $\mathbf{A} \to \mathbf{A} + \nabla\chi$, the electric scalar potential $\phi \to \phi - \partial\chi/\partial t$, and $\theta \to \theta + (e^*/\hbar)\chi$, where χ is an arbitrary scalar quantity (see also eqns.(2.2.19) and (2.2.20)). Therefore, eqns.(2.13.12) and (2.13.14) may now be written in a gauge invariant form

$$J = J_0 \sin(\varphi' + \frac{e^*}{\hbar}\int_2^1 A \cdot dL) \tag{2.13.16}$$

and

$$\frac{\partial}{\partial t}(\varphi' + \frac{e^*}{\hbar}\int_2^1 \mathbf{A}\cdot d\mathbf{L}) = \frac{e^*V}{\hbar} . \qquad (2.13.17)$$

Eqns.(2.13.16) and (2.13.17) present the two basic equations of the general theory of Josephson junction.

By noting that $\frac{e^*V}{\hbar} = -\frac{e^*}{\hbar}\int_2^1 \mathbf{E}\cdot d\mathbf{L}$ at constant temperature, and $\mathbf{E} = -\nabla\phi - \partial\mathbf{A}/\partial t$, one can get

$$\frac{\partial\varphi'}{\partial t} = \frac{e^*}{\hbar}\int_2^1 \nabla\phi\cdot d\mathbf{L} . \qquad (2.13.18)$$

To derive an equation for describing the spatial variation of the phase difference, one may consider **Figure** 2.13.2 where one lets P_1, P_2 and Q_1, Q_2 be two pairs of points, the members of each pair being adjacent to each other but on opposite sides of the barrier.

From eqn.(2.13.1), one can obtain

$$\varphi'(P) - \varphi'(Q) = \int_{\Gamma_1+\Gamma_2} \frac{e^*}{\hbar}(\mathbf{A} + \frac{m^*}{e^{*2}n_s^*}\mathbf{J}_s)\cdot d\mathbf{L} . \qquad (2.13.19)$$

By choosing a suitable integration contour on which $\mathbf{J}_s$ is either zero or is made to be small by considering that the dominant part of the current flows parallel to the barrier surface, shielding the field from the interior of the bulk superconductors, the second term on the r.h.s.

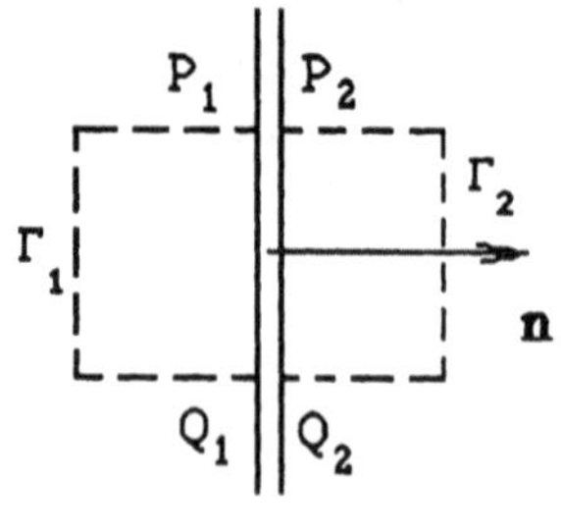

Figure 2.13.2 Contours of integration Γ_1 and Γ_2 used to derive the magnetic field dependence of the phase difference.

of eqn.(2.13.19) becomes negligibly small in comparison with the first term, and thus one can get, by adding the line integrals from P_2 to P_1 and Q_1 to Q_2 to both sides and replacing the line integral of **A** around the closed loop by the surface integral of **B**,

$$\varphi(P) - \varphi(Q) = \frac{e^*}{\hbar}\Phi_s(\Gamma) \qquad (2.13.20)$$

where $\Phi_s(\Gamma)$ is the magnetic flux enclosed by the contour Γ (= $\Gamma_1 + \Gamma_2 + P_2P_1 + Q_1Q_2$) and φ is the gauge-invariant generalization of the relative pair phase defined by

$$\varphi(P) = \varphi'(P) + \frac{e^*}{\hbar}\int_{P_2}^{P_1} \mathbf{A}\cdot d\mathbf{L} \qquad (2.13.21)$$

where $\varphi'(P) = \theta(P_2) - \theta(P_1)$.

If P and Q are close to each other, one may write

$$\varphi(P) - \varphi(Q) = \nabla\varphi \cdot QP \quad \text{and} \quad \Phi_s(\Gamma) = \mathbf{B}\cdot(d\mathbf{n}\times QP) = (\mathbf{B}\times d\mathbf{n})\cdot QP \qquad (2.13.22)$$

from which one arrives at

$$\nabla\varphi = \frac{e^*\delta}{\hbar}\mathbf{B}\times\mathbf{n} \qquad (2.13.23)$$

with $\delta = d + \lambda_{L1} + \lambda_{L2}$. Here, d is the thickness of the barrier, and λ_{L1} and λ_{L2} are the effective London penetration depths of the two superconductors forming the junction. **B** represents the actual magnetic field in the plane of the junction including both externally applied magnetic field and the field induced by the currents flowing in the junction. **n** is the unit normal vector of the junction surface and directed from superconductor 1 to superconductor 2.

Eqns.(2.13.16), (2.13.17) and (2.13.23) form a set of constitutive relations for the Josephson junction composed of thick superconductors. These constitutive equations relate the current density to the phase and phase to the electromagnetic fields in the junction. The electrodynamic equations of Josephson junctions are then obtained by combining Maxwell's equations with the constitutive current-phase-field relations.

In the case of a junction that lies in the x-y plane with **n** along the z-axis direction, one has from Maxwell's equations

$$\frac{\partial B_y}{\partial x} - \frac{\partial B_x}{\partial y} = \mu_o J_z + \mu_o c_s \frac{\partial V}{\partial t} \tag{2.13.24}$$

where $c_s \frac{\partial V}{\partial t}$ is the displacement current. Expressing B_y and B_x from eqn.(2.13.23), V from eqn.(2.13.17), J_z from eqn.(2.13.16) and substituting them into eqn.(2.13.24), we obtain a single differential equation in φ

$$\frac{\partial^2 \varphi}{\partial x^2} + \frac{\partial^2 \varphi}{\partial y^2} - \frac{1}{v^2}\frac{\partial^2 \varphi}{\partial t^2} = \frac{1}{\lambda_J{}^2} \sin \varphi \tag{2.13.25}$$

where v is the phase velocity (Swihart (1961)) given by

$$v = \frac{1}{\sqrt{\mu_o c_s \delta}} = c\sqrt{\frac{d}{\varepsilon_r \delta}} \tag{2.13.26}$$

with $\varepsilon_r = \varepsilon/\varepsilon_o$ being the effective relative dielectric constant, c the velocity of light in vacuum, and λ_J is the Josephson penetration depth defined by

$$\lambda_J = \sqrt{\frac{\hbar}{e^* \delta \mu_o J_o}} \ . \tag{2.13.27}$$

For typical junctions λ_J is of the order of a millimeter.

If we further include the current due to the tunnelling of normal electrons, we may write approximately (Solymar (1972))

$$J_z = J_o \sin\varphi + g_o V \tag{2.13.28}$$

with $1/g_o$ characterizing the resistance of the junction, and eqn.(2.13.25) becomes

$$\frac{\partial^2\varphi}{\partial x^2}+\frac{\partial^2\varphi}{\partial y^2}-\frac{1}{v^2}\frac{\partial^2\varphi}{\partial t^2}-\frac{\beta}{v^2}\frac{\partial\varphi}{\partial t}=\frac{1}{\lambda_J{}^2}\sin\varphi \qquad (2.13.29)$$

where $\beta = g_o/c_s$.

This differential equation together with proper boundary conditions (see Barone et al. (1982)) has no general analytical solution. It is, however, possible to obtain solutions in some special cases (see, for instance, Solymar (1972) and Barone and Paterno (1982)). In what follows, some illustrative examples will be given for explaining the basic phenomena of Josephson junctions and their uses in superconducting quantum interference devices (SQUIDs). The possible applications of the SQUIDs in various research areas are also discussed briefly.

2.13.4 **AC Josephson effect**

It has been shown from the basic Josephson relations (2.13.12) and (2.13.14) that a finite supercurrent of maximum value J_0 can flow with zero voltage drop (the dc Josephson effect). We may now consider the case that, if we apply a dc voltage V_0, the argument of the sine function in eqn.(2.13.12) becomes $\varphi_0' + (e^*/\hbar)V_0t$, and we find

$$J = J_0\sin(\varphi_0' + \frac{e^*V_0}{\hbar}t) \qquad (2.13.30)$$

which means that, if a constant voltage is maintained at the junction, an alternating current will flow across the junction. The frequency of the alternating current is $\omega = e^*V_0/\hbar$ (quantitatively $\omega/2\pi V_0 \approx 484$ MHz/μV). Such an effect is called the ac Josephson effect.

If we now applies a voltage at a very high frequency ω_a in addition to the dc voltage V_0 by setting

$$V = V_0 + V_1\cos(\omega_a t) \qquad (2.13.31)$$

we can obtain from eqn.(2.13.15) in the absence of applied magnetic fields

$$\varphi'(t) = \varphi_0' + \omega_J t + \frac{V_1e^*}{\omega_a\hbar}\sin(\omega_a t) \qquad (2.13.32)$$

where $\omega_J = e^*V_o/\hbar$ is called the Josephson frequency.
In the case of $V_1 << V_o$, with the use of the approximation $\sin(x+\Delta x) \approx \sin x + \Delta x \cos x$, we can get

$$J = J_o \left[\sin(\varphi_o' + \omega_J t) + \frac{V_1 e^*}{\omega_a \hbar} \sin(\omega_a t) \cos(\varphi_o' + \omega_J t) \right]. \tag{2.13.33}$$

The first term is zero on average, but the second term is not if $\omega_a = \omega_J$. This means that there should be a dc current component if the ac voltage has just the Josephson frequency (Shapiro (1963)).

In general, the Josephson current density may be expressed as

$$J = J_o \sin[\varphi_o' + \omega_J t + \frac{V_1 e^*}{\omega_a \hbar} \sin(\omega_a t)]$$

$$= J_o \sum_{k=-\infty}^{\infty} J_k\left(\frac{V_1 e^*}{\omega_a \hbar}\right) \sin[(\omega_J + k\omega_a)t + \varphi_o'] \tag{2.13.34}$$

where J_k is the ordinary Bessel function of order k. It can be seen that the current is phase modulated. If the Josephson frequency ω_J is equal to an integer multiple $n\omega_a$ ($n = 1, 2, \ldots$) of the microwave frequency of the ac voltage, the Josephson current density contains a time-independent (dc) current component given by

$$J_{DC} = (-1)^n J_o J_n\left(\frac{V_1 e^*}{\omega_a \hbar}\right) \sin\varphi_o' \tag{2.13.35}$$

The condition $\omega_J = n\omega_a$ corresponds to the condition

$$V_n = \frac{n\hbar\omega_a}{e^*}, \tag{2.13.36}$$

which may be observed experimentally.

Such an ac Josephson effect has, in fact, been used in the detection of electromagnetic radiation (see, for instance, Richards (1977) and Claeson (1983)). For more detailed analysis of the ac Josephson effect, the reader is referred to the work of Barone and Paterno (1982).

2.13.5 Josephson junctions in magnetic fields

Here, we shall study the influence of applied magnetic fields on the behavior of Josephson junctions. Let us now look at the case where there is a constant (both temporally and spatially) magnetic field, $\mathbf{B}=(B_x, B_y, 0)$ in the plane of the junction lying in the x-y plane. In this case, eqn.(2.13.23) may be integrated, with the approximation of ignoring the self-magnetic field generated by the supercurrent as shown in eqn.(2.13.24), to give

$$\varphi = \frac{e^*\delta}{\hbar}(B_y x - B_x y) + \alpha \tag{2.13.37}$$

where α is an integration constant. The total current across the junction can now be expressed as

$$I_z = \int J_o \sin\varphi \, dS = \mathrm{Im}\int J_o \exp\{i[\frac{e^*\delta}{\hbar}(B_y x - B_x y) + \alpha]\} dS \tag{2.13.38}$$

where Im denotes the imaginary part and the integration is taken over the area of the junction. The maximum supercurrent through the insulator is given by the maximum value of this expression with respect to changes in α, which is

$$I_o = \left| \int J_o \exp\{i[\frac{e^*\delta}{\hbar}(B_y x - B_x y)]\} dS \right| . \tag{2.13.39}$$

This integral may be calculated if the spatial variation of J_o is known. For most cases of interest J_o may be taken as a constant; eqn.(2.13.39) is then reduced to

$$I_o = J_o a_x a_y \left|\frac{\sin u_1 a_x}{u_1 a_x}\right| \left|\frac{\sin u_2 a_y}{u_2 a_y}\right| \tag{2.13.40}$$

with

$$u_1 = \frac{\pi e^*\delta}{h} B_y , \qquad u_2 = \frac{\pi e^*\delta}{h} B_x \tag{2.13.41}$$

in the case of a rectangular junction of dimensions a_x and a_y. The quantity $J_o a_x a_y$ is the maximum supercurrent in the absence of a magnetic field.

If the applied magnetic field is parallel with one of the edges of the junction (say $B_x = 0$), we get

$$I_o = J_o a_x a_y \left| \frac{\sin(\pi\Phi_J/\Phi_o)}{\pi\Phi_J/\Phi_o} \right| \tag{2.13.42}$$

where Φ_J denotes the magnetic flux enclosed, i.e. $\Phi_J = \delta a_x B$. Thus whenever the magnetic flux is an integral multiple of Φ_o, the flux quantum, no supercurrent will flow across the junction. This effect was first observed by Rowell (1963). **Figure** 2.13.3 shows quantitatively the dependence of the normalized maximum supercurrent on the normalized magnetic flux, which has been well confirmed experimentally by Langenberg et al (1966).

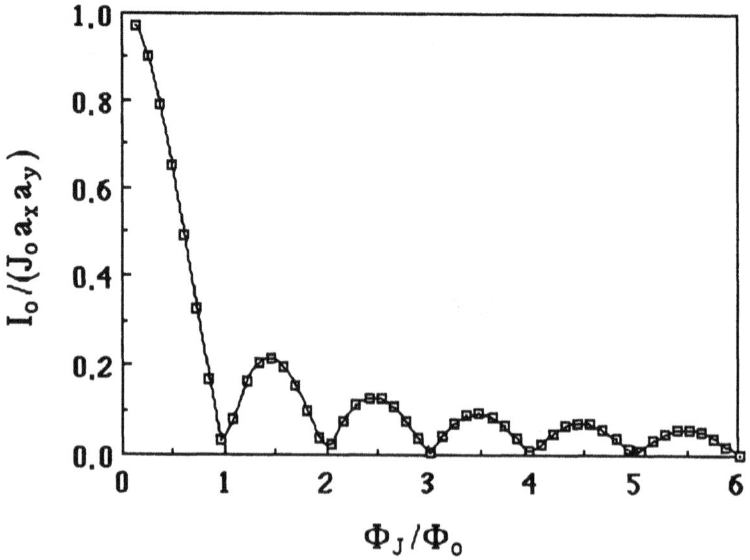

Figure 2.13.3 The normalized maximum supercurrent as a function of magnetic flux.

If we now consider the case of no externally applied magnetic field, we should use, by including the self-magnetic field effect, eqn.(2.13.29), which, in the time-independent case and by assuming a one-dimensional geometry, becomes

$$\frac{\partial^2\varphi}{\partial x^2} = \frac{1}{\lambda_J^2}\sin\varphi \, . \tag{2.13.43}$$

For small φ (valid for small magnetic fields (Josephson (1969)), eqn.(2.13.43) may be written as

$$\frac{\partial^2\varphi}{\partial x^2} = \frac{1}{\lambda_J^2}\varphi \tag{2.13.44}$$

which has the solution of the form $\varphi \sim \exp(-x/\lambda_J)$, in analogy to the Meissner effect. It can be seen that if the Josephson penetration depth λ_J is small, the current (also the magnetic field) is confined to the edge of the junction and the maximum supercurrent is drastically reduced. If, however, λ_J is large in comparison with the dimensions of the junction, φ can be regarded a constant and the total current can be calculated by simply multiplying the current density by the area of the junction. It should be noticed that the exclusion of the magnetic field from the junction (in analogy with type II superconductors) will be destroyed by sufficiently large fields and quantized flux lines will enter the barrier. Such cases have to be studied by solving eqn.(2.13.43) without assuming that φ is small (see Solymar (1972)).

It has been shown in eqn.(2.13.42) that, for a single Josephson junction, I_0 cannot be made to depend very sensitively on applied magnetic field since the effective junction width cannot be increased above the Josephson penetration depth λ_J. The situation is, however, quite different for double junctions. Let us now consider a combination of two identical Josephson junctions in parallel as shown in **Figure** 2.13.4. For simplicity, we shall ignore the effect of self-inductance in the following discussion.

The total current through the double junction is equal to the sum of the supercurrents through the individual contacts and is a function of φ_a and Φ_s (the externally applied magnetic flux enclosed by the loop)

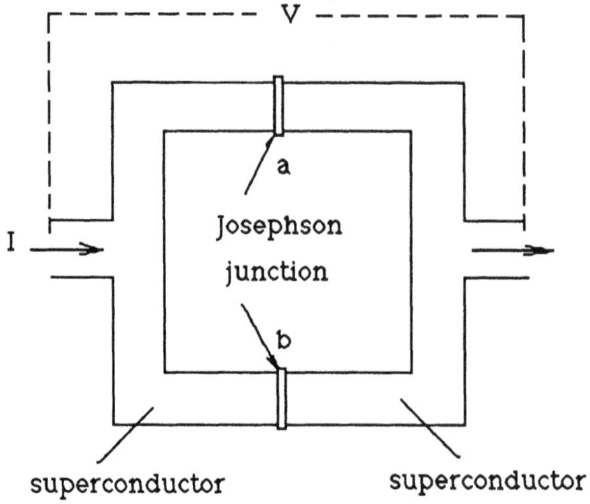

Figure 2.13.4 A double Josephson junction configuration.

$$I(\Phi_s, \varphi_a) = I_0(\sin \varphi_a + \sin \varphi_b) = I_0[\sin \varphi_a + \sin (\varphi_a + \frac{e^*}{\hbar}\Phi_s)]$$

$$= 2I_0 \cos(\frac{e^*}{2\hbar}\Phi_s) \sin (\varphi_a + \frac{e^*}{2\hbar}\Phi_s) . \qquad (2.13.45)$$

The maximum current that can flow through the junctions in the superconducting loop without any voltage appearing across the junctions may be obtained by maximizing the above expression with respect to φ_a at a given value of Φ_s. The result is

$$I_{max}(\Phi_s) = 2I_0 \left|\cos(\frac{e^*}{2\hbar}\Phi_s)\right| = 2I_0 \left|\cos(\frac{\pi\Phi_s}{\Phi_0})\right| \qquad (2.13.46)$$

which defines a critical current for the double junction in the superconducting loop. One of the significant features of this I_{max} vs Φ_s dependence is that $I_{max} = 0$ when $\Phi_s = (n+1/2)\Phi_0$ and $I_{max} = 2I_0$ when $\Phi_s = n\Phi_0$. In practice, it is hard to observe experimentally the result $I_{max}=0$, which might partly be due to an asymmetry of the junctions and to the self-induced flux (see, for instance, de Bruyn Ouboter (1977)).

Eqn.(2.13.46) shows that I_{max} is a periodic function of Φ_s, the flux through the loop, with the period being equal to one flux quantum Φ_o, in contrast to the case of the single junction critical current from eqn.(2.13.42) which is periodic in Φ_J, the flux through the junction. Since the double junction loop area can be made arbitrarily large, the critical current of the junction pair becomes a very sensitive measure of applied field.

We may now consider the situation when the current exceeds the maximum value possible with zero voltage. Note that eqn.(2.13.45) is still valid for the dc supercurrent flowing in the circuit, only φ_a is now related to the voltage. Adding the normal current term V/R_n, we may write the total current with the use of eqn.(2.13.18)

$$I = 2I_o \cos(\frac{e^*}{2\hbar}\Phi_s) \sin(\varphi_a + \frac{e^*}{2\hbar}\Phi_s) + \frac{\hbar}{e^*R_n}\frac{\partial\varphi_a}{\partial t} \quad . \tag{2.13.47}$$

Following McCumber (1968), we may introduce the dimensionless variables

$$\tau = \frac{2e^*R_nI_o\cos(\frac{e^*}{2\hbar}\Phi_s)}{\hbar}\,t\,, \qquad \zeta = \frac{I}{2I_o\cos(\frac{e^*}{2\hbar}\Phi_s)}, \tag{2.13.48}$$

$$\eta(\tau) = \frac{V}{2R_nI_o\cos(\frac{e^*}{2\hbar}\Phi_s)} = \frac{df}{d\tau} \tag{2.13.49}$$

yielding

$$\zeta = \frac{df}{d\tau} + \sin f \tag{2.13.50}$$

with $f = \varphi_a + \frac{e^*}{2\hbar}\Phi_s$ by noting that Φ_s is independent of time.

In the case of $|\zeta| \geq 1$, eqn.(2.13.50) may be written in the form

$$\frac{df}{\zeta - \sin f} = d\tau \ . \tag{2.13.51}$$

Using the x = tan f/2 substitution, eqn.(2.13.51) may be integrated to give

$$f = 2\tan^{-1}\{\frac{1}{\zeta} + \sqrt{1 - \frac{1}{\zeta^2}}\tan[\frac{\tau}{2}\sqrt{\zeta^2 - 1}\,]\} \,. \qquad (2.13.52)$$

We can then get

$$\eta(\tau) = \frac{df}{d\tau} = \frac{\zeta^2 - 1}{\zeta + \sin[\tau\sqrt{\zeta^2 - 1} + \tan^{-1}(\zeta^2 - 1)^{-1/2}]} \,. \qquad (2.13.53)$$

The average value of $\eta(\tau)$ may be obtained by integrating eqn.(2.13.53) over a period, yielding

$$\overline{\eta(\tau)} = \sqrt{\zeta^2 - 1} \qquad (2.13.54)$$

Thus, the averaged (dc) voltage $\bar{V}$ with respect to time is found as

$$\overline{V(\tau)} = R_n\sqrt{I^2 - I_{max}^2} \qquad (2.13.55)$$

where I_{max} is given by eqn.(2.13.46).

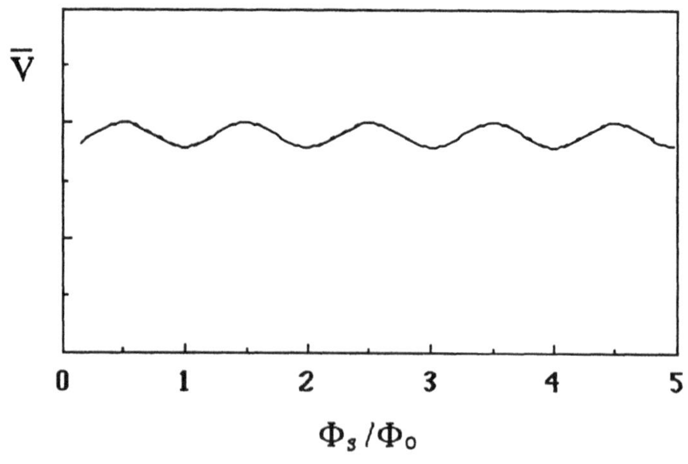

Figure 2.13.5 The dependence of dc voltage $\bar{V}$ on the magnetic flux Φ_s/Φ_0.

It follows clearly from eqn.(2.13.55) that the dc voltage as a function of magnetic field is in anti-phase with I_{max}. Whenever I_{max} is a minimum, $\bar{V}$ is maximum and vice versa. It is also shown that the dc voltage oscillates as a function of the external applied magnetic flux with the same period Φ_0 in the case where the applied constant current I exceeds the critical value I_{max} (see **Figure** 2.13.5). The experimental evidence of such phenomenon was observed by Zimmerman and Silver (1966).

2.13.6 **SQUIDs**

The dependence of the maximum supercurrent or the dc voltage on the applied magnetic field may be used for the measurement of small magnetic fields and their small changes. Such a device is the well-known dc SQUID composed of two Josephson junctions mounted on a superconducting ring. Since both I_{max} and the dc voltage across the double junctions are periodic functions of the magnetic field, we can gain information about the change in magnetic field by measuring the change in voltage (Zimmmerman and Silver (1966)). The resolution obtained by such a device with a 1 sec time constant was about 10^{-15} Tesla (Mercereau (1970)). The device may also be used to measure larger magnetic fields by counting the number of flux quanta moving in or out of the ring (Forgacs and Warnick (1967)). Details of some technical problems in such devices are referred to in the work of, for instance, Clarke (1983) and Solymar (1972).

The another type of device is the r.f. (radio-frequency) SQUID. An r.f. SQUID consists basically of a single Josephson junction mounted on a superconducting ring of inductance L (typically 10^{-9} H), shunted by a resistance and a capacitance for eliminating hysteresis of current-voltage character (Ziemmerman et al. (1970), Mercereau (1970) and Clarke (1989)) as shown schematically in **Figure** 2.13.6. For radio frequency operation of the device, the ring is inductively coupled to a coil of a LC-resonant circuit that is excited by a sinusoidal current I_{rf} at its resonant frequency (typically 20 or 30 MHz). The r.f. voltage, V_T developed across the tank circuit is amplified and detected with a diode (for example). If one plots V_T (vertically) versus I_{rf} (horizontally), one obtains a series of "steps" and "risers" (see **Figure** 2.13.7). On the steps, V_T is nearly independent of I_{rf}. If the external magnetic flux Φ_s is

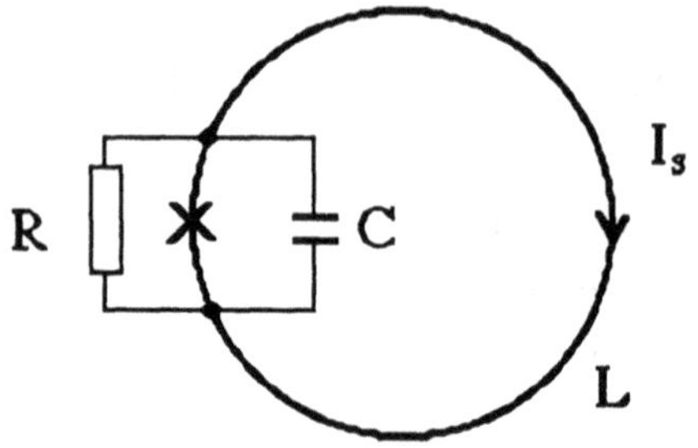

Figure 2.13.6 The r.f. SQUID.

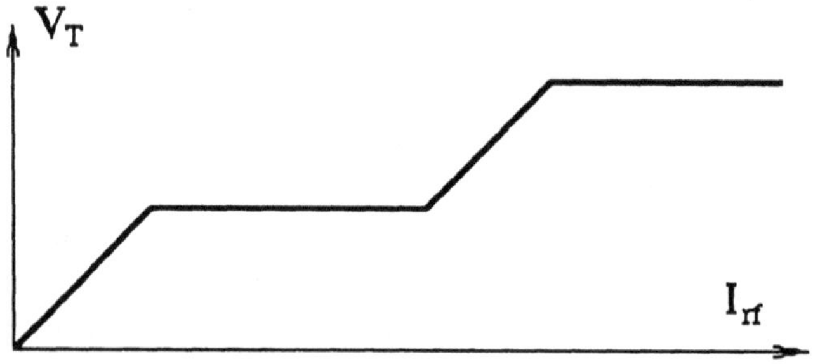

Figure 2.13.7 The sketch of V_T vs. I_{rf} in the absence of thermal noise.

slowly changed, the voltage at which the steps appear oscillates as a periodic function of Φ_s with period Φ_0. The measurement of magnetic field may then proceed in the same manner as for the dc SQUID.

Some possible applications of SQUIDs may be listed as follows.

(1) SQUIDs in biomedical applications. The measurement of the magnetic signals from various parts of human body can yield new informations about the organs that generate electric currents, not available to surface electrodes, and also about organs which contain foreign ferromagnetic particles. Signals measured from the heart are called magnetocardiograms (MCG) (Cohen (1975)), those from neural activity within the brain are called magnetoencephalograms (MEG) (Cohen (1968) and Hari and Ilmoniemi (1986)), those from muscle action are called magnetomyogram (MMG) (Cohen and Givler (1972)), and those due to eye movements are called magnetooculograms (MOG).

Magnetic signals have also been detected from blood flow, injured tissues, fetuses in utero, and the magnetite dust in stomach and lung (Robinson (1981)).

The first observation of the magnetocardiogram was reported by Baule and McFee (1963), where a peak field of about 50 pT was observed. This achievement marks the birth of experimental biomagnetism, which is the area of studying the weak magnetic fields associated with biological activity. The interest in finding theoretical models for presenting data was then motivated by the desire to decrease the effort in taking data and to establish a format that conveys a more intuitive appreciation of the actual heart action. One operationally simple approach is to assume that the observed fields originate from a magnetic dipole, called the "magnetic heart vector", which is located at the center of the heart and whose magnitude and direction vary through the cardiac cycle (Wynn et al. (1975)). More sophisticated models based on current dipole and higher magnetic multipoles were also proposed (Horacek (1973) and Grynszpan and Geselowitz (1973)). Later the observation of a neuromagnetic field produced by neural activity was made by Cohen (1968), where the strongest MEG signals are typically about 2 pT. Using a SQUID magnetometer, Cohen and Givler (1972) had also found that human skeletal muscles at the elbow produce both dc and ac magnetic fields when contracted. The fields are found to be quite large near the skin, having an amplitude of about 20 pT. More discussions about the recent advances in biomagnetism may be found in the Proceedings of the international conference on biomagnetism edited by Atsumi et al (1988).

(2) SQUIDs in geophysics. They may be used for studying the geological history of the earth, for determining the properties of the earth deep below the surface and for measuring the magnetic properties of rocks under various temperatures and pressures (Goubau (1980)). For instance, seawater motion, seismic stresses and electric conductivity of the earth's crust may be studied by measuring the fluctuating magnetic gradients generated respectively by seawater oscillation across the earth's magnetic field (Podney (1975)), by piezomagnetic effects (Stacey (1964)), and by the distortion of magnetic fields of electric currents flowing in the ionosphere (Cagniard (1953)). Information from the measurement of magnetic fields caused by piezomagnetism of rocks may also be useful in earthquake prediction

since there is evidence (Clarke (1977)) that the magnetic field along a fault line changes over periods of a few days prior to an earthquake.

(3) SQUIDs in nuclear gyros which may be used in navigation system. The principle of the operation of a nuclear gyroscope is as follows. For a species of nuclei with a magnetic moment **m** in a magnetic field **B**, the nuclear moments precess about the direction of the field at the Larmor frequency ω. If the frame of reference containing the nuclear free precession detection system were fixed in an inertial frame the precession frequency measured by an observer in that frame would be the Larmor frequency. If, however, the whole apparatus experiences a rotation about an axis parallel to the axis of the magnetic field then the measured precessional frequency will be shifted by the rotation frequency. Using the exceptional sensitivity of SQUID in place of conventional NMR free precession detection methods the nuclear sample can be made much smaller or operated in lower ambient field (Adams et al. (1980)).

(4) SQUIDs in submarine communications. At extremely low frequencies (from 30 to 3000 Hz), SQUIDs may permit submarines to operate much deeper than those using conventional communication systems in which receiving antennas have to be placed within a few meters of the ocean surface for maximum signal-to-noise ratio (SNR) (Davis and Nisenoff (1977)).

(5) SQUIDs in radiation detection at millimeter and submillimeter frequencies. They may be used, for instance, as receivers and detectors of millimeter and far infrared radiation from plasma diagnostics to analyze electron and ion temperature and to measure plasma drift velocities and the density of the plasma (Pedersen (1980) and Ulrich and Tutter (1980)).

(6) SQUIDs to opto-magnetic investigation for studying basic interactions in solids. They have the advantage of being capable of measuring the magnetic interactions in crystal-compounds without destroying the sensitive electronic structure by an external field of the measuring apparatus (Heidrich and Mataew (1980)).

Other applications of SQUIDs are, for instance, tracking of magnetic objects, such as a moving magnetic dipole (Wynn et al. (1975)), their use as computer elements (Wolf (1977)) and their use as high resolution voltmeters (Clarke (1966)).

2.14 Micromechanism of superconductivity

In above sections, we have presented the macroscopic electrodynamics of superconductors on the assumption that there exist some sort of electrons in the superconductor which behave as superelectrons, with the mysterious property that they can move through the material medium without resistance of any kind. In this section, we shall introduce the micromechanism which leads to the formation of such superelectrons and discuss briefly their collective behavior with the introduction of the well-known BCS microscopic theory (Bardeen, Cooper and Schrieffer (1957)).

2.14.1 **Isotope effect**

Since the discovery of superconductivity by H. Kamerlingh Onnes in 1911, it took many decades for scientists to understand this fascinating phenomenon. A crucial step towards the emergence of the BCS microscopic theory of superconductivity is the isotope effect discovered by Reynolds et al (1950) and Maxwell (1950), which shows that the superconducting transition temperature T_c and the critical field H_c depend on the mass of ions forming the crystal lattice of the material in the form

$$T_c \propto M^{-1/2} \qquad \text{and} \qquad H_c \propto M^{-1/2}. \tag{2.14.1}$$

This isotope effect implies that superconductivity is not a purely electronic phenomenon since the mass of the ions manifests itself only when lattice vibrations are taken into account. On the basis of this fact, Fröhlich (1950) and Bardeen (1950, 1951) demonstrated independently that the electrons, while residing in the crystal lattice, are capable of attracting one another. The physical origin of such an attraction may be seen by considering that if one puts an electron into a polarizable lattice, the positively charged lattice deforms to lower the electrostatic energy by crowding positive charge nearer the electron's position. If now there is a second electron to be introduced, the concentration of positive charge around the first electron makes that region also favorable to the second electron, which leads to an effective attractive force between the two electrons in addition to their direct mutually repulsive Coulomb interaction.

After the discovery of the attraction between electrons, it seems to become immediately possible to work out the theory of superconductivity. However, again it took several years for the formation of the concept of pairing of electrons. In 1956, it was Cooper who first studied theoretically what happens when two electrons with kinetic energies E_1 and E_2 are added to a metal at absolute zero. He concluded that, in the presence of even very weak attractive interaction, the Fermi sea of single electrons is unstable and any small perturbation that moves two electrons above E_F, where scattering is possible, will lower the system energy by pairing (Cooper (1956)). Because of the importance of the pairing concept in the theory of superconductivity, we shall present here some analyses on the formation of the bound Cooper pair of electrons.

2.14.2 Cooper pair of electrons

According to Cooper when two electrons with kinetic energies E_1 and E_2 are added to a metal at absolute zero, both energies E_1 and E_2 of the electrons must be above E_F in order not to violate Pauli's exclusion principle since all the eigenstates with kinetic energies up to E_F (Fermi energy) are occupied at absolute zero. The lowest values of E_1 and E_2, which are above E_F, lie within an energy about $\hbar\omega_c$ of the Fermi energy, where ω_c denotes the cutoff frequency. Consider now the scattering between the electron states resulting from interaction between the electrons. In the electron-phonon scattering mechanism, the electron occupying state $\mathbf{k}_1$ emits a phonon (i.e. vibrates the lattice) which is absorbed by the electron occupying the state $\mathbf{k}_2$. The total momentum $\mathbf{K} = \mathbf{k}_1 + \mathbf{k}_2$ in the final state is the same as in the initial state (i.e. the total momentum is conserved in the scattering event). It is, however, not necessary to require conservation of energy owing to Heisenberg's uncertainty principle. The largest number of allowed scattering processes, yielding the maximum lowering of the energy as we shall see later, is obtained by pairing electrons with equal and opposite momenta, i.e. $\mathbf{K} = 0$ for a translationally invariant system carrying no current.

The wave function of such a pair of electrons with opposite momenta and with a weak interaction can thus be expressed, according to perturbation theory (see, for instance, Schiff (1968)), as

$$\Psi(\mathbf{x}_1, \mathbf{x}_2) = \sum_{\mathbf{k}} a(\mathbf{k})\exp(i\hbar\mathbf{k}\cdot\mathbf{r}) \qquad (2.14.2)$$

with $\mathbf{r} = \mathbf{x}_1 - \mathbf{x}_2$ and $\mathbf{k} = (\mathbf{k}_1 - \mathbf{k}_2)/2$, where $a(\mathbf{k})$ is the probability amplitude of finding an electron in the state $\hbar\mathbf{k}$ and another electron in the state $-\hbar\mathbf{k}$.

The Schrödinger equation for the two electrons with the interaction potential $V(\mathbf{r})$ reads

$$-\frac{\hbar^2}{2m}[\nabla_1^2 + \nabla_2^2]\Psi + V(\mathbf{r})\Psi = (E + 2E_F)\Psi \qquad (2.14.3)$$

where the energy E of the pair is measured from the state where the two electrons are at the Fermi level. The equation for $a(\mathbf{k})$ can easily be found to be

$$\frac{\hbar^2}{m}k^2 a(\mathbf{k}) + \sum_{\mathbf{k}'} a(\mathbf{k}')V_{\mathbf{kk}'} = (E + 2E_F)a(\mathbf{k}) \qquad (2.14.4a)$$

with

$$a(\mathbf{k}) = 0 \qquad \text{for } k < k_F \qquad (2.14.4b)$$

where $V_{\mathbf{kk}'}$ is the scattering matrix element defined by

$$V_{\mathbf{kk}'} = \langle \hbar\mathbf{k}' | V | \hbar\mathbf{k} \rangle = \frac{1}{\Omega}\int_{\Omega} V(\mathbf{r})\exp[i\hbar(\mathbf{k}-\mathbf{k}')\cdot\mathbf{r}]d\mathbf{r} \qquad (2.14.4c)$$

with Ω being the volume of the system. The condition (2.14.4b) expresses the fact that states $k < k_F$ are already occupied (Pauli's exclusion principle).

Cooper pointed out that bound states exist for $E < 0$ provided that V is attractive, whatever the magnitude of this interaction. To see this, one may consider a simple model in which $V_{\mathbf{kk}'}$ is assumed to be a constant, i.e.

$$V_{\mathbf{kk}'} = -V \quad \text{for } E_F < \frac{\hbar^2k^2}{2m} < E_F + \hbar\omega_c \text{ and } E_F < \frac{\hbar^2k'^2}{2m} < E_F + \hbar\omega_c \qquad (2.14.5)$$

with $V > 0$ and $V_{\mathbf{kk}'} \equiv 0$ everywhere else. Thus, eqn.(2.14.4a) becomes

$$\left(\frac{\hbar^2}{m}k^2 - E - 2E_F\right)a(\mathbf{k}) = \{\sum_{\mathbf{k}'} a(\mathbf{k}')\}V \tag{2.14.6}$$

where the summation over $\mathbf{k}'$ is restricted to the band $\hbar\omega_c$ above E_F. Dividing each term of eqn.(2.14.6) by $(\frac{\hbar^2}{m}k^2 - E - 2E_F)$ and then summing over $\mathbf{k}$ yields

$$\frac{1}{V} = \sum_{\mathbf{k}} \frac{1}{(\hbar^2/m)k^2 - E - 2E_F} \; . \tag{2.14.7}$$

The discrete summation over $\mathbf{k}$ may be replaced by an integral over the energy with the range bounded by $E_F < \frac{\hbar^2 k^2}{2m} < E_F + \hbar\omega_c$, or $0 < \varepsilon(k) < \hbar\omega_c$ by introducing a new definition of the energy

$$\varepsilon(k) = \frac{\hbar^2 k^2}{2m} - E_F \tag{2.14.8}$$

so that the integral reads simply

$$\frac{1}{V} = \int_0^{\hbar\omega_c} \frac{N(\varepsilon)}{2\varepsilon - E} d\varepsilon \tag{2.14.9}$$

in which $N(\varepsilon)$ denotes the density of state (per unit energy and unit volume) by

$$N(\varepsilon) = \frac{4\pi k^2}{(2\pi)^3}\frac{dk}{d\varepsilon} \; . \tag{2.14.10}$$

If $\hbar\omega_c << E_F$, the density of the states $N(\varepsilon)$ is nearly constant and equal to its value at the Fermi surface $N(0)$ over the range of integration so that eqn.(2.14.9) may be integrated easily to give

$$E = \frac{2\hbar\omega_c}{1 - \exp[\frac{2}{VN(0)}]} . \qquad (2.14.11)$$

For weak interaction having VN(0) << 1, one has approximately

$$E = -\,2\hbar\omega_c \exp[-\frac{2}{VN(0)}] \qquad (2.14.12)$$

which states that an allowed energy state exists with $E < 0$.

The calculation thus shows that, in the presence of even very weak attractive interaction, the Fermi sea of single electrons is unstable and any small perturbation that moves two electrons above E_F, where scattering is possible, will lower the system energy by pairing.

2.14.3 **BCS theory of superconductivity**

With the concept of pairing of electrons and the electron-phonon interaction mechanism, which provides the necessary attractive interaction strong enough to overcome the repulsive screened Coulomb interaction between electrons, Bardeen, Cooper and Schrieffer (1957) were able to give the first microscopic theory of superconductivity, now known as the BCS theory.

In the BCS theory, it is assumed that the ground state can be expressed wholly in terms of paired electrons having the same total momentum, which are in a singlet state such that if the state $\mathbf{k}\uparrow$ is occupied, so is $-\mathbf{k}\downarrow$; and similarly if $\mathbf{k}\uparrow$ is vacant, $-\mathbf{k}\downarrow$ is also vacant. The total reduced Hamiltonian treated by BCS for electron pairs with zero net momentum is taken to be of the following form:

$$H_{red} = 2\sum_{k>k_F} \varepsilon_k b_k^* b_k + 2\sum_{k<k_F} |\varepsilon_k| b_k b_k^* + \sum_{\mathbf{kk'}} V_{\mathbf{kk'}} b_{k'}^* b_k \qquad (2.14.13)$$

where $\varepsilon_k = (\hbar^2 k^2/2m) - E_F$ is the kinetic energy of a Bloch state measured relative to the Fermi level. The first two terms on the r.h.s. of eqn.(2.14.13) denote the kinetic energy of the electron pairs relative to the normal ground state and the last term is the potential energy relative to the normal state, the reduced potential energy. The factors of 2 appear in the expression since there are two electrons in the pair, i.e. there are two possible spins for each $\mathbf{k}$. The summation puts a spin-up electron in each **k** state and a spin-down electron in the corresponding -**k** state with the result that each state for $k < k_F$ contains both spin-up electron and spin-down electrons.

The operators b_k and b_k^* used in the formalism of second quantization denote respectively the pair operators defined by

$$b_{\mathbf{k}} = c_{-\mathbf{k}\downarrow}c_{\mathbf{k}\uparrow} \qquad \text{and} \qquad b_{\mathbf{k}}^* = c_{\mathbf{k}\uparrow}^* c_{-\mathbf{k}\downarrow}^* \tag{2.14.14}$$

where $c_{\mathbf{k}\uparrow}^*$ is the single-electron creation operator, which places an electron in state **k** with spin up, and $c_{\mathbf{k}\uparrow}$ is the single-electron annihilation operator, which causes the elimination of an electron (details about the operator algebra is referred to in the work of, for instance, Schiff (1968)).

To find the distribution of pair occupancy in the superconducting state, BCS introduced a ground state function with the use of a Hartree-like approximation in which the probability that a specific configuration of pairs occurs in the wave function is given by a product of occupancy probabilities for the individual pair states, i.e.,

$$|\Psi\rangle = \prod_{\mathbf{k}} [u_{\mathbf{k}} + v_{\mathbf{k}} b_{\mathbf{k}}^*]\ |0\rangle \tag{2.14.15}$$

where $|0\rangle$ denotes the vacuum state with no electrons present, $v_{\mathbf{k}}^2$ is the probability of pair occupancy, and $u_{\mathbf{k}}^2 = 1 - v_{\mathbf{k}}^2$ (for all **k**) is the probability of pair vacancy, so that the wave function (2.14.15) is correctly normalized to unity.

The probability that the pair (**k**↑, -**k**↓) is occupied reads

$$\langle\Psi|c_{\mathbf{k}\uparrow}^* c_{\mathbf{k}\uparrow}|\Psi\rangle = v_{\mathbf{k}}^2 \tag{2.14.16}$$

and the probability that both pairs **k** and **k**' are occupied is

$$\langle\Psi|c_{\mathbf{k}\uparrow}^* c_{\mathbf{k}\uparrow} c_{\mathbf{k}'\uparrow}^* c_{\mathbf{k}'\uparrow}|\Psi\rangle = v_{\mathbf{k}}^2 v_{\mathbf{k}'}^2 \tag{2.14.17}$$

and the probability that **k** is occupied and **k**' unoccupied is

$$\langle\Psi|c_{\mathbf{k}\uparrow}^* c_{\mathbf{k}\uparrow}(1 - c_{\mathbf{k}'\uparrow}^* c_{\mathbf{k}'\uparrow})|\Psi\rangle = v_{\mathbf{k}}^2 u_{\mathbf{k}'}^2 \ . \tag{2.14.18}$$

It is assumed here that only those many body states in which the two members of a pair with the pair state of spin-antiparallel (**k**↑, -**k**↓) are either both occupied or both unoccupied are chosen. It is also assumed that the probability amplitude v_k^2, for the occupation of a particular pair of state (**k**↑, -**k**↓), does not depend on the occupation of the other states. Essentially, the basic approximation of the BCS theory of superconductivity rests on their assumption that it is the two-body correlations that are responsible for the qualitative features of superconductivity and that of the two-body correlations there is a very strong preference for singlet zero momentum pairs (so strong that one can get an adequate description of superconductivity by treating these correlations alone (Blatt (1964))). The ground wave function (2.14.15) chosen by BCS has been found to be very satisfied from the empirical point of view since the BCS theory is in excellent agreement with experiment. However, there are some questions yet to be studied from the theoretical point of view, especially for the new high-T_c superconductors.

The energy of the superconducting ground state relative to the normal ground state is the expectation value of the reduced Hamiltonian

$$W = \langle\Psi|H_{red}|\Psi\rangle \,. \tag{2.14.19}$$

Substituting (2.14.13) and (2.14.15) into (2.14.19), one obtains

$$W = 2\sum_{k>k_F} \varepsilon_k v_k^2 + 2\sum_{k<k_F} |\varepsilon_k| u_k^2 + \sum_{kk'} V_{kk'} u_k v_k u_{k'} v_{k'} \,. \tag{2.14.20}$$

In the equilibrium state, the energy W is minimized with respect to v_k^2 and with the restriction condition $u_k^2 = 1 - v_k^2$. The resulting probability of occupancy reads

$$v_k^2 = \frac{1}{2}\left[1 - \frac{\varepsilon_k}{\sqrt{\varepsilon_k^2 + \Delta_k^2}}\right] \tag{2.14.21}$$

in which Δ_k is called the gap parameter defined by

$$\Delta_{\mathbf{k}} = -\sum_{\mathbf{k}'} V_{\mathbf{k}\mathbf{k}'} u_{\mathbf{k}'} v_{\mathbf{k}'} \quad . \tag{2.14.22}$$

Substitution of eqn.(2.14.21) into eqn.(2.14.22) gives

$$\Delta_{\mathbf{k}} = -\sum_{\mathbf{k}'} V_{\mathbf{k}\mathbf{k}'} \frac{\Delta_{\mathbf{k}'}}{2E_{\mathbf{k}'}} \tag{2.14.23}$$

with $E_{\mathbf{k}}=\sqrt{\varepsilon_{\mathbf{k}}^2 + \Delta_{\mathbf{k}}^2}$.
Eqn.(2.14.23) is a nonlinear integral equation since $\Delta_{\mathbf{k}'}$ is present in the denominator in $E_{\mathbf{k}'}$. This integral equation has obviously the trivial solution $\Delta_{\mathbf{k}} = 0$ for all $\mathbf{k}$ leading to normal state. The condition for superconductivity is, thus, whether the integral equation possesses a nontrivial solution leading to a lower energy.

For general interactions this equation has to be solved numerically. In the simple BCS model, an assumption is made that the matrix element $V_{\mathbf{k}\mathbf{k}'}$ can be replaced by a constant average matrix element

$$V_{\mathbf{k}\mathbf{k}'} = \begin{cases} -V & \text{for } |\varepsilon_{\mathbf{k}}| \text{ and } |\varepsilon_{\mathbf{k}'}| < \hbar\omega_c \\ 0 & \text{otherwise} \end{cases} \tag{2.14.24}$$

with $V > 0$ (meaning the attractive interaction between electrons in a pair). Here, the cutoff frequency ω_c is taken to be the Debye phonon frequency ω_D in the electron-phonon mechanism.

For this simple interaction, $\Delta_{\mathbf{k}}$ is zero if $|\varepsilon_{\mathbf{k}}| > \hbar\omega_c$ and it is a constant Δ if $|\varepsilon_{\mathbf{k}}| < \hbar\omega_c$; and it is determined from the following equation

$$\frac{1}{V} = \sum_{\mathbf{k}'} \frac{1}{2\sqrt{\varepsilon_{\mathbf{k}'}^2 + \Delta^2}} \tag{2.14.25}$$

which has a solution only if V is positive. Replacing the summation over $\mathbf{k}'$ by an integral over the corresponding energy range and by noting the fact that the density of states is nearly constant close to the Fermi surface, so $N(\varepsilon) \approx N(0)$ over the range $|\varepsilon_{\mathbf{k}}| < \hbar\omega_c$, eqn.(2.14.25) becomes

$$\frac{1}{N(0)V} = \int_0^{\hbar\omega_c} \frac{d\varepsilon}{\sqrt{\varepsilon^2 + \Delta^2}} . \tag{2.14.26}$$

Solving for Δ, we obtain

$$\Delta = \hbar\omega_c/\sinh[1/N(0)V] . \tag{2.14.27}$$

If N(0)V << 1 for the weak coupling between electrons and phonons as in most elemental materials, eqn.(2.14.27) is reduced to

$$\Delta = 2\hbar\omega_c \exp[-1/N(0)V] \tag{2.14.28}$$

which has a typical value of about 1 meV.

The ground state energy relative to the normal ground state at the absolute zero can then be found from eqns.(2.14.20), (2.14.21), (2.14.24) and (2.14.28)

$$W_s - W_n = -\frac{1}{2}N(0)\Delta^2[1 - \exp(-\frac{2}{N(0)V})] \tag{2.14.29a}$$

$$\approx -\frac{1}{2}N(0)\Delta^2 \tag{2.14.29b}$$

where the last form holds in the weak coupling limit.

By equating eqn.(2.14.29) with the thermodynamic condensation energy at T = 0 (see eqn.(2.7.8)), we find

$$\frac{\mu_o H_o^2}{2} = (g_n - g_s)_{T=0} = \frac{1}{2}N(0)\Delta^2 . \tag{2.14.30}$$

It is shown that although the BCS theory of the ground state has all electrons paired, only those in a narrow energy range of order Δ participate in the condensation. Those further below the Fermi level are described mathematically as pairs with no loss of veracity but they are too far from the surface to be scattered by the electron-phonon interaction, and hence do not participate in the reduction of the system energy. This shows that the electron pairs are not true Bose particles otherwise the ground state would be formed by placing all the

Bosons in the lowest state. The effect of Pauli's exclusion principle acting on the individual electrons forming the pair is shown from the fact that no **k**↑ and -**k**↓ state may be occupied by more than one pair at a time, which then makes the probability v_k^2 of the occupancy of the pair state (**k**↑, -**k**↓) less than 1 as expressed by eqn.(2.14.21).

To study the physical properties of the system, we have to treat the excitation states of the system which deviates from its ground state due to electrons being excited out of the ground state by thermal lattice vibrations or by incident photons. To find the excited states of the BCS reduced Hamiltonian, we consider adding an electron to the system in the state **k**↑ with its mate -**k**↓ being empty. The only effect of this process is to block the pair state (**k**↑, -**k**↓) from participating in the pairing interaction due to Pauli's exclusion principle. The state of the system, with an excitation in state $\mathbf{k}_1$ and all other states occupied by pairs, reads

$$|\Psi_{k_1}\rangle = c^*_{k_1\uparrow} \prod_{k \neq k_1} [u_k + v_k b^*_k] \, |0\rangle \tag{2.14.31}$$

where the creation operator $c^*_{k_1\uparrow}$ gives sure occupancy in state $\mathbf{k}_1\uparrow$, but the state $-\mathbf{k}_1\downarrow$ remains empty. The kinetic energy of state (2.14.31) relative to the ground state is reduced by the absence of the pair occupying $(\mathbf{k}_1\uparrow, -\mathbf{k}_1\downarrow)$ with probability v_k^2 and increased by the sure occupancy of $\mathbf{k}_1\uparrow$ by the excitation, and the reduced potential energy is changed by eliminating the pair scattering from and into $(\mathbf{k}_1\uparrow, -\mathbf{k}_1\downarrow)$. The total change of energy from the ground state to the excited state (2.14.31) is

$$W_{k_1\uparrow} - W_o = \varepsilon_{k_1}(1 - 2v_{k_1}^2) + 2u_{k_1}v_{k_1}\Delta_{k_1} = E_{k_1} \tag{2.14.32}$$

which shows that the parameter E_{k_1} is just the energy required to create an excitation (a quasi-particle) in state $\mathbf{k}_1\uparrow$. A plot of E_k as a function of k is given in **Figure** 2.14.1.

It is shown that the excitation energy has a minimum value Δ, which means that excitations cannot be created with an arbitrarily small amount of energy as in the case of the normal metal. To break a pair, two excitations must be created and, thus, the minimum energy required is at least 2Δ, Δ for removing an electron from one state and Δ

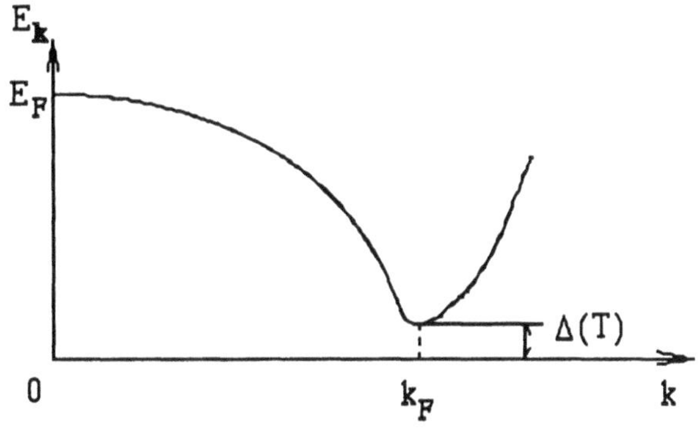

Figure 2.14.1 Temperature-dependent gap parameter Δ in the excitation spectrum.

for placing it in another state. The occurrence of thermally excited particles at finite temperature will result in increase of energy and entropy. At any given temperature, the number of excitations at thermal equilibrium is determined by minimizing the free energy of the system.

In the BCS theory, the free energy of the system at finite temperature is taken to be of the following form (Bardeen et al. (1957))

$$F = 2\sum_{k} |\varepsilon_k|[f_k + (1 - 2f_k)v_k^2(|\varepsilon_k|)] + \sum_{kk'} V_{kk'}u_k v_k u_{k'} v_{k'}(1 - 2f_k)(1 - 2f_{k'}) - TS \tag{2.14.33}$$

where the entropy S is

$$S = -2k_B \sum_{k} \{f_k \ln f_k + (1 - f_k)\ln(1 - f_k)\} . \tag{2.14.34}$$

By minimizing the free energy of the system with respect to the probability of pair occupancy v_k^2, one can find the same expression (2.14.21) for v_k^2 but with the new expression for the gap parameter

$$\Delta = -V\{\sum_{k'} u_{k'}v_{k'}(1 - 2f_k)\} \tag{2.14.35}$$

where one has used the assumption given by eqn.(2.14.24).

By minimizing F with respect to the probability of excitation occupancy f_k, one finds the expression for the probability of excitation occupancy f_k by

$$f_k = \frac{1}{1 + \exp(\frac{E_k}{k_B T})} \tag{2.14.36}$$

which is the usual Fermi-Dirac distribution function. Thus the excitations act completely like a set of independent fermions whose energies are given by the dispersion law $E_k = \sqrt{\varepsilon_k^2 + \Delta^2}$.

By substituting eqns.(2.14.21) and (2.14.36) into eqn.(2.14.35), a self-consistent condition for the determination of the gap parameter Δ_k is found

$$\frac{1}{V} = \sum_{k'} \frac{1}{2E_{k'}}[1 - \frac{2}{1 + \exp(\frac{E_{k'}}{k_B T})}] \tag{2.14.37}$$

which may, further, be expressed in the following integral form, by assuming a constant density of state N(0),

$$\frac{1}{N(0)V} = \int_0^{\hbar\omega_c} \tanh[\frac{\sqrt{\varepsilon^2 + \Delta^2}}{2k_B T}] \frac{d\varepsilon}{\sqrt{\varepsilon^2 + \Delta^2}} . \tag{2.14.38}$$

This equation is an implicit relation for the determination of the temperature-dependent gap parameter $\Delta(T)$. As temperature T increases from zero, Δ decreases and vanishes at the transition temperature T_c. Thus T_c is given by

$$\frac{1}{N(0)V} = \int_0^{\hbar\omega_c} \tanh[\frac{\varepsilon}{2k_B T_c}] \frac{d\varepsilon}{\varepsilon} . \tag{2.14.39}$$

In the weak-coupling limit ($N(0)V \ll 1$) this gives

$$k_BT_c = 1.14\hbar\omega_c \exp[-\frac{1}{N(0)V}] \,. \tag{2.14.40}$$

Since N(0)V is expected to be independent of isotopic mass M, we see that the transition temperature T_c is proportional to $\hbar\omega_c$, which is consistent with the isotope effect since $\omega_c \propto M^{1/2}$ in the electron-phonon interaction mechanism.

By comparing eqn.(2.14.40) with eqn.(2.14.28), one finds the following relation for the energy gap at T=0

$$2\Delta(0) = 3.52k_BT_c \tag{2.14.41}$$

where the actual values of the constant, 3.52, is found to be in reasonably good agreement with experiment for weak-coupling superconductors. In general, it lies within about 30% of 3.52.

In general, the temperature-dependent energy gap Δ(T) may be calculated numerically from eqn.(2.14.38). Near the transition temperature T_c, the gap parameter can be found from the following approximate relation

$$\Delta(T) \approx 3.2k_BT_c\sqrt{1 - T/T_c} \tag{2.14.42}$$

which has the form suggested by Buckingham (1956), and is consistent with Landau's theory of second-order phase transition.

So far, nothing has been said about the electrodynamic properties of superconductors, such as the disappearance of resistance and the Meissner effect. Both the ground state and the excited states which we have discussed have a perfectly isotropic distribution of electrons in momentum space, so that there are as many electrons travelling one way as the other, and no macroscopic net current flows. It is, however, possible to visualize a situation in which each Cooper pair, instead of having zero total momentum, has a resultant momentum $\hbar\mathbf{K}$, which is the same for all pairs. In this case the pair wave function is then of the form

$$\Psi_K = \Psi\exp(i\hbar\mathbf{K}\cdot\mathbf{R}) \tag{2.14.43}$$

with $\mathbf{K} = \mathbf{k}_1 + \mathbf{k}_2$ and $\mathbf{R} = (\mathbf{x}_1 + \mathbf{x}_2)/2$ being the position of the centre of mass of the pair.

In this picture the current is carried by pairs of electrons which have a total momentum **K**. When a current is carried by an ordinary conductor, such as a normal metal or a semiconductor, resistance is inevitably present because the current carriers (either electrons or holes) can be scattered with a change in momentum so that their free acceleration in the direction of the electric field is hindered. This scattering may be due to impurity atoms, lattice defects or thermal vibrations. In the case of a superconductor, the electrons which make up a Cooper pair are constantly scattering each other, but since the total momentum remains constant in such a process there is no change in the current flowing. The only scattering process which can reduce the current flow is one in which the total momentum of a pair in the direction of the current changes, and this can only happen if the pair is broken up. This de-pairing requires a minimum amount of energy 2Δ so the scattering can only happen if this energy can be supplied from somewhere. Thus, up to a certain critical value, there is a supercurrent, which flows without resistance. Macroscopically, we have shown that the electrodynamic behavior of the superconductors may be described by an ensemble-average wave function of the form of $\psi = |\psi| \exp(i\theta)$, where θ is the phase of the electron pairs. More detail discussions about this point may be found in the work of Blatt (1964).

After the BCS microscopic theory appeared, alternative and more general formulations were developed, among which an important and useful method of Green's functions and Feynman diagrams was introduced to the theory of superconductivity by Gor'kov (1958). Other topics, such as the microscopic theory of superconducting tunnel junctions and the possible micromechanisms other than the electron-phonon mechanism responsible for the superconductivity, were also studied intensively. In particular, researches into revealing the micromechanism and developing theories for the newly discovered high-T_c oxide superconductors are still in progress. Readers who are interested in those specific topics are referred to the works of, for instance, Josephson (1962, 1969), Barone and Paterno (1982), Khurana (1987) and Lundqvist et al. (1987).

Chapter 3

Electromagneto-mechanical interactions in superconductors

In many engineering applications of superconductivity, superconductors may not only conduct electric currents but they may also be subject to thermal and mechanical loadings. Since the discovery of hard superconductors, the mechanical behavior of superconducting materials has been of interest and concern to researchers for many years because of the large mechanical forces expected in future applications of superconducting devices, such as magnets for plasma confinement and energy storage, superconducting generators, electrical transmission lines, magnetic levitating trains, electromagnetic-propulsion ships, or still unimagined new devices. In particular, the latest generation of superconductors uses ceramic materials which are brittle and much harder to work with than metals, and a considerable amount of work is needed to improve the mechanical behavior of the new superconductors. Furthermore, AC currents or changing magnetic fields may lead to losses and generation of heat in superconductors, which had in fact limited most of the past engineering applications of superconductivity to static cases. Also dynamic effects in nonequilibrium superconductors can introduce interesting acoustic phenomena in superconductors, which are of importance for potential technological applications. Obviously, the possibility for wider applications of superconductivity hinges on the success in improving our understanding of the superconductivity and its related electrodynamic and mechanical problems. This chapter is, therefore, devoted to the introduction of some theoretical models developed recently for the study of the interaction between the mechanics and electromagnetics of superconductors.

3.1 Macroscopic theory of elastic superconductors

We shall start with the introduction of a continuum model which may describe phenomenologically the macroscopic electromagnetoelastic behavior of superconductors. We shall assume that the superconductor is in thermodynamic equilibrium and its deformation is elastic. No dissipative processes, such as plasticity, normal electric and heat conductions, are being considered in this model.

We begin with the following virtual variation equation for the elastic superconductor in thermodynamic equilibrium at a constant temperature

$$\int_V \delta F \, dV = \int_{\partial V} \mathbf{t}^{(n)} \cdot \delta \mathbf{x} \, dS + \int_V (\mathbf{f}^{me} + \mathbf{f}^{em}) \cdot \delta \mathbf{x} \, dV - \int_V \mathbf{J} \cdot \delta \mathbf{A} dV \qquad (3.1.1)$$

where ρ is the mass density of the superconductor, V the volume of the superconductor, $\mathbf{J}$ the supercurrent density, $\mathbf{t}^{(n)}$ the surface traction, $\mathbf{f}^{me}$ the mechanical body force and $\mathbf{f}^{em}$ the electromagnetic body force defined by

$$\mathbf{f}^{em} = \mathbf{J} \times \mathbf{B} \, . \qquad (3.1.2)$$

The thermodynamic function of free energy density F may be expressed by

$$F = F_n + F_{sn} \qquad (3.1.3)$$

where F_n denotes the free energy density of normal state, being only a functional of the Lagrangian strain E_{KL} defined by eqn.(1.10.6) at a given temperature T. F_{sn} denotes the free energy density difference between superconducting state and normal state, which, near superconducting transition temperature, may be expressed by

$$F_{sn} = \alpha(E_{KL})|\psi|^2 + \frac{1}{2}\beta(E_{KL})|\psi|^4 + \frac{1}{2m^*_{kj}}[(-i\hbar\frac{\partial}{\partial x_k} - e^* A_k)\psi][(i\hbar\frac{\partial}{\partial x_j} - e^* A_j)\psi^*] \qquad (3.1.4)$$

where the conventional Einstein's summation rule has been used. Here, $\mathbf{A}$ is the magnetic vector potential (defined by $\mathbf{B} = \nabla \times \mathbf{A}$). The second term on the r.h.s. of eqn.(3.1.4) represents presumably the kinetic energy of the superelectrons with its macroscopic quantum generalization. The necessity of such a macroscopic quantum generalization was confirmed by Josephson-junction experiments. The anisotropic behavior of the superconductors has been taken into account by introducing the effective mass tensor **m*** (Tilley (1965) and Bulaevskii et al. (1988)), which may, in general, be a function of the elastic strain tensor E_{KL}. However, to simplify the formulation for practical applications, we may assume that the effective mass is a material constant with negligible effects from the elastic lattice deformation by noting the fact that m* plays essentially the role of the effective mass of one Cooper pair of superconducting electrons rather than the mass density of Cooper pair superelectrons (Gor'kov and Melik-Barkhudarov (1964) and Zimmerman and Mercereau (1965)). Nevertheless, there would be, in principle, no difficulty in considering mathematically more general cases where such dependence may be taken into account if it were necessary.

After carrying out the variation procedure for eqn.(3.1.1), we can derive the following set of field equations and boundary conditions for the elastic superconductor at equilibrium state.

The mechanical equilibrium equation is

$$t_{ji,j} + f_i^{me} + f_i^{em} = 0 \qquad \text{in V} \tag{3.1.5}$$

where t_{ij} is the Cauchy stress tensor given by

$$t_{ji} = \frac{\partial F}{\partial E_{KL}} x_{i,K} x_{j,L} \qquad \text{in V.} \tag{3.1.6}$$

with $t_{[ji]} \equiv t_{ji} - t_{ij} = 0$, which means that the Cauchy stress tensor t_{ij} is a symmetric tensor.

Eqn.(3.1.5) may also be written as

$$(t_{ji} + t_{ji}^{em})_{,j} + f_i^{me} = 0 \qquad \text{in V} \tag{3.1.7}$$

with the introduction of a Maxwell stress tensor $\mathbf{t}^{em}$ defined by

$$t_{ji}^{em} = \frac{1}{\mu_o}(B_j B_i - \frac{1}{2} B_k B_k \delta_{ji}) . \tag{3.1.8}$$

The generalized anisotropic Ginzburg-Landau equation reads

$$\alpha\psi + \beta|\psi|^2\psi + \frac{1}{2m^*_{kj}}(-i\hbar\frac{\partial}{\partial x_k} - e^*A_k)(-i\hbar\frac{\partial}{\partial x_j} - e^*A_j)\psi = 0 \qquad \text{in V.} \tag{3.1.9}$$

and the equation for the superconducting current density is

$$J_k = \frac{e^*\hbar}{i2m^*_{kj}}(\psi^*\frac{\partial\psi}{\partial x_j} - \psi\frac{\partial\psi^*}{\partial x_j}) - \frac{e^{*2}}{m^*_{kj}}|\psi|^2 A_j \qquad \text{in V.} \tag{3.1.10}$$

The boundary conditions for the superconductor are

$$t_{ji}n_j = t_i^{o(n)} \qquad \text{on } \partial_T V \tag{3.1.11}$$

$$\mathbf{u} = \mathbf{u}^o \qquad \text{on } \partial_u V \tag{3.1.12}$$

$$\frac{1}{2m^*_{kj}}(-i\hbar\frac{\partial\psi}{\partial x_j} - e^*\psi A_j)n_k = 0 \qquad \text{on } \partial V \tag{3.1.13}$$

where $\mathbf{t}^{o(n)}$ denotes the prescribed mechanical surface traction on $\partial_T V$, and $\mathbf{u}^o$ denotes the prescribed displacement on $\partial_u V$ ($\equiv \partial V - \partial_T V$). The boundary condition (3.1.13) is used for the system of superconductor-vacuum (or superconductor-insulator). For the system of superconductor-normal-conductor, the boundary condition (3.1.13) has to be modified (see section 2.8.2). For isotropic superconductors, eqns.(3.1.9), (3.1.10) and (3.1.13) are reduced to the following form

$$\alpha\psi + \beta|\psi|^2\psi + \frac{1}{2m^*}(-i\hbar\nabla - e^*\mathbf{A})^2\psi = 0 \qquad \text{in V} \tag{3.1.14}$$

and

$$\mathbf{J} = \frac{e^*\hbar}{i2m^*}(\psi^*\nabla\psi - \psi\nabla\psi^*) - \frac{e^{*2}}{m^*}|\psi|^2\mathbf{A} \qquad \text{in V} \tag{3.1.15}$$

and

$$(-i\hbar\nabla\psi - e^*\psi\mathbf{A})\cdot\mathbf{n} = 0 \qquad \text{on } \partial V. \tag{3.1.16}$$

A simple dynamic modification of the theory may be eligible for studying elastic superconductors in weak magnetic fields at the magneto-quasistatic approximation. In certain problems, if AC losses in the superconductor are negligible and elastic deformation of the superconductor is small and there is no global motion of the superconductor, then, the mechanical equilibrium equation (3.1.5) may be modified to be the equation of motion by simply adding an inertia force term $\rho_0 \partial^2 u_i / \partial t^2$ on the right-hand side of eqn.(3.1.5). The generalized G-L equations and boundary conditions require no modification since one has ignored the effect of finite relaxation time in the superelectron and there is no surface concentration of mass. The equilibrium theory presented here for elastic superconductors is general and can ensure a consistent treatment of different-order-effects. The resulting field equations are, however, highly non-linear and coupled. In order to obtain solutions of practical interest for applications, some simplifications are, therefore, needed.

3.2 Effects of elastic deformation on superconducting properties

In this section, we shall study the effects of elastic deformation on the superconducting properties of an elastic superconductors by using the macroscopic theory for elastic superconductors given in section 3.1. We shall suppose that the superconductor is in thermodynamic equilibrium and is in a uniform temperature field constant in time. Possible applied electromagnetic fields are static. Furthermore, the superconductor is assumed to be homogeneous, and its elastic deformation is small. In such a case, the mechanical constitutive equation (3.1.6) of the elastic superconductor at a constant temperature T may be written approxi-mately as

$$t_{ij} = C^{n}_{ijkl}\varepsilon_{kl} + \alpha'(a_{ij} + a'_{ijkl}\varepsilon_{kl})|\psi|^2 + \frac{\beta'}{2}(b_{ij} + b'_{ijkl}\varepsilon_{kl})|\psi|^4 \qquad (3.2.1)$$

where ε_{kl} is the infinitesimal strain tensor defined by eqn.(1.10.8b) and C^{n}_{ijkl} the elastic moduli of the superconductor in normal state. The phenomenological coefficients α', a_{ij}, a'_{ijkl} and β', b_{ij} and b'_{ijkl} are defined from the expansion coefficients of α and β by

$$\alpha(T, \varepsilon) = \alpha'(T)[1 + a_{ij}(T)\varepsilon_{ij} + \frac{1}{2}a^*_{ijkl}(T)\varepsilon_{ij}\varepsilon_{kl}] \tag{3.2.2}$$

$$\beta(T, \varepsilon) = \beta'(T)[1 + b_{ij}(T)\varepsilon_{ij} + \frac{1}{2}b^*_{ijkl}(T)\varepsilon_{ij}\varepsilon_{kl}] . \tag{3.2.3}$$

It is noticed that the coefficients a'_{ijkl} and b'_{ijkl} in eqn.(3.2.1) are, in general, different from the expansion constants a^*_{ijkl} and b^*_{ijkl}, which is due to the influence of the deformation gradient **F** (defined by eqn.(1.10.2)) in the expansion calculation. For isotropic superconductors, one has

$$a_{ij} = a_o\delta_{ij} \tag{3.2.4}$$

$$a^*_{ijkl} = a^*_o\,\delta_{ij}\delta_{kl} + a^*_1(\delta_{ik}\delta_{jl} + \delta_{il}\delta_{jk}) \tag{3.2.5}$$

$$b_{ij} = b_o\delta_{ij} \tag{3.2.6}$$

$$b^*_{ijkl} = b^*_o\,\delta_{ij}\delta_{kl} + b^*_1(\delta_{ik}\delta_{jl} + \delta_{il}\delta_{jk}) \tag{3.2.7}$$

and

$$a'_{ijkl} = a^*_o\,\delta_{ij}\delta_{kl} + (a_o + a^*_1)(\delta_{ik}\delta_{jl} + \delta_{il}\delta_{jk}) \tag{3.2.8}$$

$$b'_{ijkl} = b^*_o\,\delta_{ij}\delta_{kl} + (b_o + b^*_1)(\delta_{ik}\delta_{jl} + \delta_{il}\delta_{jk}) \tag{3.2.9}$$

and

$$C^n_{ijkl} = \frac{2G\nu}{1 - 2\nu}\delta_{ij}\delta_{kl} + G(\delta_{ik}\delta_{jl} + \delta_{il}\delta_{jk}) \tag{3.2.10}$$

where G and ν are respectively the elastic shear modulus and Poisson's ratio.

We shall now look at some specific cases. The simplest one is the case in which we consider a superconductor in equilibrium state and in the absence of applied magnetic field. We consider that the density of the superelectron is uniform by ignoring the effect of elastic deformation at the zeroth-order approximation. Thus, eqn.(3.1.14) becomes simply

$$\alpha\psi + \beta|\psi|^2\psi = 0 \tag{3.2.11}$$

which has a non-zero solution $|\psi_0|^2 = -\alpha/\beta$, describing the usual superconducting state with perfect Meissner effect (neglecting surface effects). The effect of elastic deformation on the superelectron density may be taken into account by using a perturbation (iterative) method. At the first-order approximation, the effect may be accounted for by using eqns.(3.2.2) and (3.2.3) for $|\psi_0|^2 = -\alpha/\beta$, where the elastic strain is obtained from the zeroth-order approximate solution of the mechanical equilibrium equation for the superconductor under purely mechanical loadings.

3.2.1 **Strain effect on transition temperature T_c**

By noting the fact that the phenomenological coefficient $\alpha(T, \varepsilon)$ must change its sign at the transition temperature T_c, one may also make an expansion of α in the following form

$$\alpha(T, \varepsilon) = \alpha_0(\varepsilon)(T - T_c) + o(|T-T_c|^3) \tag{3.2.12}$$

with α_0 being a real and positive quantity. Here, the expansion with respect to temperature is made near the superconducting transition temperature T_c defined at the actual configuration in the absence of applied magnetic field. In general, the transition temperature T_c of the superconductor is also strain-dependent. However, such a dependence of T_c on elastic strain of the superconductor cannot be derived simply from the macroscopic model. Empirically, it may be assumed that the strain-dependent transition temperature T_c may be described up to the second-order approximation of the elastic strain by

$$T_c(\varepsilon) = T_c^0(1 + \Delta_{ij}\,\varepsilon_{ij} + \frac{1}{2}\Xi_{ijkl}\,\varepsilon_{ij}\,\varepsilon_{kl}) \tag{3.2.13}$$

where T_c^0 is defined as the "ideal" superconducting transition temperature for the superconductor in a natural reference state where the material is stress-free, temperature field is uniform, and no magnetic field is applied. Here, Δ_{ij} and Ξ_{ijkl} are dimensionless material constants.

Also, we may make an expansion of $\alpha_0(\varepsilon)$ by

$$\alpha_o(\varepsilon) = \alpha_{\varepsilon o} + \alpha_{\varepsilon ij}\varepsilon_{ij} + \frac{1}{2}\alpha_{\varepsilon\varepsilon ijkl}\,\varepsilon_{ij}\,\varepsilon_{kl}\,. \tag{3.2.14}$$

Up to the second-order approximation with respect to the elastic strain, we may find the following relations between the coefficients defined in eqn.(3.2.2) and eqns.(3.2.13) and (3.2.14):

$$\alpha'(T) = \alpha_{\varepsilon o}(T - T_c^o) \tag{3.2.15}$$

$$a_{ij}(T) = \frac{1}{\alpha_{\varepsilon o}}\,\alpha_{\varepsilon ij} - \frac{T_c^o}{T - T_c^o}\Delta_{ij} \tag{3.2.16}$$

$$a^*_{ijkl}(T) = \frac{1}{\alpha_{\varepsilon o}}\,\alpha_{\varepsilon\varepsilon ijkl} - \frac{T_c^o}{T - T_c^o}\Xi_{ijkl} - \frac{2T_c^o}{\alpha_{\varepsilon o}(T - T_c^o)}\alpha_{\varepsilon ij}\Delta_{kl}\,. \tag{3.2.17}$$

For an isotropic superconductor, one has $\Delta_{ij} = \Delta_o\delta_{ij}$, where the constant Δ_o may be either positive or negative depending on the materials. For instance, Smith and Chu (1967) observed a linear hydrostatic pressure dependence of T_c for Al, Cd, Zn, Sn, In and Pb with $\Delta_o > 0$, which means that the transition temperature decreases with increasing pressure. Borges et al. (1987), however, found that the transition temperatures of some high temperature superconductors, such as $YBa_2Cu_3O_x$ and $RBa_2Cu_3O_x$ (R=Gd, Er and Yb), are enhanced by pressure, which implies a negative value of Δ_o. For relatively large deformation, it may be necessary to include the term ε^2 by considering the second-order approximation (Welch (1980)).

Using elastic stress-strain relations, one may also expressed the strain-dependence of the transition temperature T_c in terms of the stress-dependence of T_c. Experimental studies of the pressure effects on superconductivity has been made since 1926 when Sizoo and Kamerlingh Onnes (1926) first reported a small decrease in T_c of In and Sn when subjected to pressure generated by liquid helium. A good many experimental data for some conventional superconductors have been summarized in the work of Narlikar and Ekbote (1983). Here, in **Figure** 3.2.1, we show the experimental result on the pressure dependence of the resistance of the high-T_c oxide superconductor $EuBa_2(Cu_{0.99}Zn_{0.01})_3O_7$ (Phillips (1989)).

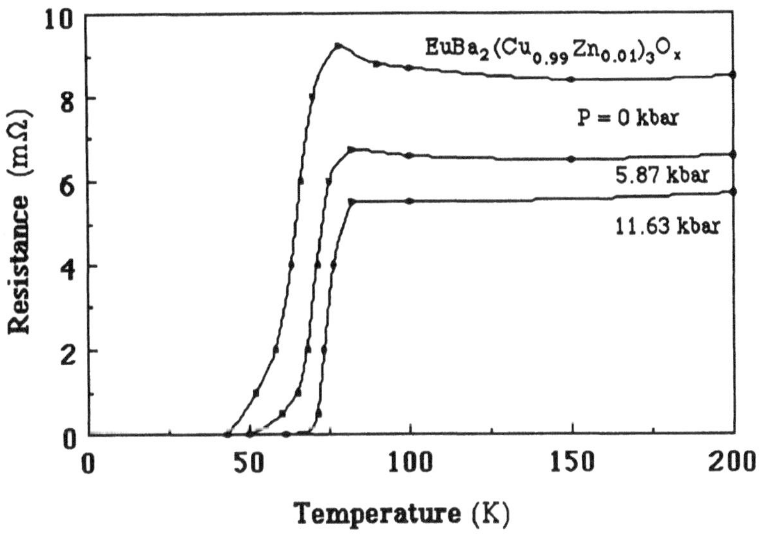

Figure 3.2.1 Resistance of $EuBa_2(Cu_{0.99}Zn_{0.01})_3O_7$ at three pressures.

3.2.2 **Strain effect on thermodynamic critical field H_c**

With the use of eqns.(3.2.2), (3.2.3) and (3.2.15), the thermodynamic critical field H_c defined in eqn.(2.8.10) can be expressed, to the first-order approximation with respect to the elastic strain, by

$$H_c = H_{co}\,[1 + \frac{1}{2}(2a_{ij} - b_{ij})\varepsilon_{ij}] \tag{3.2.18}$$

with H_{co} being defined as the thermodynamic critical field of the superconductor at zero strain and near transition temperature by

$$H_{co} = \sqrt{\frac{\alpha_{\varepsilon o}^2 (T - T_c^0)^2}{\mu_o \beta_o}} \tag{3.2.19}$$

with $\beta_o = \beta'(T_c^0) > 0$.

3.2.3 **Strain effect on penetration depth λ**

In a weak magnetic field ($H < H_{c1}$) for a homogeneous isotropic superconducting material, to the first order in **B**, $|\psi|^2$ may be replaced by $|\psi_o|^2$; eqn.(3.1.15) then becomes (for $|\psi_o|^2$ being independent of **x**)

$$J = \frac{e^*\hbar}{m^*}|\psi_o|^2 \nabla\theta - \frac{e^{*2}}{m^*}|\psi_o|^2 A \tag{3.2.20}$$

where θ is the phase of ψ (i.e. $\psi = |\psi| \exp(i\theta)$). By taking the curl of eqn.(3.2.20), one derives immediately the second London equation (2.2.9) describing the Meissner effect with the penetration depth λ defined by eqn.(2.8.21), which is a constant for a given temperature in classical London theory.

By now taking into account elastic deformation, we may find that the penetration depth λ also depends on elastic strain. As a first-order approximation with respect to the elastic strain, the penetration depth λ may be expressed, near the transition temperature, by

$$\lambda = \lambda_{\varepsilon o} [1 + \frac{1}{2}(b_o - a_o)\varepsilon_{kk}] \tag{3.2.21}$$

with $\lambda_{\varepsilon o}$ being defined as the penetration depth of the superconductor at zero strain

$$\lambda_{\varepsilon o} = \sqrt{\frac{m^*\beta_o}{\mu_o e^{*2}\alpha_{\varepsilon o}|T - T_c^o|}} \; . \tag{3.2.22}$$

3.2.4 **Strain effect on coherence length ξ and G-L parameter κ**

Similarly, at the first-order approximation with respect to the elastic strain, we may also find the strain-dependent coherence length ξ, near the transition temperature, by

$$\xi = \xi_{\varepsilon o}(1 - \frac{1}{2} a_o \varepsilon_{kk}) \tag{3.2.23}$$

with $\xi_{\varepsilon o}$ being defined as the G-L coherence length of the superconductor at zero strain

$$\xi_{\varepsilon o} = \sqrt{\frac{\hbar^2}{2m^*\alpha_{\varepsilon o}|T - T_c^o|}} \; . \tag{3.2.24}$$

The strain-dependent G-L parameter κ may also be found, at the first-order approximation with respect to the elastic strain, as

$$\kappa = \kappa_{\varepsilon o}(1 + \frac{1}{2} b_o \varepsilon_{kk}) \tag{3.2.25}$$

with $\kappa_{\varepsilon o}$ being defined as the G-L parameter of the superconductor at zero strain

$$\kappa_{\varepsilon o} = \frac{m^*}{e^* \hbar} \sqrt{\frac{2\beta_o}{\mu_o}} \; . \tag{3.2.26}$$

The upper critical field H_{c2} can be found, at the first order approximation with respect to the elastic strain, as

$$H_{c2} = H_{co2}(1 + a_o \varepsilon_{kk}) \tag{3.2.27}$$

with H_{co2} being defined as the upper critical field of the superconductor at zero strain by

$$H_{co2} = \sqrt{2}\, \kappa_{\varepsilon o} H_{co} \tag{3.2.28}$$

with $\kappa_{\varepsilon o}$ and H_{co} defined respectively by eqn.(3.2.26) and eqn.(3.2.19).

It is shown that, at the first-order approximation, the superconducting properties of an isotropic superconducting solid, such as the transition temperature T_c, the penetration depth λ, the coherence length ξ, the G-L parameter κ as well as the critical field H_{c2} are only influenced by the dilatation ($\varepsilon_{ii} = \Delta V/V$) of the superconductive body. Thus, the effect of shearing strain on the superconducting properties of isotropic superconductors can only be taken into account by considering the second-order approximation with respect to the elastic strain.

3.3 Elastic behavior of superconductor at superconducting state

In the above section, some studies have been made on the problem of how superconductive properties of the material can be influenced by mechanical deformation. We are now going to consider the problem of how the elastic behavior of the superconductor is influenced by superconducting-normal phase transition. We shall assume here that the superconductor considered is a homogeneous medium.

3.3.1 **Elastic superconductor in weak magnetic field**

First, we consider the case of a superconductor in a weak magnetic field ($B < \mu_0 H_{c1}$) and in a constant temperature field. Using a perturbation method, we may write, at the first-order approximation with respective to the elastic strain,

$$|\psi|^2 = -\frac{\alpha'}{\beta'}[1 + (a_{ij} - b_{ij})\varepsilon_{ij}], \qquad |\psi|^4 = \frac{\alpha'^2}{\beta'^2}[1 + 2(a_{ij} - b_{ij})\varepsilon_{ij}] \tag{3.3.1}$$

where α', β', a_{ij} and b_{ij} are material coefficients defined in eqns.(3.2.2) and (3.2.3). From eqn. (3.3.1), we may write

$$\alpha' a_{ij}\frac{\partial|\psi|^2}{\partial x_j} + \frac{1}{2}\beta' b_{ij}\frac{\partial|\psi|^4}{\partial x_j} = -\frac{\alpha'^2}{\beta'}(a_{ij} - b_{ij})(a_{kl} - b_{kl})\frac{\partial\varepsilon_{kl}}{\partial x_j} \tag{3.3.2}$$

where we have assumed that the deformation of the superconductor may be non-uniform, but the temperature field is uniform.

Furthermore, from eqn.(3.2.1), we can get

$$t_{ij,j} = C^n_{ijkl}\varepsilon_{kl,j} + \alpha' a'_{ijkl}\varepsilon_{kl,j}|\psi|^2 + \frac{\beta'}{2}b'_{ijkl}\varepsilon_{kl,j}|\psi|^4 + \alpha' a_{ij}\frac{\partial|\psi|^2}{\partial x_j} + \frac{1}{2}\beta' b_{ij}\frac{\partial|\psi|^4}{\partial x_j}. \tag{3.3.3}$$

Thus, by eqn.(3.3.2), we may find

$$t_{ij,j} = \{C^n_{ijkl} - \frac{\alpha'^2}{\beta'}[a'_{ijkl} - \frac{1}{2}b'_{ijkl} + (a_{ij} - b_{ij})(a_{kl} - b_{kl})]\}\varepsilon_{kl,j}\,. \tag{3.3.4}$$

from which, we may write

$$t_{ij} = C^s_{ijkl}\,\varepsilon_{kl} \tag{3.3.5}$$

where C^s_{ijkl} may be called the effective elastic modulus tensor of the superconductor in superconducting state, defined by

$$C^{s}_{ijkl} = C^{n}_{ijkl} - \mu_o H^2_{co}[a'_{ijkl} - \frac{1}{2} b'_{ijkl} + (a_{ij} - b_{ij})(a_{kl} - b_{kl})] \tag{3.3.6}$$

Here, H_{co} is the thermodynamic critical magnetic field of the superconductor at zero strain, defined by eqn.(3.2.19). The relation (3.3.6) is in agreement with the result derived by Labusch (1968) in a somewhat different way. However, a little modification between the coefficients of a'_{ijkl} and a^*_{ijkl} and of b'_{ijkl} and b^*_{ijkl} is found here as shown by eqns.(3.2.5)-(3.2.9).

3.3.2 **Elastic superconductor in strong magnetic field**

Next, we consider the case of the superconductor in a strong applied magnetic field B^e which is close to $\mu_0 H_{c2}$. In such a case, a mixed state will be present in the superconducting solid which splits into some fine-scale mixture of superconducting and normal regions whose boundaries lie parallel to the applied field, and the arrangement being such as to give the maximum boundary area relative to the volume of normal material since the surface energy associated with the boundary between a normal and superconducting region is negative for type II superconductors. The structure of such a mixed state is, in general, on a very fine scale with a periodicity less than 10^{-4} cm. Thus, for macroscopic strain fields slowly varying compared with $|\psi|^2$ in the mixed state, one may introduce the local averages $<t_{kl}>$, $<\varepsilon_{kl}>$, $<|\psi|^2>$ and $<|\psi|^4>$ over a small macroscopic volume. We then have, at the zero-order approximation, the classical results (see eqns.(2.8.66) and (2.8.67))

$$<|\psi|^2> = -\frac{\alpha}{\beta} \cdot \frac{2\kappa^2}{\beta_A(2\kappa^2-1)}(1 - \frac{B^e}{\mu_o H_{c2}}) \tag{3.3.7a}$$

$$<|\psi|^4> = \beta_A <|\psi|^2>^2 \tag{3.3.7b}$$

with β_A = 1.16 for the triangular flux line lattice, which is known as the most thermodynamically stable lattice among all other possible periodic lattices (Essmann and Träuble (1967)).

Thus, with the use of eqns.(3.2.2), (3.2.3), (3.2.25) and (3.2.27), we can derive, at the first order approximation with respect to the elastic strain,

$$<|\psi|^2> = -\frac{\alpha'}{\beta'}\cdot\frac{2\kappa_{\varepsilon o}^2}{\beta_A(2\kappa_{\varepsilon o}^2-1)}(1-\frac{B^e}{\mu_o H_{co2}})[1+(\frac{H_{co2}}{H_{co2}-B^e/\mu_o}a_{ij}-\frac{2\kappa_{\varepsilon o}^2}{2\kappa_{\varepsilon o}^2-1}b_{ij})<\varepsilon_{ij}>] \tag{3.3.8}$$

and

$$\alpha' a_{ij}\frac{\partial<|\psi|^2>}{\partial x_j}+\frac{1}{2}\beta' b_{ij}\frac{\partial<|\psi|^4>}{\partial x_j} = -\frac{\alpha'^2}{\beta'}\cdot\frac{2\kappa_{\varepsilon o}^2}{\beta_A(2\kappa_{\varepsilon o}^2-1)}[a_{ij}-\frac{2\kappa_{\varepsilon o}^2}{2\kappa_{\varepsilon o}^2-1}(1-\frac{B^e}{\mu_o H_{co2}})b_{ij}]$$

$$\times[a_{kl}-\frac{2\kappa_{\varepsilon o}^2}{2\kappa_{\varepsilon o}^2-1}(1-\frac{B^e}{\mu_o H_{co2}})b_{kl}]\frac{\partial<\varepsilon_{kl}>}{\partial x_j}. \tag{3.3.9}$$

From eqn.(3.2.1), we can then obtain

$$<t_{ij}>_{,j} = C^n_{ijkl}<\varepsilon_{kl}>_{,j} - \frac{\alpha'^2}{\beta'}\frac{2\kappa_{\varepsilon o}^2}{\beta_A(2\kappa_{\varepsilon o}^2-1)}\{(1-\frac{B^e}{\mu_o H_{co2}})(a'_{ijkl}-\frac{\kappa_{\varepsilon o}^2}{2\kappa_{\varepsilon o}^2-1}b'_{ijkl})$$

$$+[a_{ij}-\frac{2\kappa_{\varepsilon o}^2}{2\kappa_{\varepsilon o}^2-1}(1-\frac{B^e}{\mu_o H_{co2}})b_{ij}][a_{kl}-\frac{2\kappa_{\varepsilon o}^2}{2\kappa_{\varepsilon o}^2-1}(1-\frac{B^e}{\mu_o H_{co2}})b_{kl}]\}<\varepsilon_{kl}>_{,j} \tag{3.3.10}$$

from which we may define an effective elastic modulus tensor of the superconductor in a strong magnetic field and in superconducting state by

$$C^s_{ijkl} = C^n_{ijkl} - \frac{2\mu_o H_{co}^2\kappa_{\varepsilon o}^2}{\beta_A(2\kappa_{\varepsilon o}^2-1)}\{(1-\frac{B^e}{\mu_o H_{co2}})(a'_{ijkl}-\frac{\kappa_{\varepsilon o}^2}{2\kappa_{\varepsilon o}^2-1}b'_{ijkl})$$

$$+[a_{ij}-\frac{2\kappa_{\varepsilon o}^2}{2\kappa_{\varepsilon o}^2-1}(1-\frac{B^e}{\mu_o H_{co2}})b_{ij}][a_{kl}-\frac{2\kappa_{\varepsilon o}^2}{2\kappa_{\varepsilon o}^2-1}(1-\frac{B^e}{\mu_o H_{co2}})b_{kl}]\}. \tag{3.3.11}$$

Eqn.(3.3.11) indicates a jump of the elastic moduli at $B^e = \mu_o H_{co2}$, which is given by

$$C^n_{ijkl} - C^s_{ijkl} = \frac{2\mu_o H_{co}^2\kappa_{\varepsilon o}^2}{\beta_A(2\kappa_{\varepsilon o}^2 - 1)}a_{ij}a_{kl}\,. \tag{3.3.12}$$

The relation (3.3.12) is also in agreement with the result obtained by Labusch (1968), with, however, some modifications in eqn.(3.3.12). It is seen that most of calculations given above have been restricted to the first order approximation. However, it is, in principle, not difficult to generalize such calculations to higher order approximations according to specific requirements.

Some experiments in studying elastic properties of superconductors have been made by a number of researchers. Bourne et al. (1987) performed Young's modulus measurements of the $La_{2-x}Sr_xCuO_4$ superconducting compound by using a resonant vibration technique, where changes in Young's modulus were determined directly from the changes in vibration frequency. It was observed that, at the superconducting transition temperature (T_c = 35 ^oK), there was an anomaly in Young's modulus. The anomalous singularities in the elastic shear modulus and Young's modulus of polycrystalline $La_{1.85}Sr_{0.15}CuO_4$ were also observed in the vicinity of their superconducting transitions by Xiang et al. (1988). Other observations on various anomalies in elastic behavior of, for instance, Y-Ba-Cu-O superconducting materials at varying low temperatures were also made by Ledbetter et al. (1987) and Bhattachary et al. (1988a, b) etc. Further efforts are needed to correlate experimental observations to the theoretical models when sufficient experimental data are available.

3.4 Ferromagnetism of magnetic solids

Recently, exciting experimental and theoretical activities in the study of the interaction between superconductivity and magnetism have been prompted by the discovery of magnetic superconductors (Matsubara and Kotani (1984) and Kakani and Upadhyaya (1988)). In order to study the interaction of superconductivity and magnetism and also due to its intrinsic theoretical and practical interest, some basic phenomena concerning ferromagnetism and their theoretical modelling will be introduced first in this section.

3.4.1 Saturation magnetization and Curie temperature

The distinguishing feature of macroscopic behavior of ferromagnetic materials in an external magnetic (intensity) field **H** may be seen from the shape of its magnetization curve M(H) as shown in **Figure** 3.4.1 (Stoletov (1873a, b)). It is shown that starting from the initial state

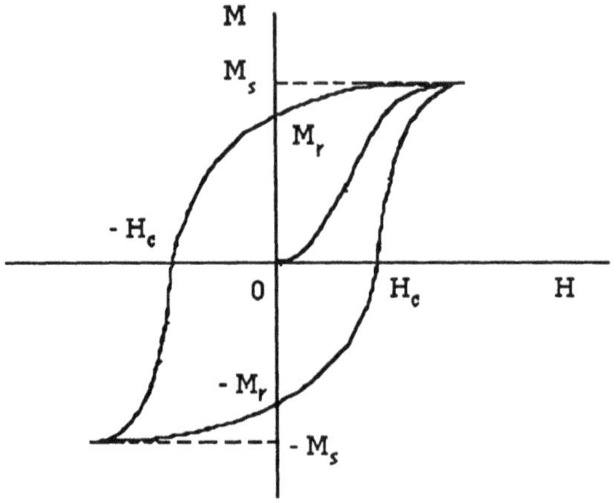

Figure 3.4.1 Magnetic hysteresis loop.

(H=M=0), the magnetization of the ferromagnetic sample increases sharply with increase in the field and eventually reaches a constant value M_s called the saturation magnetization. Some values of $\mu_0 M_s$ for typical ferromagnetic materials are, for instance, 0.79 Tesla for supermalloy, 0.69 Tesla for nickel, 1.79 Tesla for cobalt and 2.15 Tesla for iron (Chikazumi (1964)).

It is noted that, since $\mathbf{B} = \mu_0(\mathbf{H} + \mathbf{M})$, the magnetic induction field cannot in any event reach a saturation value because the component $\mu_0 H$ increases continuously with increasing H. The saturation magnetization of a ferromagnetic body is a structure-insensitive property. However, the approach to the saturation is a structure-sensitive process. The magnitude of the saturation magnetization depends on temperature. It decreases with the increase of temperature and vanishes at a certain (critical) temperature T_{cu}, called the Curie temperature. Above this temperature, the ferromagnet no longer retains its characteristic magnetic properties, and it becomes paramagnetic. Some typical values of the Curie temperature T_{cu} are, for instance, 673 °K for supermalloy, 631 °K for nickel, 1400 °K for cobalt and 1043 °K for iron (Chikazumi (1964)) and Brailsford (1966)).

3.4.2 **Magnetic hysteresis and its mathematical modelling**

During the process of magnetization, if we decrease the field H, we find that the magnetization curve does not retrace its path, and M does not decrease as rapidly as it originally increased as shown in **Figure** 3.4.1. Such a phenomenon is known as hysteresis behavior. If we continue decreasing H until H = 0, we find that $M = M_r$ ($\neq 0$). This value M_r is called the remnant magnetization of the sample, with a value somewhat lower than M_s. Typical values of $\mu_0 M_r$ are, for instance, 0.5 Tesla for supermalloy, 1 Tesla for cobalt steel, and 0.5 Tesla for iron. In order to reduce the magnetization of a ferromagnet to zero, a reverse magnetic field must be applied so that M = 0 when $H = - H_c$. This value H_c is called the coercive force, which has typical values of, for instance, 0.48 $A{\cdot}m^{-1}$ for supermalloy, 18500 $A{\cdot}m^{-1}$ for cobalt steel and 48000 $A{\cdot}m^{-1}$ for iron. The remnant magnetization and the coercive force are both structure-sensitive properties. The coercive force is often used as the most important single criterion for determining whether a ferromagnetic material is soft or hard. Although the classification is not sharp, materials with the maximum H_c values (coercivity) less than 400 $A{\cdot}m^{-1}$ are definitely considered soft, and those with coercitivity values greater than 8000 $A{\cdot}m^{-1}$ are labelled hard. For materials with coercivity values between 400 and 8000 $A{\cdot}m^{-1}$, other criteria, such as the energy product $\mu_0 M_r H_c$, must be considered to provide a clear classification.

As shown in **Figure** 3.4.1, further strengthening of the negative field may finally bring the sample to magnetization in the opposite direction ($- M_s$). Returning backward, we reach $M = - M_r$ at H=0, then M = 0 at $H = + H_c$, and finally the saturation $M = M_s$ again. Thus, over a complete cycle of H, a closed curve is traced out. This closed curve is called the maximum loop of magnetic hysteresis. If, in a cycle, the maximum values of the magnetization are lower than M_s but the same, the loop will describe a similar symmetrical partial cycle within the maximum loop. If the maximum values of the magnetization arc not equal, the loop will have the shape of unsymmetrical partial cycle.

To describe such macroscopic behavior of ferromagnetic materials, we may introduce here a simple mathematical model (Potter and Schmulian (1971)) without going into details about the structure of ferromagnetic domains and their orientational configurations influenced by external magnetic fields. This mathematical model may be used in practical computation of electromagnetic field problems for isotropic ferromagnetic materials (see, for instance, Oohira (1989)). In this model, the magnetic hysteresis loop, shown by **Figure** 3.4.1, is represented by the following hyperbolic tangent formula:

$$\frac{M(H;\,\alpha)}{M_s} = \operatorname{sgn}\alpha - \alpha\{1 + \tanh[\frac{H_c - H\operatorname{sgn}\alpha}{H_c}\tanh^{-1}(\frac{M_r}{M_s})]\} \tag{3.4.1}$$

where M_s, M_r and H_c are respectively the saturation magnetization, the remnant magnetization and the coercive force of the material. The function sgnα is defined by sgnα = 1 if $\alpha > 0$ and sgnα = -1 if $\alpha < 0$, where α is a curve parameter. This parameter characterizes the location of hysteresis loops and it varies between -1 and +1, with $\alpha > 0$ for increasing H and $\alpha < 0$ for decreasing H, and with $\alpha = \pm 1$ corresponding to the maximum loop. The determination of the curve parameter α is by initial conditions and subsequently by external magnetic fields.

The virgin magnetization curve is represented by

$$\alpha = \pm\frac{1}{1 + M_r/M_s} \tag{3.4.2}$$

where the positive sign is taken when the loop is in the region of $H > 0$ and $M > 0$, and the negative sign is taken when the loop is in the region of $H < 0$ and $M < 0$.

If the magnetization is given by M(H; α) and H has an extremum at $H = H_m$, then a new magnetization curve M(H; α') is selected for H being reversed, where

$$\alpha' = -\frac{2\operatorname{sgn}\alpha - \alpha\{1 + \tanh[(1 - \frac{H_m\operatorname{sgn}\alpha}{H_c})\tanh^{-1}(\frac{M_r}{M_s})]\}}{1 + \tanh[(1 + \frac{H_m\operatorname{sgn}\alpha}{H_c})\tanh^{-1}(\frac{M_r}{M_s})]}\,. \tag{3.4.3}$$

It should, however, be noticed that the presented mathematical model does not take into account reversible susceptibilities. Thus, calculated magnetization values could be unstable if small fluctuations in field intensity are considered.

3.4.3 Weiss theory of molecular field

The existence of the saturation magnetization M_s reached in a comparatively weak field constitutes a very striking external manifestation of the magnetic properties of ferromagnetic materials. Though magnetic saturation may also exist in a paramagnet, except in a temperature close to 0 °K, it is not observed in practice even for the highest fields attainable (Vonsovskii (1974)). The fundamental internal property of a ferromagnetic body was studied by Weiss (1907) with the first postulation of the presence of temperature-dependent spontaneous magnetization within the temperature interval from 0 °K to the Curie temperature T_{cu}. This spontaneous magnetization is independent of an external magnetic field, and it disappears completely above the Curie temperature. Furthermore, due to the fact that a ferromagnetic body may have zero net magnetic moment in the absence of external magnetic field, a second postulation was made by considering the existence of small regions of spontaneous magnetization, called the magnetic domains, each of which is magnetized to the saturation m_s for a given temperature. The net magnetic moment of a ferromagnetic body is then the vector sum of the magnetic moments of each domain. Since the direction of magnetization of each domain need not be parallel, certain domain configurations may, therefore, lead to a zero net magnetic moment.

In accordance with the above two postulations, the theory of ferromagnetism may be divided into two parts: (1) the theory of spontaneous magnetization (in domains) which explains the actual nature of ferromagnetism, and (2) the theory of the appearance and modification of ferromagnetic domains which explains the macroscopic behavior of a ferromagnetic body in an external magnetic field.

In what follows, we shall discuss the first part of the theory of ferromagnetism on the spontaneous magnetization in a single ferromagnetic (macroscopic) domain resulting from an ordered alignment of the atomic magnetic moments caused by strong interaction among these moments. As to the second part of the theory of ferromagnetism, interested readers are referred to some more specialized work of, for instance, Chikazumi (1964), Vonsovskii (1974) and Chen (1977).

The first successful theory of the temperature dependence of the spontaneous magnetization was given by Weiss in 1907. It was postulated that the strong interaction which tends to align the atomic dipole moments parallel in a ferromagnetic material may be considered as equivalent to some internal magnetic field H_m, which is usually called the "effective" molecular field. Weiss assumed that this molecular field H_m was proportional to the magnetization M (the spontaneous magnetization) so that

$$H_m = \lambda_w M \tag{3.4.4}$$

where λ_w is a constant, called the molecular field constant.

In the presence of an externally applied magnetic intensity field H, the actual field acting on a given dipole moment is then

$$H_T = H + \lambda_w M \tag{3.4.5}$$

where the demagnetizing and Lorentz (dipole-dipole) fields are omitted, since their effect is small compared to the molecular field. The potential energy of a magnetic dipole with the moment **m** in the field H_T is given by

$$U' = - \mathbf{m} \cdot (\mu_o H_T). \tag{3.4.6}$$

Now, we consider a medium containing N atoms per unit volume, each with a permanent moment **m**, which has a component parallel to the field, $g\mu_B M_J$. The quantity M_J may take the values of J, (J-1),..., -(J-1), -J. Here, J denotes the total angular momentum quantum number of the atom (which includes the total orbital contribution and total spin contribution of the electronic system per atom), g is the gyromagnetic factor (or the spectroscopic splitting factor), and μ_B is known as the Bohr magneton. In thermal equilibrium, the average magnetization of the medium parallel to the field H_T can, thus, be calculated by the principle of statistical mechanics as (Morrish (1965))

$$M = N \frac{\sum_{-J}^{+J} g\mu_B M_J \exp(g\mu_B M_J \mu_o H_T / k_B T)}{\sum_{-J}^{+J} \exp(g\mu_B M_J \mu_o H_T / k_B T)} = Ng\mu_B J\, B_J(x) \tag{3.4.7}$$

with

$$x = \frac{g\mu_B J\mu_o}{k_B T}(H + \lambda_w M). \tag{3.4.8}$$

where k_B is Boltzmann's constant. The factor k_BT characterizes the order of magnitude of the energy of thermal motion, its typical value at room temperature being $k_BT = 1.38 \times 10^{-23} \times 300 = 4.1 \times 10^{-21}$ J. The function $B_J(x)$ is usually called the Brillouin function, defined by

$$B_J(x) = \frac{2J+1}{2J}\coth(\frac{2J+1}{2J}x) - \frac{1}{2J}\coth(\frac{x}{2J}) \tag{3.4.9}$$

which has the limit behavior as

$$B_J(x) \to \coth(x) - \frac{1}{x} = L(x) \qquad \text{for} \quad J \to \infty \tag{3.4.10}$$

where L(x) is known as the Langevin function.

To study the spontaneous magnetization, we set H = 0, and rewrite eqns.(3.4.7) and (3.4.8) in the following form

$$\frac{M(T)}{M(0)} = B_J(x) \tag{3.4.11}$$

and

$$\frac{M(T)}{M(0)} = \frac{k_B T}{N\lambda_w\mu_o g^2\mu_B{}^2J^2}x \tag{3.4.12}$$

with $M(0) = Ng\mu_BJ$. The set of eqns.(3.4.11) and (3.4.12) determines the temperature-dependent behavior of the spontaneous magnetization.

For small x, the Brillouin function may be approximately expressed as

$$B_J(x) = \frac{J+1}{3J}x - \frac{J+1}{3J}\frac{2J^2+2J+1}{30J^2}x^3 \qquad \text{for } x \ll 1 \tag{3.4.13}$$

With $x \to 0$ and $T \to T_{cu}$, we have

$$\frac{J+1}{3J} = \frac{k_B T_{cu}}{N\lambda_w\mu_o g^2\mu_B{}^2J^2} \tag{3.4.14}$$

from which we can find that the Curie temperature T_{cu} is

$$T_{cu} = \frac{N\lambda_w \mu_0 g^2 \mu_B{}^2 J(J+1)}{3k_B} \; . \tag{3.4.15}$$

By this equation, eqn.(3.4.12) may be expressed as

$$\frac{M(T)}{M(0)} = \frac{J+1}{3J}\left(\frac{T}{T_{cu}}\right)x \; . \tag{3.4.16}$$

In particular, when T is near the Curie temperature T_{cu}, x is small. We may substitute eqn.(3.4.16) into eqn.(3.4.13). The result, by noting eqn. (3.4.11), is approximately

$$\left[\frac{M(T)}{M(0)}\right]^2 = \frac{10(J+1)^2}{3[J^2 + (J+1)^2]}\left(1 - \frac{T}{T_{cu}}\right) \tag{3.4.17}$$

which shows that the spontaneous magnetization disappears continuously but has an infinite slope at the Curie temperature. Experimental values of the spontaneous magnetization of iron, cobalt and nickel as a function of temperature have been found to be fitted fairly well by the curve of $J = 1/2$ and $g \approx 2$, particularly for higher temperatures. This implies that the magnetization originates mainly from the electron spin, although there is a small contribution from orbital motions of the electrons (Tyler (1931) and Barnett (1915)). At lower temperatures close to absolute zero, the Weiss theory is not in agreement with experiment. This problem was attacked by Bloch (1930) with the introduction of a spin wave theory.

For temperatures above the Curie temperature T_{cu}, the spontaneous magnetization is zero. However, the application of a magnetic field H will produce a magnetization. In the case of H small, the magnetization may be given by

$$M = \frac{Ng\mu_B(J+1)}{3}\,\frac{g\mu_B J\mu_0}{k_B T}(H + \lambda_w M) \tag{3.4.18}$$

from which we can find, by noting eqn.(3.4.15), the susceptibility χ by

$$\chi = \frac{M}{H} = \frac{C}{T - T_{cu}} \tag{3.4.19}$$

where

$$C = \frac{N\mu_0 g^2 \mu_B{}^2 J(J+1)}{3k_B} \ . \tag{3.4.20}$$

Eqn.(3.4.19) shows the fact that above the Curie temperature T_{cu}, the ferromagnet becomes a paramagnet with the susceptibility being inversely proportional to T-T_{cu}. This fact was discovered experimentally by Curie, and it is now called the Curie-Weiss law.

It is seen that the Weiss theory of molecular field is essentially a self-consistent theory of mean field, in which the magnetization is determined by the self-consistent equation (3.4.7). The origin of the postulated molecular field was, however, not explained in the Weiss theory. The first explanation of the origin of the molecular field was given by Heisenberg (1928) in terms of a positive exchange interaction between spins in neighboring atoms. According to his theory, the force which makes spins line up is an exchange force of quantum mechanical nature. The exchange energy of a given atom i with its neighbors reads

$$W_i^{ex} = -2\sum_j J_{ij}\mathbf{S}_i \cdot \mathbf{S}_j \tag{3.4.21}$$

where $\mathbf{S}_i$ and $\mathbf{S}_j$ are the total spins of atoms i and j, and J_{ij} the exchange integral which is a measure of the strength of the interaction and is related to the overlap of the charge distributions of the two atoms. Although this quantum mechanical result has no classical analogue, it is electrostatic in origin. This exchange integral J_{ij} may be either positive or negative. For a hydrogen molecule, it is negative, which results in the antiparallelism of the spins of the two electrons (i and j) in the 1s ground state. It was Heisenberg's postulation that the exchange integral is positive for a ferromagnet so that its stable state may have parallel spins. For a crystal in which the exchange integral is isotropic, we have for the total exchange interaction by summing eqn.(3.4.21) over all the atoms in the crystal

$$W^{ex} = -2J_{ex}\sum_{ij}{}' \mathbf{S}_i \cdot \mathbf{S}_j \ . \tag{3.4.22}$$

For the exchange integral differing appreciably from zero only for atoms i and j which are nearest neighbors in the crystal lattice, an approximation of the nearest interaction model may be used, in which the exchange energy is given by eqn.(3.4.22) with the sum over only the pairs of nearest neighbors. A simple relation between the molecular field constant λ_w and the exchange integral J_{ex} may be found with the aid of an Ising model (Ising (1925)) which makes use of the nearest neighbor interaction and assumes further that the spin system is quantized along a certain direction, usually the direction of the applied field **H**. The result is

$$\lambda_w = \frac{2zJ_{ex}}{N\mu_0 g^2 \mu_B{}^2} \tag{3.4.23}$$

where z denotes the number of the nearest neighbors.

A continuum representation of the exchange energy may also be deduced (see, for instance, Brown (1963) and Vonsovskii (1974)), which shows that, for a rigid ferromagnet, the exchange energy may be expressed in terms of the volume integral $\frac{1}{2}\int \alpha_{ij} M_{k,i} M_{k,j} dV$, where **M** is the magnetization vector and α_{ij} the exchange modulus tensor. This fact will be used later in our modelling of ferromagnetic superconductors.

3.4.4 **Landau theory of ferromagnetic phase transition**

It has been found that, in most cases, the transition from the ferromagnetic state to the paramagnetic state is a phase transition of second order (Vonsovskii (1974)). Thus, a thermodynamic theory may be used with the introduction of a thermodynamic potential density $\Phi'(\mathbf{M},\mathbf{H})$ in the presence of an external field **H** by the relation

$$\frac{\partial \Phi'}{\partial \mathbf{H}} = -\mathbf{B} = -\mu_0 \mathbf{H} - \mu_0 \mathbf{M} \tag{3.4.24}$$

from which we may get

$$\Phi'(\mathbf{M}, \mathbf{H}) = \Phi'(\mathbf{M}, 0) - \frac{\mu_0}{2} H^2 - \mu_0 \mathbf{M}\cdot\mathbf{H} \,. \tag{3.4.25}$$

Neglecting the effects of magnetic anisotropy and inhomogeneity, we may expand $\Phi'(\mathbf{M}, 0)$ in accordance with the general Landau theory of second-order phase transitions (Landau et al. (1984)), and write

$$\Phi'(M, H) = \Phi'_0 + aM^2 + bM^4 - \mu_0 MH - \frac{\mu_0}{2}H^2 \tag{3.4.26}$$

near the Curie temperature (point) T_{cu}, where the magnetization M is small and acts as an order parameter. Here, Φ'_0, a and b are functions only of temperature and deformation of the material. For simplicity, the effect of deformation will, however, not be discussed in this section. Thus, the quantity a(T) may be expressed near the Curie point by

$$a(T) = a_0(T - T_{cu}) \tag{3.4.27}$$

where a_0 is a positive quantity independent of temperature. The quntity b is positive at $T = T_{cu}$ as well as in the vicinity of T_{cu} and may be taken approximately as $b = b(T_{cu})$ near the Curie temperature T_{cu}.

In thermodynamic equilibrium, the thermodynamic potential Φ' must be a minimum for any given field H. Thus, differentiation of (3.4.26) with respect to M at constant H gives the following equilibrium equation

$$2a_0(T - T_{cu})M + 4bM^3 = \mu_0 H \tag{3.4.28}$$

which is a basic relation between the field and the magnetization in a ferromagnet. In the case of H = 0 and below the Curie temperature, we may find the non-zero spontaneous magnetization

$$M^2 = \frac{a_0}{2b}(T_{cu} - T) \tag{3.4.29}$$

which is in accordance with the result (3.4.17) obtained by Weiss theory. A comparison of (3.4.29) and (3.4.17) with J = 1/2 shows that

$$\frac{a_0}{2b} = \frac{3M(0)}{T_{cu}} . \tag{3.4.30}$$

Above the Curie temperature, there is no spontaneous magnetization, and if $H \neq 0$, we may find the susceptibility $\chi = (\partial M/\partial H)_{H=0}$ near the Curie temperature. This can be done by differentiating eqn.(3.4.28) with respect to H, i.e.

$$[2a_o(T - T_{cu}) + 12bM^2]\frac{\partial M}{\partial H} = \mu_o . \tag{3.4.31}$$

Therefore, for $T > T_{cu}$, where $M = 0$ at $H = 0$, the susceptibility is

$$\chi = \frac{\mu_o}{2a_o(T - T_{cu})} \tag{3.4.32}$$

which is, in fact, the Curie-Weiss law. By a comparison of eqn.(3.4.32) and (3.4.19), we find

$$a_o = \frac{\mu_o}{2C} . \tag{3.4.33}$$

Below the Curie temperature, for $H = 0$, the magnetization M is given by eqn.(3.4.29), so that in this case eqn.(3.4.31) gives

$$\chi = \frac{\mu_o}{4a_o(T_{cu} - T)} \tag{3.4.34}$$

which is the initial susceptibility of the paraprocess since $M \neq 0$ for $H=0$. The entropy per unit volume near the Curie point may be found by

$$s = -\frac{\partial \Phi'}{\partial T} = s_o - a_o M^2 \tag{3.4.35}$$

where $s_o = -\partial \Phi'/\partial T$ is the part of entropy independent of M. The term $(\partial \Phi'/\partial M)\cdot(\partial M/\partial T)$ drops out because of the condition $\partial \Phi'/\partial M = 0$. Eqn.(3.4.35) shows that the entropy varies continuously at the Curie point since M is continuous there. The continuity of the entropy at the transition temperature $T = T_{cu}$ indicates the absence of a latent heat of transition. Furthermore, we can find that, at the ferromagnetic phase transition, there is an abrupt change in the specific heat,

$$C_{vferr} - C_{vpara} = vT_{cu}\frac{a_o^2}{2b} \tag{3.4.36}$$

which implies that at T_{cu} the transition from the ferromagnetic state to the paramagnetic state is a phase transition of second order. It is seen that, to a certain degree, the phenomenological thermodynamic theory verifies Weiss´ theory of molecular field for T near T_{cu}.

3.5 Macroscopic theory of magnetoelastic superconductors

3.5.1 Reentrant superconductivity of ferromagnetic superconductors

Recent discovery of magnetic superconductors has prompted exciting experimental and theoretical activities in the study of the interaction between superconductivity and magnetism (Matsubara and Kotani (1984) and Kakani and Upadhyaya (1988)). The earliest theoretical investigation on the possibility of the existence of ferromagnetic superconductors was made by Ginzburg (1957), who showed that it is almost impossible for superconductivity to coexist with ferromagnetic order. In 1977, however, after almost 20 years of continuous efforts in the investigation of magnetic superconductors, Matthias' group at LaJolla (Fertiq et al. (1977)) and Fischer's group at Geneva (Ishikawa and Fischer (1977)) independently found the coexistence of superconductivity and long-range magnetic order in some ternary rare earth compounds, such as $RERh_4B_4$ and $REMo_6S_8$, which have an ordered RE sublattice plus weak exchange interaction between the conduction electron spins and the RE magnetic moments.

Experiments show that some rare earth ternary compounds, such as $ErRh_4B_4$, $REMo_6S_8$ and $HoMo_6S_8$ are ferromagnetic superconductors in which the onset of ferromagnetism causes them to exhibit reentrant superconductivity behavior; i.e. they become superconducting below the upper critical temperature T_{c1} (8.7 °K for $ErRh_4B_4$ and 1.2 °K for $HoMo_6S_8$) and return to the normal state at the lower critical temperature T_{c2} (0.93 °K for $ErRh_4B_4$ and 0.64 °K for $HoMo_6S_8$) near the Curie temperature T_M (0.98 °K for $ErRh_4B_4$ and 0.67 °K for $HoMo_6S_8$) due to the onset of the ferromagnetic order (Izyumov and Skryabin (1980)). The superconducting to normal ferromagnetic state transition at T_{c2} is observed to be a first-order transition. In particular, in a narrow temperature range just above T_{c2}, the magnetic order may occur, coexisting with the superconductivity for $ErRh_4B_4$ (Lynn et al. (1981) and Matsubara and Kotani (1984)). However, studies by Woolf et al. (1979) show that there is no range in which superconductivity and ferromagnetism coexist for the compound $HoMo_6S_8$ (see also Izyumov and Skryabin (1980)).

Typical ac magnetic susceptibility χ_{ac} and electrical resistance vs temperature data for $ErRh_4B_4$ are shown in **Figure** 3.5.1, which shows clearly the thermal hysteresis at T_{c2} for both properties. It can be seen that the susceptibility suddenly increases at T_{c2} and then falls with temperature. This shows that the transition from superconducting to normal state at T_{c2} is accompanied by a transition to a magnetically ordered state.

The failure of the resistance below T_{c2} to attain its full normal state value may be due to the presence of filaments of another phase though other explanations such as superconducting fluctuations were suggested (Fertig et al. (1977) and Shenoy et al. (1980)). The specific heat C_v vs temperature T for $ErRh_4B_4$ is also plotted in **Figure** 3.5.2.

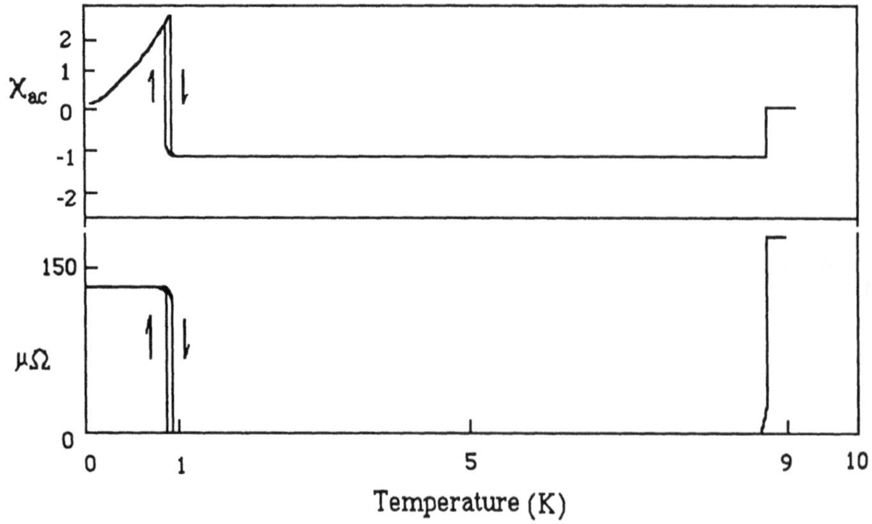

Figure 3.5.1 Typical ac magnetic susceptibility χ_{ac} and electrical resistance vs temperature data for $ErRh_4B_4$.

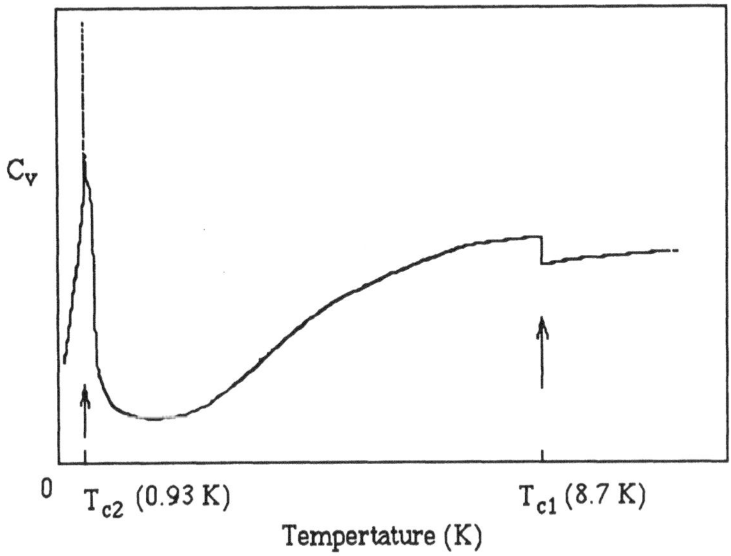

Figure 3.5.2 Specific heat C_v vs temperature T for $ErRh_4B_4$.

It is seen that there is a jump in the specific heat at T_{c1} = 8.7 °K, and a spike-shaped feature in the specific heat at a lower temperature T_{c2}=0.93 °K, indicating a first order transition (Maple (1983)).

Thermal expansion measurements by Ott et al. (1978) revealed a cusp at a slightly higher temperature (0.98 °K), which is typical of a magnetic ordering transition, indicating that superconductivity and long-range ferromagnetism coexist in a small region. In addition, experimental data from the measurements of electrical resistance vs temperature (Ott et al (1978)) showed that applied magnetic field has the effect of lowering T_{c1} and raising T_{c2} and changing the transition at T_{c2} from first order to second order. Anisotropic superconducting properties of the system $ErRh_4B_4$ have also been revealed by Crabtree et al. (1982) in their measurements of upper critical field H_{c2} vs temperature of a single crystal specimen of $ErRh_4B_4$.

Attempts at the development of microscopic theories from first principles and phenomenological modelling based on Ginzburg-Landau free energy expressions have both been made to investigate the novel phenomena of ferromagnetic superconductors (see, for instance, Krey

(1972), Blount and Varma (1979) and Kakani and Upadhyaya (1988)). However, there is no a general phenomenological theory which has uniformly taken into account the interactions between magnetism, superconductivity and mechanical deformation presented in ferromagnetic superconductors, though there is experimental evidence that indicates the influence of the mechanical deformation on the magnetic and superconducting properties of the materials (Vonsovskii (1974), Koch and Easton (1977) and Welch (1980)). In the next section, we shall, therefore, present a macroscopic continuum theory for the study of elastic ferromagnetic superconductors in which the magnetoelastic-superconducting interaction may be taken into account.

3.5.2 **Continuum theory of elastic ferromagnetic superconductors**

Ferromagnetism and superconductivity by their very nature are both quantum-mechanical effects which manifest themselves at the macroscopic scale. The phenomena by which ferromagnetic superconductors exhibit reentrant superconductivity represent their unusual macroscopic behavior due to the interaction of ferromagnetism and superconductivity. In addition, the ferromagnetism and superconductivity have also been found to be dependent on the mechanical deformation of the superconductors as shown by various experiments (Vonsovskii (1974) and Welch (1980)), which implies that the lattice deformation in the material may also contribute to the very nature of the ferromagnetic superconductors.

In order to study phenomenologically the novel macroscopic behavior of the ferromagnetic superconductor, we shall introduce here a continuum model which can, in general, take account of the magneto-mechanical-superconducting interaction in the ferromagnetic superconductors. In this model, the existence of the superconducting phase is described by an order parameter ψ, and the effect of Heisenberg's exchange forces leading to the magnetic order in the ferromagnetic superconductor is described by including the magnetization gradient as internal parameters in the free energy expression for the material. Furthermore, we shall assume that the material is in thermodynamic equilibrium and its deformation is elastic. Some nonequilibrium phenomena will, however, be studied later in this chapter.

We shall start from the following virtual variation equation for the elastic ferromagnetic superconductor in thermodynamic equilibrium at a constant temperature

$$\int_V \delta F \, dV = \int_{\partial V} \mathbf{t}^{(n)} \cdot \delta \mathbf{x} \, dS + \int_V \mathbf{f} \cdot \delta \mathbf{x} \, dV + \int_V \mu_o \rho \mathbf{H} \cdot \delta \mu \, dV$$

$$- \int_V \mathbf{J}_s \cdot \delta \mathbf{A} dV + \int_{\partial V} \rho \gamma \mu \cdot \delta \mu \, dS - \int_V \rho \lambda \mu \cdot \delta \mu dV \qquad (3.5.1)$$

where ρ is the mass density of the superconductor, $\mu = \mathbf{M}/\rho$ the magnetization per unit mass, $\mathbf{J}_s$ the supercurrent density, $\mathbf{f}$ the body force being the sum of the mechanical body force $\mathbf{f}^{me}$ and the electromagnetic body force $\mathbf{f}^{em}$ defined by

$$\mathbf{f}^{em} = \mu_o \mathbf{J}_s \times \mathbf{H} + \mu_o (\mathbf{M} \cdot \nabla) \mathbf{H} \qquad (3.5.2)$$

with μ_o being the magnetic permeability in vacuum. $\mathbf{t}^{(n)}$ is the surface force being the sum of a prescribed surface traction $\mathbf{t}^{o(n)}$ and the magnetic pressure $0.5\mu_o(\mathbf{M} \cdot \mathbf{n})^2 \mathbf{n}$ with $\mathbf{n}$ being the unit outward normal vector on the boundary surface ∂V of the superconductor.

The free energy density of the ferromagnetic superconductor F may be expressed, by noting the fact that the internal part of the free energy density is an objective scalar valued function (invariant with respect to rigid body rotations), as

$$F = F_n(E_{KL}, G_L, R_{KL}) + F_{sn} \qquad (3.5.3)$$

where F_n denotes the free energy density of normal state and F_{sn} the free energy density difference between the superconducting state and the normal state.

To describe the superconducting state of the material and its transition to normal state, we may introduce the order parameter ψ in the free energy function, the actual value of which is determined by thermal equilibrium conditions according to the general Landau theory of phase transition. Furthermore, we shall assume that the superconducting pairing in the ferromagnetic superconductor is also of s-type so that the order parameter ψ is a scalar complex function with the physical meaning that $\psi\psi^*=|\psi|^2$ denotes the density of the Cooper pairs of superconducting electrons. Thus, we may write F_{sn} in the following form:

$$F_{sn}= F'(E_{KL}, G_L, R_{KL}, |\psi|) + \frac{1}{2m^*_{kj}}[(-i\hbar\frac{\partial}{\partial x_k} - e^*A_k)\psi][(i\hbar\frac{\partial}{\partial x_j} - e^*A_j)\psi^*] \qquad (3.5.4)$$

with $F' = 0$ when $|\psi| = 0$. Here, the conventional Einstein's summation rule has been used. The second term on the r.h.s. of eqn.(3.5.4) represents essentially the kinetic energy of the superelectrons with its macroscopic quantum generalization. $\mathbf{A}$ is the magnetic vector potential defined from $\mathbf{B} = \nabla \times \mathbf{A}$ with $\mathbf{B}$ being the magnetic induction field which is related to the magnetization vector $\mathbf{M}$ by

$$\mathbf{B} = \mu_o(\mathbf{H} + \mathbf{M}). \qquad (3.5.5)$$

The anisotropic behavior of the superconductor has been taken into account in eqn.(3.5.4) by introducing the effective mass tensor $\mathbf{m}^*$ (Ginzburg (1952), Tilley (1965) and Bulaevskii et al. (1988)), which may, in general, be a function of the elastic deformation, the magnetization and the magnetization gradient. To simplify the formulation for practical applications, we may assume that the effective mass is a material constant with negligible effects from the elastic lattice deformation and the spontaneous magnetization, by noting the fact that m^* essentially plays the role of the effective mass of a Cooper pair with the effective charge e^* ($\approx 2e$) of superconducting electrons rather than the mass density of the Cooper pair superelectrons.

Near superconducting transition temperatures, F' may be expanded approximately into the following form with respect to the order parameter ψ

$$F' = \alpha(E_{KL}, G_L, R_{KL})|\psi|^2 + \frac{1}{2}\beta(E_{KL}, G_L, R_{KL})|\psi|^4 \qquad (3.5.6)$$

where E_{KL} is the Lagrangian strain tensor defined by eqn.(1.10.6), and G_L and R_{KL} are respectively defined by

$$G_L = \mu_i\, x_{i,L} \qquad \text{and} \qquad R_{KL} = \mu_{i,K}\, x_{i,L}\,. \qquad (3.5.7)$$

It should, however, be noticed that the validity of the expansion (3.5.6) is influenced by the fluctuation of the order parameter near critical transition temperatures. The classical G-L theory with the expansion of eqn.(3.5.6) may be adequate for conventional bulk superconductors where the fluctuation is negligibly small, but it may fail for others in which the fluctuation plays an important role. The ferromagnetic superconductor might be the case in which the magnetic fluctuation and the superconducting fluctuation are important to explain their unusual behaviors (Sakurai (1978) and Shenoy et al. (1980)). We are not going to discuss this problem further, but leave it for future study.

In eqn.(3.5.1), the saturation condition of magnetization $\mu\cdot\mu = \mu_s^2$ with μ_s being assumed to be a constant at a constant temperature (i.e. $\mu\cdot\delta\mu=0$) has been taken into account by introducing the Lagrangian multipliers γ and λ (Brown (1966)). In the case where the magnetization saturation condition is not imposed, we may set γ and λ to zero (Alblas (1979)). Such a generalization may be necessary for studying the ferromagnetic superconductor when it is in a state near the transition region of ferromagnetic normal state and superconducting state, where the exchange interaction among magnetic moments would be weakened considerably with the onset of superconductivity; so that the magnitude of the magnetization $|\mu|$ may not be a constant but may vary from place to place, depending on the structure of superconducting and ferromagnetic ordering in the material (Krey (1972)).

After carrying out the variation procedure for eqn.(3.5.1), we can arrive at the following set of equilibrium field equations and boundary conditions for the ferromagnetic superconductor.

The mechanical equilibrium equation reads

$$t_{ji,j} + f_i = 0 \qquad \text{in V} \tag{3.5.8}$$

where t_{ij} is the Cauchy stress tensor given by

$$t_{ji} = \frac{\partial F}{\partial E_{KL}} x_{i,K} x_{j,L} + \frac{\partial F}{\partial G_L} \mu_i x_{j,L} + \frac{\partial F}{\partial R_{KL}} \mu_{i,K} x_{j,L} \qquad \text{in V.} \tag{3.5.9}$$

Here, V denotes the volume of the superconductive body.

Eqn.(3.5.8) may also be written as

$$(t_{ji} + t^{em}_{ji})_{,j} + f^{me}_i = 0 \qquad \text{in V} \tag{3.5.10}$$

in which $\mathbf{t}^{em}$ is the Maxwell stress tensor defined by

$$t^{em}_{ji} = B_j H_i - \frac{1}{2}\mu_o H_k H_k \delta_{ji} \,. \tag{3.5.11}$$

The magnetic equilibrium equation is

$$\mu_o H_i - \frac{1}{\rho}\frac{\partial F}{\partial G_L} x_{i,L} + \frac{1}{\rho}\left(\frac{\partial F}{\partial R_{KL}} x_{i,L} x_{j,K}\right)_{,j} - \lambda\mu_i = 0 \qquad \text{in V.} \tag{3.5.12}$$

The first three terms of eqn.(3.5.12) are regarded as the i component of an effective field $\mu_o\mathbf{H}^{eff}$ which for equilibrium must be in (or opposite to) the direction of μ due to the fact that the total torque per unit volume must be zero. Eqn.(3.5.12) shows that this equilibrium condition is satisfied by noting $\rho\mu \times \mu_o\mathbf{H}^{eff} = \rho\mu \times \lambda\mu = 0$ (Brown (1966) and Maugin (1988)). The antisymmetric part of the Cauchy stress tensor, when taking account of the effect of exchange force, becomes

$$t_{[ji]} = \frac{\partial F}{\partial G_L}\mu_{[i} x_{j],L} + \frac{\partial F}{\partial R_{KL}}\mu_{[i,K} x_{j],L} \tag{3.5.13}$$

where the notation $t_{[ji]} = t_{ji} - t_{ij}$ has been used.

When exchange forces are omitted, the terms containing $\mu_{i,K}$ are absent, and eqn.(3.5.13) is reduced to $t_{[ji]} = \mu_o H_{[j} M_{i]}$ by noting eqn.(3.5.12), which is, in fact, the classical result for the nonsymmetry of the Cauchy stress tensor due to the magnetic body couple $\mu_o\mathbf{M} \times \mathbf{H}$.

The generalized G-L equation may be found as

$$\frac{\partial F'}{\partial \psi^*} + \frac{1}{2m^*_{kj}}\left(-i\hbar\frac{\partial}{\partial x_k} - e^* A_k\right)\left(-i\hbar\frac{\partial}{\partial x_j} - e^* A_j\right)\psi = 0 \qquad \text{in V} \tag{3.5.14}$$

which, in the case where F' is taken in the expansion form of eqn.(3.5.6), becomes

$$\alpha\psi + \beta|\psi|^2\psi + \frac{1}{2m^*_{kj}}\left(-i\hbar\frac{\partial}{\partial x_k} - e^* A_k\right)\left(-i\hbar\frac{\partial}{\partial x_j} - e^* A_j\right)\psi = 0 \qquad \text{in V.} \tag{3.5.15}$$

The equation for the superconducting current density reads

$$J_{sk} = \frac{e^*\hbar}{i2m^*_{kj}}\left(\psi^*\frac{\partial\psi}{\partial x_j} - \psi\frac{\partial\psi^*}{\partial x_j}\right) - \frac{e^{*2}}{m^*_{kj}}|\psi|^2 A_j \qquad \text{in V.} \tag{3.5.16}$$

In particular, for isotropic superconductors, we have

$$\alpha\psi + \beta|\psi|^2\psi + \frac{1}{2m^*}(-i\hbar\nabla - e^*\mathbf{A})^2\psi = 0 \qquad \text{in V} \tag{3.5.17}$$

and the supercurrent density

$$\mathbf{J}_s = \frac{e^*\hbar}{i2m^*}(\psi^*\nabla\psi - \psi\nabla\psi^*) - \frac{e^{*2}}{m^*}|\psi|^2\,\mathbf{A} \qquad \text{in V.} \tag{3.5.18}$$

The boundary conditions are

$$t_{ji}n_j = t_i^{o(n)} + \frac{\mu_o}{2}(\mathbf{M}\cdot\mathbf{n})^2 n_i \qquad \text{on } \partial_T V \tag{3.5.19}$$

$$\mathbf{u} = \mathbf{u}^o \qquad \text{on } \partial_u V \tag{3.5.20}$$

$$\frac{1}{2m^*_{kj}}\left(-i\hbar\frac{\partial\psi}{\partial x_j} - e^*\psi A_j\right)n_k = 0 \qquad \text{on } \partial V \tag{3.5.21}$$

$$\frac{1}{\rho}\frac{\partial F}{\partial R_{KL}}x_{i,L}x_{j,K}\,n_j = \gamma\mu_i \qquad \text{on } \partial V \tag{3.5.22}$$

where $\partial_T V$ and $\partial_u V$ denote respectively the boundaries of the superconductor with the traction prescribed and with the displacement specified ($\partial V = \partial_T V + \partial_u V$). The Lagrangian multiply γ is either to be taken as zero in the case of magnetization not saturated, or is to be determined from the saturation condition (Alblas (1979)).

From eqn.(3.5.22), we may also derive

$$\mu_{[k}\frac{\partial F}{\partial R_{KL}}x_{i],L}x_{j,K}\,n_j = 0 \qquad \text{on } \partial V. \tag{3.5.23}$$

Equations (3.5.22) and (3.5.23) are in accordance with the result given by, for instance, Alblas (1979) that $\mathbf{Q} = \gamma\mathbf{M}$ on ∂V and $\mu \times \mathbf{Q} = 0$ ∂V

provided that we introduce the so-called "surface exchange contact force" **Q** defined by

$$Q_i = \frac{\partial F}{\partial R_{KL}} x_{i,L} x_{j,K}\, n_j \, . \qquad (3.5.24)$$

Detail discussions about such a "stress" model by using a lattice continuum combined with a spin continuum will, however, not be given here. Interested readers are referred to the work of, for instance, Maugin (1988) and Tiersten (1964). The boundary condition (3.5.21) is used for the system of superconductor-vacuum (or superconductor-insulator). For the system of superconductor-normal conductor, the boundary condition (3.5.21) has to be modified as stated in section 2.8.

Similarly to non-magnetic elastic superconductors, a simple dynamic modification of the theory may be possible for studying the elastic ferromagnetic superconductor in weak magnetic fields at the magneto-quasistatic approximation. If the magnetic dissipation and AC losses in the ferromagnetic superconductor may be ignored in a problem in which the elastic deformation of the superconductor is also supposed to be small and there is no global motion of the superconductor, then the mechanical equilibrium equation may be modified to be the mechanical equation of motion by simply adding an inertia force term $\rho_o \partial^2 u_i/\partial t^2$ on the right-hand side of eqn.(3.5.8), and the magnetic equation of motion may be written as $\gamma_o \mu \times \mathbf{H}^{eff} = \partial \mu/\partial t$ with γ_o being a material constant. The G-L equations and boundary conditions require no modification since we ignore the effect of finite relaxation time of superelectrons and there is no surface concentration of mass or magnetic moment (Brown (1966)). It should, however, be noticed that a general study of dynamic phenomena has to be necessarily based on a nonequilibrium thermomechanics analysis taking account of various dissipative mechanisms in the superconductor as we shall show later.

3.5.3 **Magnetoelastic properties of ferromagnetic superconductors**

In what follows, we shall study the magnetoelastic properties of ferromagnetic superconductors. We shall show how the elastic, the magnetostrictive and the magnetic anisotropic properties could be influenced by the superconducting phase transition. For simplicity, we shall assume that the elastic deformation of the ferromagnetic superconductor is infinitesimal, and the material is of centrosymmetry, so that the piezomagnetic term and all odd power terms of the magnetization and of the magnetization gradient are absent in the energy expression due to the space-time symmetry requirement (Kanamori (1963) and Landau et al. (1984)). We also assume that the temperature of the superconductor is uniform and near the superconducting transition temperature, where the magnetization, the magnetization gradient and the order parameter ψ are all small so that the free energy density may be expanded in the following form:

$$F = \frac{1}{2}C_{ijkl}\varepsilon_{ij}\varepsilon_{kl} + \frac{1}{2}\Gamma_{ij}M_iM_j + B_{ijkl}\varepsilon_{ij}M_kM_l + \frac{1}{4}\kappa_{ijkl}M_iM_jM_kM_l + \frac{1}{2}\Lambda_{ij}M_{p,i}M_{p,j}$$

$$+ \alpha|\psi|^2 + \frac{1}{2}\beta|\psi|^4 + \frac{1}{2m^*_{kj}}[(-i\hbar\frac{\partial}{\partial x_k} - e^*A_k)\psi][(i\hbar\frac{\partial}{\partial x_j} - e^*A_j)\psi^*] \qquad (3.5.25)$$

in which the phenomenological coefficients α and β may be expanded respectively as

$$\alpha = \alpha'(T)(1 + a^e_{ij}(T)\varepsilon_{ij} + \frac{1}{2}a^{e2}_{ijkl}(T)\varepsilon_{ij}\varepsilon_{kl} + \frac{1}{2}a^M_{ij}(T)M_iM_j + a^{eM}_{ijkl}(T)\varepsilon_{ij}M_kM_l$$

$$+ \frac{1}{2}a^{ex}_{ij}(T)M_{p,i}M_{p,j} + \frac{1}{4}a^{M2}_{ijkl}(T)M_iM_jM_kM_l) \qquad (3.5.26)$$

and

$$\beta = \beta'(T)(1 + b^e_{ij}(T)\varepsilon_{ij} + \frac{1}{2}b^{e2}_{ijkl}(T)\varepsilon_{ij}\varepsilon_{kl} + \frac{1}{2}b^M_{ij}(T)M_iM_j + b^{eM}_{ijkl}(T)\varepsilon_{ij}M_kM_l$$

$$+ \frac{1}{2}b^{ex}_{ij}(T)M_{p,i}M_{p,j} + \frac{1}{4}b^{M2}_{ijkl}(T)M_iM_jM_kM_l) \qquad (3.5.27)$$

where ε_{ij} denotes the infinitesimal strain tensor. The terms on the r.h.s. of eqn.(3.5.25) represent respectively the elastic energy, the second order magnetocrystalline energy, the magnetostrictive energy, the fourth order magnetocrystalline energy, the exchange energy and the superconducting energy. Here, we have excluded higher order terms of, for instance, the exchange-strictive effect etc., the effects of which are supposed to be negligibly small in comparison with those terms retained. In addition, e^* and m^*_{ij} are both assumed to be material constants characterizing respectively the effective charge and

the effective mass of a Cooper pair of superconducting electrons. Those material coefficient tensors evidently satisfy symmetry relations. In particular, for material with cubic symmetry, the coefficient tensors will have particularly simple forms (Maugin (1988)). It is worth mention that a consistent treatment of the free energy expression as well as the field equations from their Lagrangian form to Eulerian expressions may be necessary for studying problems of high order effects and of highly anisotropic effects (Zhou and Hsieh (1985)).
To the same degree of approximation, we can find that the Cauchy stress may be written explicitly from eqn.(3.5.9)

$$t_{ji} = C_{jikl}\epsilon_{kl} + \Gamma_{jikl}M_kM_l + \kappa_{jklp}M_kM_lM_pM_i + \Lambda_{kl}M_{i,k}M_{j,l}$$
$$+ \alpha'(a^e_{ji} + a^{e2}_{jikl}\epsilon_{kl} + a'^M_{jikl}M_kM_l + a^{M2}_{jpkl}M_pM_kM_lM_i + a^{ex}_{kl}M_{i,k}M_{j,l})|\psi|^2$$
$$+\frac{\beta'}{2}(b^e_{ji} + b^{e2}_{jikl}\epsilon_{kl} + b'^M_{jikl}M_kM_l + b^{M2}_{jpkl}M_pM_kM_lM_i + b^{ex}_{kl}M_{i,k}M_{j,l})|\psi|^4 \quad (3.5.28)$$

where

$$\Gamma_{jikl} = \Gamma_{jk}\delta_{il} + B_{jikl} + 2B_{pqkj}\delta_{il}\epsilon_{pq} \quad (3.5.29a)$$

$$a'^M_{jikl} = a^M_{jk}\delta_{il} + a^{eM}_{jikl} + 2a^{eM}_{pqkj}\delta_{il}\epsilon_{pq} \quad (3.5.29b)$$

$$b'^M_{jikl} = b^M_{jk}\delta_{il} + b^{eM}_{jikl} + 2b^{eM}_{pqkj}\delta_{il}\epsilon_{pq} \quad (3.5.29c)$$

with δ_{il} being the Kronecker delta.

The magnetic equilibrium equation (3.5.12), in the case that the magnetization saturation condition is not imposed, becomes

$$\mu_oH_i + (\Lambda_{jk} + \alpha'a^{ex}_{jk}|\psi|^2 + \frac{\beta'}{2}b^{ex}_{jk}|\psi|^4)M_{i,kj} = \{\Gamma_{ij} + 2B_{klij}\epsilon_{kl} + \alpha'(a^M_{ij} + 2a^{eM}_{klij}\epsilon_{kl})|\psi|^2$$
$$+\frac{\beta'}{2}(b^M_{ij} + 2b^{eM}_{klij}\epsilon_{kl})|\psi|^4\}M_j$$
$$+ (\kappa_{ijkl} + \alpha'a^{M2}_{ijkl}|\psi|^2 + \frac{\beta'}{2}b^{M2}_{ijkl}|\psi|^4)M_jM_kM_l$$
$$(3.5.30)$$

By eqn.(3.5.30), the Cauchy stress may be expressed as

$$t_{ji} = C_{jikl}\varepsilon_{kl} + B_{jikl}M_k M_l + \mu_o H_j M_i + \Lambda_{kl}(M_i M_{j,l})_{,k}$$
$$+ \alpha'[a^e_{ji} + a^{e2}_{jikl}\varepsilon_{kl} + a^{eM}_{jikl}M_k M_l + a^{ex}_{kl}(M_i M_{j,l})_{,k}]|\psi|^2$$
$$+\frac{\beta'}{2}[b^e_{ji} + b^{e2}_{jikl}\varepsilon_{kl} + b^{eM}_{jikl}M_k M_l + b^{ex}_{kl}(M_i M_{j,l})_{,k}]|\psi|^4 . \qquad (3.5.31)$$

The G-L equation is given by eqn.(3.5.15) with the phenomenological coefficients α and β given by eqns.(3.5.26) and (3.5.27).

Let us now consider the following case, in which the temperature of the superconductor is supposed to be near the transition temperature T_{c2}. Before the appearance of ferromagnetic ordering in the superconductor, the density of the superconducting electrons may be assumed to be uniform in the superconductor (neglecting surface effects) in the absence of applied magnetic field (or for applied field much smaller than the critical field of the superconductor), and can be obtained approximately from the G-L eqn.(3.5.15)

$$|\psi|^2 = -\frac{\alpha}{\beta} . \qquad (3.5.32)$$

By slowly decreasing the temperature, the ferromagnetic order structure appears and the magnetization increases gradually while the superconductivity is being suppressed. At certain states in the coexistence of ferromagnetism and superconductivity, the exchange interaction is small and we may assume that the magnetization is small so that the effects of the fourth order term of the magnetization and of the magnetization gradient may be ignored in the analysis. Thus, as the first approximation of the perturbation, we may write

$$|\psi|^2 = -\frac{\alpha'}{\beta'}[1 + (a^e_{kl} - b^e_{kl})\varepsilon_{kl} + \frac{1}{2}(a^M_{kl} - b^M_{kl})M_k M_l] \qquad (3.5.33a)$$

and

$$|\psi|^4 = \frac{\alpha'^2}{\beta'^2}[1 + 2(a^e_{kl} - b^e_{kl})\varepsilon_{kl} + (a^M_{kl} - b^M_{kl})M_k M_l] . \qquad (3.5.33b)$$

We then have

$$\alpha' a^e_{ji}\frac{\partial|\psi|^2}{\partial x_j} + \frac{\beta'}{2} b^e_{ji}\frac{\partial|\psi|^4}{\partial x_j} = -\frac{\alpha'^2}{\beta'}(a^e_{ji} - b^e_{ji})(a^e_{kl} - b^e_{kl})\varepsilon_{kl,j} - \frac{\alpha'^2}{2\beta'}(a^e_{ji} - b^e_{ji})(a^M_{kl} - b^M_{kl})(M_k M_l)_{,j} \tag{3.5.34}$$

From eqns.(3.5.31) and (3.5.34) and by neglecting magnetization gradient term, we may derive finally

$$t_{ji,j} = C^{eff}_{jikl}\varepsilon_{kl,j} + B^{eff}_{jikl}(M_k M_l)_{,j} + \mu_o(H_j M_i)_{,j} \tag{3.5.35}$$

where we have introduced the effective elastic coefficients of the superconductor at superconducting state, defined by

$$C^{eff}_{jikl} = C_{jikl} - \frac{\alpha'^2}{\beta'}[a^{e2}_{jikl} - \frac{1}{2} b^{e2}_{jikl} + (a^e_{ji} - b^e_{ji})(a^e_{kl} - b^e_{kl})] \tag{3.5.36}$$

and the effective magnetostrictive coefficients of the superconductor at the superconducting state, defined by

$$B^{eff}_{jikl} = B_{jikl} - \frac{\alpha'^2}{2\beta'}[a^{eM}_{jikl} - \frac{1}{2} b^{eM}_{jikl} + (a^e_{ji} - b^e_{ji})(a^M_{kl} - b^M_{kl})] \tag{3.5.37}$$

where C_{jikl} and B_{jikl} are respectively the elastic moduli and the magnetostrictive coefficients in the normal state.

From eqn.(3.5.30), we can also get, by ignoring the term of magnetization gradient and the third order term of magnetization,

$$\mu_o H_{i,p} = \Gamma^{eff}_{ij} M_{j,p} + 2B^{eff}_{klij}(\varepsilon_{kl} M_j)_{,p} \tag{3.5.38}$$

where Γ^{eff}_{ij} denote the effective magnetic anisotropic coefficients of the superconductor in the superconducting state, defined by

$$\Gamma^{eff}_{ij} = \Gamma_{ij} - \frac{\alpha'^2}{\beta'}(a^M_{ij} - \frac{1}{2} b^M_{ij}) \tag{3.5.39}$$

where Γ_{ij} are the magnetic anisotropic coefficients in normal state.

It is seen that eqn.(3.5.36) shows the similar result for the effective elastic moduli given by eqn.(3.3.6) for an elastic superconductor at superconducting state. However, eqns.(3.5.37) and (3.5.39) provide additional information on the magnetoelastic effect in elastic ferromagnetic superconductors. If we now consider those effective material coefficients being defined to represent the macroscopic magnetoelastic properties of the ferromagnetic superconductor in its superconducting state, we find the change of magnetoelastic properties of the ferromagnetic superconductor from its superconducting state to normal ferromagnetic state, which is dependent on the superconducting properties of the material. In particular, we find that the change of the magnetostrictive coefficients is, to the first approximation, directly related to the magneto-superconducting coefficients a_{ij}^{M} and b_{ij}^{M} and the magnetostrictive-superconducting coefficients a_{jikl}^{eM} and b_{jikl}^{eM}.

Thus, the indication of the change of magnetostrictive coefficients of the ferromagnetic superconductor from its normal state to the coexistence state of ferromagnetism and superconductivity may provide an additional means (such as using mechanical waves) of revealing unusual behavior of the ferromagnetic superconductor without introducing much disturbance to the superconductivity, in comparison with the conventional specific heat, thermal expansion and magnetic measurements (Fischer and Peter (1973)). The investigation of the magnetoelastic effects in normal materials has been of interest for many researchers (Maugin (1984), Yamomoto and Miya (1987), and Zhou and Hsieh (1988)), the results of which may be used and generalized for the study of possiblly interesting phenomena in ferromagnetic superconductors.

3.5.4 Coexistence of superconductivity and ferromagnetism

In this section, we shall study the conditions for the possible coexistence of superconducting and magnetic phases in the ferromagnetic superconductor using the proposed phenomenological theory. For simplicity, we shall ignore here the effect of elastic deformation of the superconductor, and only consider a homogeneous and isotropic superconductor. Let us first consider possible solutions for $|\psi|$ and M being independent of coordinates for the homogeneous superconductor in the absence of externally applied magnetic field. In such a case, we have the following set of equilibrium equations at the first order approximation

$$\Gamma M_i + \alpha_m|\psi|^2 M_i + \kappa_m M^2 M_i = 0 \tag{3.5.40}$$

$$\alpha'\psi + \beta'|\psi|^2\psi + \frac{1}{2}\alpha_m M^2\psi = 0. \tag{3.5.41}$$

At the same order of the approximation, the corresponding free energy density function may be written as

$$F = \alpha'|\psi|^2 + \frac{1}{2}\beta'|\psi|^4 + \frac{1}{2}\Gamma M^2 + \frac{1}{4}\kappa_m M^4 + \frac{1}{2}\alpha_m M^2|\psi|^2 \tag{3.5.42}$$

where the phenomenological coefficients Γ, α_m, κ_m, α' and β' are only possible functions of temperature T. In particular, β', κ_m and α_m are positive values. The positivity of α_m reflects the antagonistic character of order parameters to the problem of the coexistence of superconductivity and ferromagnetism: that is for a fixed magnetization **M** the superconducting transition temperature diminishes according to the condition $\alpha' \to \alpha' + \alpha_m M^2/2$; but for a given value of ψ, the magnetic ordering temperature diminishes according to the analogous condition $\Gamma \to \Gamma + \alpha_m|\psi|^2/2$ (Izyumov and Skryabin (1980)).

The set of eqns.(3.5.40) and (3.5.41) may have the following solutions:

$$M = 0 \qquad \text{and} \qquad |\psi| = 0 \tag{3.5.43}$$

for the material at normal state or paramagnetic state;

$$M = 0 \qquad \text{and} \qquad |\psi|^2 = -\frac{\alpha'}{\beta'} \tag{3.5.44}$$

for the material at purely superconducting state;

$$M^2 = -\frac{\Gamma}{\kappa_m} \qquad \text{and} \qquad |\psi| = 0 \tag{3.5.45}$$

for the material with purely magnetic phase;

$$M^2 = \frac{-\beta'\Gamma + \alpha'\alpha_m}{\beta'\kappa_m - \alpha_m^2/2} \quad \text{and} \quad |\psi|^2 = \frac{-\alpha'\kappa_m + \alpha_m\Gamma/2}{\beta'\kappa_m - \alpha_m^2/2} \tag{3.5.46}$$

for the material with coexistence of superconducting and magnetic phases (MS phase).

Using of the above solutions, we can find the corresponding free energy densities respectively as

$$F_s = -\frac{\alpha'^2}{2\beta'} \quad \text{for purely superconducting phase} \tag{3.5.47}$$

and

$$F_m = -\frac{\Gamma^2}{4\kappa_m} \quad \text{for purely magnetic phase} \tag{3.5.48}$$

and

$$F_{ms} = -\frac{\alpha'^2}{2\beta'} - \frac{1}{4\beta'}\frac{[\beta'\Gamma - \alpha'\alpha_m]^2}{\beta'\kappa_m - \alpha_m^2/2} \tag{3.5.49a}$$

$$= -\frac{\Gamma^2}{4\kappa_m} - \frac{1}{2\kappa_m}\frac{[\alpha'\kappa_m - \alpha_m\Gamma/2]^2}{\beta'\kappa_m - \alpha_m^2/2} \tag{3.5.49b}$$

for the coexistence of superconducting and magnetic phases. It can be seen from eqn.(3.5.49) that in order to have a stable state of the coexistence of the superconducting and magnetic phases, F_{ms} must be less than the free energy densities of the purely magnetic and superconducting states, which results in the following condition

$$\beta'\kappa_m > \alpha_m^2/2 . \tag{3.5.50}$$

It is shown that the coexistence of superconductivity and ferromagnetism is only possible when the coupling constant α_m connecting the two order parameters ψ and M in the homogeneous magnetic superconductor system is quite weak so that the condition (3.5.50) is satisfied. In such a case, phase transitions between all phases are second order. If the condition (3.5.50) is reversed, no MS phase arises and there is a first-order phase transition between the superconducting and magnetic phases (Izyumov and Skryabin (1980)).

We shall now consider the possible coexistence state of superconducting and ferromagnetic phases with spatially varying order parameters. To study such problems, we start with the following set of equilibrium equations

$$-\Lambda_o\nabla^2 M_i - \alpha^{ex}|\psi|^2\nabla^2 M_i + \Gamma M_i + \alpha_m|\psi|^2 M_i + \kappa_m M^2 M_i = 0 \qquad (3.5.51)$$

$$-\frac{\hbar^2}{2m^*}\nabla^2\psi + \alpha'\psi + \beta'|\psi|^2\psi + \frac{1}{2}\alpha_m M^2\psi + \frac{1}{2}\alpha^{ex}\psi M_{i,j}M_{i,j} = 0 \qquad (3.5.52)$$

where the effect of spatial variation of the order parameters is taken into account by including the derivative terms of the order parameters. In addition, we also include the term ($\alpha^{ex} \neq 0$) accounting for the effect of coupling between superconducting electrons and exchange interaction of localized magnetic moments, which was originally ignored in the work of Izyumov and Skryabin (1980). At the same order of the approximation, the corresponding free energy density function may be expressed as

$$F = \alpha'|\psi|^2 + \frac{1}{2}\beta'|\psi|^4 + \frac{1}{2}\Gamma M^2 + \frac{1}{4}\kappa_m M^4 + \frac{1}{2}\alpha_m M^2|\psi|^2$$

$$+ \frac{1}{2}\Lambda_o M_{i,j}M_{i,j} + \frac{1}{2}\alpha^{ex}|\psi|^2 M_{i,j}M_{i,j} + \frac{\hbar^2}{2m^*}|\nabla\psi|^2 \qquad (3.5.53)$$

where the phenomenological coefficients Γ, α_m, κ_m, α', Λ_o, α^{ex} and β' are only possible functions of temperature T.

This set of equations (3.5.51) and (3.5.52) admits of the following type of solution

$$|\psi| = \text{const.} \qquad \text{and} \qquad \mathbf{M} = (M\cos(\mathbf{Q}\cdot\mathbf{x}), M\sin(\mathbf{Q}\cdot\mathbf{x}), 0). \qquad (3.5.54)$$

which characterizes the inhomogeneous state, known as cryptoferromagnetic (spiral) state, that was originally proposed by Anderson and Suhl (1959), and studied by Suhl (1978) and by Izyumov and Skryabin (1980) neglecting the direct coupling effect between superconducting electrons and exchange interaction of localized magnetic moments ($\alpha^{ex} = 0$).

To see the contribution of the coupling of superconducting electrons and exchange interaction of localized magnetic moments, we may obtain from eqns.(3.5.51), (3.5.52) and (3.5.54) for the MS phase

$$\Gamma + \Lambda_o Q^2 + \kappa_m M^2 + (\alpha_m + \alpha^{ex} Q^2)|\psi|^2 = 0 \tag{3.5.55}$$

$$\alpha' + \beta' |\psi|^2 + \frac{1}{2}(\alpha_m + \alpha^{ex} Q^2) M^2 = 0 \tag{3.5.56}$$

which gives

$$M^2 = \frac{-\beta'(\Gamma + \Lambda_o Q^2) + \alpha'(\alpha_m + \alpha^{ex} Q^2)}{\beta' \kappa_m - (\alpha_m + \alpha^{ex} Q^2)^2/2} \tag{3.5.57}$$

and

$$|\psi|^2 = \frac{-\alpha' \kappa_m + (\alpha_m + \alpha^{ex} Q^2)(\Gamma + \Lambda_o Q^2)/2}{\beta' \kappa_m - (\alpha_m + \alpha^{ex} Q^2)^2/2} \; . \tag{3.5.58}$$

By substituting eqns.(3.5.57) and (3.5.58) into eqn.(3.5.53) and by noting eqn.(3.5.54), we find the corresponding free energy density for the MS phase of the cryptoferromagnetic state by

$$F_{ms} = -\frac{\alpha'^2}{2\beta'} - \frac{1}{4\beta'} \frac{[\beta'(\Gamma + \Lambda_o Q^2) - \alpha'(\alpha_m + \alpha^{ex} Q^2)]^2}{\beta' \kappa_m - (\alpha_m + \alpha^{ex} Q^2)^2/2} \tag{3.5.59a}$$

$$= -\frac{(\Gamma + \Lambda_o Q^2)^2}{4\kappa_m} - \frac{1}{2\kappa_m} \frac{[\alpha' \kappa_m - (\alpha_m + \alpha^{ex} Q^2)(\Gamma + \Lambda_o Q^2)/2]^2}{\beta' \kappa_m - (\alpha_m + \alpha^{ex} Q^2)^2/2} \tag{3.5.59b}$$

which shows that, to have a stable state of the MS phase, the following condition has to be satisfied

$$\beta' \kappa_m > \frac{1}{2}(\alpha_m + \alpha^{ex} Q^2)^2 \; . \tag{3.5.60}$$

The condition (3.5.60) implies physically the fact that the free energy of the system in the cryptoferromagnetic state with the MS phase is lower than that of the system with either the purely magnetic phase

$$M^2 = -\frac{\Gamma + \Lambda_o Q^2}{\kappa_m} \qquad \text{and} \qquad |\psi| = 0, \tag{3.5.61}$$

or the purely superconducting phase

$$M = 0 \qquad \text{and} \qquad |\psi|^2 = -\frac{\alpha'}{\beta'} . \tag{3.5.62}$$

The condition (3.5.60) is shown to differ from the condition (3.5.50) due to the presence of the coupling effect between superconducting electrons and exchange interaction of localized magnetic moments. The actual value of Q^2 is obviously limited by the condition (3.5.60). In particular, eqn.(3.5.59) indicates mathematically that there might be a possibility that free energy of the system in the cryptoferromagnetic state could be lowered indefinitely for certain values of Q^2 which makes $\beta'\kappa_m - (\alpha_m + \alpha^{ex}Q^2)^2/2 \rightarrow +0$. This might be physically inaccessible. The actual value of Q^2 may have to be determined by other conditions (see, for instance, Anderson and Suhl (1959) and Suhl (1978)). If we ignore the direct coupling effect between superconducting electrons and exchange interaction of localized magnetic moments by setting $\alpha^{ex} = 0$, we will find the same condition for the MS phase as given by eqn.(3.5.50) and find that the equilibrium value of the wave vector Q_o, which determines the helical structure and is determined by minimizing the free energy over Q for the cryptoferromagnetic state, has a zero value unless the nonlocal nature of ψ is considered (Suhl (1978)). The result presented here thus shows that the effect of coupling between superconducting electrons and exchange interaction of localized magnetic moments may be of importance for the study of the cryptoferromagnetic state as well as other types of inhomogeneous states in the ferromagnetic superconductors though the coupling may be very weak (Izyumov and Skryabin (1980)). Discussions about other types of MS phases are referred to in a review article by Kakani and Upadhyaya (1988).

The result of the coupling between superconducting electrons and exchange interaction of localized magnetic moments might also provide a possible mechanism for high-T_c superconductors, in which superconducting electrons may move along certain special paths such that the coupling of superconducting electrons and exchange interaction of localized magnetic moments contributes to the lowering of system free energy, which then requires higher temperature to cause the transition of the system from the superconducting state to the normal state.

3.6 Nonequilibrium theory of thermoelastic superconductors

Superconductors can carry d.c. currents without resistance. However, a.c. currents or changing magnetic fields will lead to losses and generation of heat in superconductors, which had, in fact, limited most of the past engineering applications of superconductivity to static cases. The suppression of superconductivity under the exertion of an electromagnetic field was generally believed to be based on consideration of the physical mechanisms of interaction between an electromagnetic field and a superconductor: absorption by Cooper pairs of electromagnetic field quanta $\hbar\omega$ with energies higher than the coupling energy of Cooper pair, acceleration of superconducting electrons to critical velocities at which their kinetic energy is larger than the coupling energy, and electromagnetic energy absorption by normal electrons which results in heating of the superconductor. However, in the 1960s, it was found experimentally that the critical fields, T_c, H_c, and I_c of a superconductor may be increased when it is acted on by an electromagnetic field with a frequency of about 10 GHz; that is, the enhancement of superconductivity, which implies a new nonequilibrium steady state characterized by stronger superconducting properties is reached (Wyatt et al. (1966), Dayem and Wiegrand (1967) and Dmitriev et al. (1986)). The nonequilibrium processes in superconductors thus generate very interesting phenomena: enhancement and suppression of superconductivity. In addition, dynamic effects in nonequilibrium superconductors also introduce interesting acoustic phenomena in superconductors (Aronov et al. (1986)). Such acoustic phenomena appear if, for instance, sound propagates through a superconductor. The acousto-electric effect is also of interest for potential technological applications. The possibility for wider applications of superconductivity obviously hinges on success in improving our understanding of nonequilibrium superconductivity as well as its related electrodynamic problems.

Based on either the microscopic BCS theory or on phenomenological thermodynamic theory, several theories have been proposed (see, for instance, Stephen (1965), Schmid (1966) and Eliashberg and Ivlev (1986)). The simplest approach to nonequilibrium superconductivity, certainly at the phenomenological level, lies with the time-dependent Ginzburg-Landau (TDGL) equations, which are supposed to be applicable to the study of nonequilibrium phenomena in superconductors with small deviations from their equilibrium state (Schmid (1966) and Cyrot (1973)). However, compared with the equilibrium theory of superconductors, the theory of nonequilibrium superconductivity is still in its early stage. In particular, there are few systematic studies concerned with the problem of mechanical-electromagnetic interaction in superconductors (Labusch (1968) and Seeger and Kronmuller (1968)).

In the following sections, we shall introduce a phenomenological continuum theory for the study of thermoelastic superconductors (Zhou and Miya (1991)), in which nonequilibrium superconductivity, electromagnetics, mechanical deformation, and heat conduction can be studied uniformly by a complete coupled set of the generalized time-dependent Ginzburg-Landau (GTDGL) equations for the superconducting order parameter, the Eulerian motion equation for matter continuum, the heat conduction equation for temperature field, and Maxwell's equations for electromagnetic fields.

3.6.1 Irreversible thermodynamics of superconductors

For simplicity, we shall consider here only the non-magnetic and isotropic superconductors, and assume that the deformation of the superconductors is infinitesimal. Furthermore, we shall limit ourselves to electromagnetic problems in which the displacement current $(\partial \mathbf{D}/\partial t)$ is small in comparison with the supercurrent inside the superconducting solid and $\partial \rho_e/\partial t$ is also small (Schmid (1966) and Carr Jr. (1983)) so that they can be neglected in Maxwell's equations. In addition, no global motion, such as rotation or translation of the superconductor is being considered. Based on these assumptions, we may now start from the following set of Maxwell's equations

$$\frac{\partial \mathbf{B}}{\partial t} = -\nabla \times \mathbf{E}, \qquad \nabla \cdot \mathbf{B} = 0, \tag{3.6.1}$$

$$\nabla \times \mathbf{H} = \mathbf{J}, \qquad \nabla \cdot \mathbf{J} = 0 \tag{3.6.2}$$

which have the form of so-called magneto-quasistatic approximation commonly used in studying low-frequency (less than 100 MHz) electromagnetic phenomena of normal conductors (see section 1.8.2). Here, $\mathbf{J}$ is the total electric current density, being the sum of the normal conducting current density $\mathbf{J}_n$ and the superconducting current density $\mathbf{J}_s$. $\mathbf{E}$, $\mathbf{H}$ and $\mathbf{B}$ are respectively the electric field, the magnetic intensity field and the magnetic induction field.

Maxwell's equations have general applicability not only for normal metals but also for superconducting materials. As with any other material, Maxwell's equations must be complemented with additional specified constitutive equations for the material, which can describe correctly observed experimental results for the material. As we have seen in chapter 1, when the electromagnetic fields are interacting with deformable bodies in a thermodynamic environment not only further field equations (such as mechanical balance equation and heat conduction equation) have to be added but also coupling terms must be introduced, which account for the interaction of the deformation of the body with the electromagnetic fields and with the temperature field. Similarly, to study phenomenologically deformable superconductors, relevant field and constitutive equations have to be supplied. In this section, we shall introduce the continuum irreversible thermodynamics for the study of the thermoelastic superconductor.

We begin with the following principle of conservation of mass

$$\int_V \rho dV = \int_{V_R} \rho_R dV_R \tag{3.6.3}$$

where ρ and ρ_R are respectively the mass density in the actual configuration and the reference configuration, and V and V_R are respectively the volume of the material body in the actual configuration and the reference configuration.

The principle of balance of momentum is

$$\frac{d}{dt}\int_V \rho \mathbf{v}\, dV = \int_{\partial V} \mathbf{t}\cdot\mathbf{n}\, dS + \int_V \mathbf{f}\, dV \tag{3.6.4}$$

which has its local form as

$$\rho\frac{d\mathbf{v}}{dt} = \nabla\cdot\mathbf{t} + \mathbf{f} \tag{3.6.5}$$

where $\mathbf{v}$ is the displacement velocity, $\mathbf{t}$ the Cauchy stress tensor, $\mathbf{f}$ the body force which is the sum of the mechanical body force $\mathbf{f}^{me}$ and the electromagnetic body force $\mathbf{f}^{em}$ which is supposed to be of the Lorentz body force form; $\mathbf{f}^{em} = \mathbf{J}\times\mathbf{B}$ for the superconducting solid in the magneto-quasistatic approximation.

The principle of balance of moment of momentum is

$$\frac{d}{dt}\int_V \mathbf{x}\times\rho\mathbf{v}\, dV = \int_{\partial V} \mathbf{x}\times(\mathbf{t}\cdot\mathbf{n})dS + \int_V \mathbf{x}\times\mathbf{f}\, dV \tag{3.6.6}$$

which gives the symmetric condition for Cauchy's stress tensor $t_{ij} = t_{ji}$ with the aid of the principle of balance of momentum (3.6.5).

The principle of energy balance may be expressed as

$$\frac{d}{dt}\int_V \left(\frac{1}{2}\rho\mathbf{v}\cdot\mathbf{v} + \rho u_n\right)dV + \frac{d}{dt}\int_V \rho u_{sn}\, dV + \int_V \mathbf{H}\cdot\frac{\partial\mathbf{B}}{\partial t}\, dV$$

$$= \int_{\partial V} \mathbf{t}^{(n)}\cdot\mathbf{v}dS + \int_V \mathbf{f}^{me}\cdot\mathbf{v}dV - \int_{\partial V} \mathbf{q}\cdot\mathbf{n}dS + \int_V \rho r dV - \int_{\partial V} (\mathbf{E}\times\mathbf{H})\cdot\mathbf{n}\, dS \tag{3.6.7}$$

where $\mathbf{t}^{(n)}$ is the surface traction, $\mathbf{q}$ the heat flux, and r the heat source. The quantity $\rho\mathbf{v}\cdot\mathbf{v}/2$ is the mechanical kinetic energy density, u_n is the specific internal energy of the material body in a normal state corresponding to the same values of its variables, and u_{sn} is the specific internal energy difference between the superconducting state and the normal state. The third integral on the left-hand side of eqn.(3.6.7) represents the increase of the electromagnetic field energy density per unit time in the superconducting solid at the magneto-quasistatic approximation. The first and the second terms on the r.h.s. of eqn.(3.6.7) represent the rate of work of external mechanical force acting on the superconducting body. The third and the fourth terms on the r.h.s. of eqn.(3.6.7) represent the heat supply per unit time respectively from the heat flowing into and across the boundary surface of the body, and from the heat source inside the body. The last term on the r.h.s. of eqn.(3.6.7) represents the electromagnetic energy supply per unit time flowing into the superconductor through its boundary surface.

By using Maxwell's equations, the conservation law of mass and the balance equation of momentum, the energy balance equation (3.6.7) can be expressed as

$$\int_V \rho(\dot{u}_n + \dot{u}_{sn})dV = \int_V (t_{ij}v_{j,i} - q_{k,k} + \rho r)dV + \int_V \mathbf{J}\cdot\mathbf{E}'dV \qquad (3.6.8)$$

where $\mathbf{E}' = \mathbf{E} + \mathbf{v} \times \mathbf{B}$ denotes the electric field defined in a reference frame, stationary with respect to the material body. The term $\mathbf{v} \times \mathbf{B}$ represents the effect of the motion of the body on the electric field. For a material body, which has only time-dependent elastic deformations but does not present global motion such as translation or rotation, the electric field induced by the deformation of the superconductor may be negligible in comparison with the electric field generated by conduction electrons. This is the case which we are considering.

The second law of thermodynamics in the case considered reads

$$\frac{d}{dt}\int_V \rho\eta\, dV = \int_{\partial V} -\frac{\mathbf{q}\cdot\mathbf{n}}{T}\, dS + \int_V \rho\,\frac{r}{T}dV + \int_V \rho\sigma\, dV \qquad (3.6.9a)$$

with the condition

$$\int_V \rho\sigma\, dV \geq 0 \qquad (3.6.9b)$$

where η is the total specific entropy and σ the specific entropy production per unit time due to irreversible processes taking place inside the system considered.

With the introduction of the specific thermodynamic function

$$\Psi = u_n + u_{sn} - \eta T \tag{3.6.10}$$

the second law of thermodynamics may be expressed, with the use of eqn.(3.6.8), by

$$\int_V \rho\sigma T dV = \int_V \{-\rho(\dot{\Psi} + \eta\dot{T}) + t_{ij}\, v_{j,i} - q_i(\log T)_{,i}\}dV + \int_V \mathbf{J}\cdot\mathbf{E}'dV \geq 0. \tag{3.6.11}$$

In the approximation of infinitesimal strain, the mass density ρ remains constant and we may introduce the following expression for the thermodynamic function $\rho\Psi$

$$\rho\Psi = F_n(T, \varepsilon) + F_{sn} \tag{3.6.12}$$

where F_n denotes the free energy density of the material in normal state corresponding to the same values of its variables, and F_{sn} denotes the free energy density difference between the superconducting state and the normal state, which may be written as

$$F_{sn} = F'_{sn} + \frac{1}{2m^*}|(-i\hbar\nabla - e^*\mathbf{A})\psi|^2 \tag{3.6.13}$$

where F'_{sn} is a function of temperature T, elastic strain ε and superconducting order parameter ψ, and has the property of $F'_{sn} = 0$ for normal state ($\psi = 0$). $\mathbf{A}$ is the magnetic vector potential ($\mathbf{B} = \nabla \times \mathbf{A}$). The second term on the r.h.s. of eqn.(3.6.13) represents presumably the kinetic energy density of the supercurrent flow, where the phenomenological coefficients e^* and m^* are specified respectively as the effective electric charge ($e^* \approx 2e$) and the effective mass ($m^* \approx 2m$) of the Cooper pair superelectrons according to the microscopic theory of Gor'kov (see a review of the Gor'kov theory by Werthamer (1969)) and the elegant experiment by Zimmerman and Mercereau (1965). We shall assume here that e^* and m^* are both material constants, independent of temperature and elastic strain, since they are characterizing essentially a pair of electrons rather than the density of pairing electrons.

Substituting eqn.(3.6.12) into eqn.(3.6.11), we can derive after some manipulations

$$\int_V \rho\sigma T\, dV = \int_V \{-\rho(\frac{\partial \Psi}{\partial T} + \eta)\dot{T} + (-\rho\frac{\partial \Psi}{\partial \varepsilon_{ij}} + t_{ij})\dot{\varepsilon}_{ij}\}dV$$

$$- \int_V \{\mathbf{J}_s - [\frac{e^*\hbar}{i2m^*}(\psi^*\nabla\psi - \psi\nabla\psi^*) - \frac{e^{*2}}{m^*}|\psi|^2\mathbf{A}]\}\cdot\dot{\mathbf{A}}\, dV$$

$$- \int_{\partial V} 2R_e\{\frac{i\hbar}{2m^*}[(-i\hbar\nabla\psi - e^*\psi\mathbf{A})\cdot\mathbf{n}]\dot{\psi}^*\}dS + \int_V (\nabla\cdot\mathbf{J}_s)\phi dV - \int_{\partial V} \phi\mathbf{J}_s\cdot\mathbf{n}dV$$

$$- \int_V 2R_e\{[\frac{\partial F'_{sn}}{\partial \psi^*} + \frac{1}{2m^*}(-i\hbar\nabla - e^*\mathbf{A})^2\psi]\dot{\psi}^*\}dV - \int_V \frac{\mathbf{q}\cdot\nabla T}{T}dV + \int_V \mathbf{J}_n\cdot\mathbf{E}dV \geq 0$$

(3.6.14)

where ψ^* denotes the complex conjugate of ψ, $R_e\{.\}$ the real part of the quantity {.}, and ϕ the electric potential. Based on the above analysis, we are now ready to formulate a set of fundamental field equations for the thermoelastic superconductor.

3.6.2 Fundamental equations of thermoelastic superconductors

According to the second law of thermodynamics, the inequality (3.6.14) must be satisfied for all independent processes. Thus, for terms on the r.h.s. of eqn.(3.6.14) with $\dot{T}$, $\dot{\varepsilon}_{ij}$ and $\dot{\mathbf{A}}$ occurring only linearly with their coefficients which are not functions of these quantities, we get the following relations

$$\eta = -\frac{\partial \Psi}{\partial T} \qquad \text{in V} \qquad (3.6.15)$$

$$t_{ij} = \rho\frac{\partial \Psi}{\partial \varepsilon_{ij}} \qquad \text{in V} \qquad (3.6.16)$$

$$\mathbf{J}_s = \frac{e^*\hbar}{i2m^*}(\psi^*\nabla\psi - \psi\nabla\psi^*) - \frac{e^{*2}}{m^*}|\psi|^2\mathbf{A} \qquad \text{in V} \qquad (3.6.17)$$

$$(-i\hbar\nabla\psi - e^*\psi\mathbf{A})\cdot\mathbf{n} = 0 \qquad \text{on } \partial V. \qquad (3.6.18)$$

The physical reason for choosing the boundary condition (3.6.18) rather than the other mathematical alternative, $\psi=0$ at the boundary surface, is that the result of $\psi=0$ at boundary is incompatible with the experimental fact that very thin films of superconductor have very nearly the same transition temperature as the bulk material (see Werthamer (1969)). By substituting eqn.(3.6.18) into eqn.(3.6.17) for the supercurrent, we see that $\mathbf{J}_s \cdot \mathbf{n} = 0$ on ∂V, which means that no supercurrent is leaving the superconductor and is required in the macroscopic model. It is worth mentioning that the boundary condition (3.6.18) has to be modified for a superconductor-normal metal junction. For such an interface, a surface energy term should be added to the energy balance equation (Bulaevskii (1988)).

Eqn.(3.6.14) now becomes

$$\int_V \rho\sigma T\, dV = - \int_V 2R_e\{[\frac{\partial F'_{sn}}{\partial \psi^*} + \frac{1}{2m^*}(-i\hbar\nabla - e^*\mathbf{A})^2\psi]\dot{\psi}^*\}dV + \int_V (\nabla\cdot\mathbf{J}_s)\phi dV$$

$$- \int_V \frac{\mathbf{q}\cdot\nabla T}{T} dV + \int_V \mathbf{J}_n\cdot\mathbf{E}\, dV \geq 0 . \qquad (3.6.19)$$

Applying the concepts of nonequilibrium thermodynamics to the superconductors, we may introduce the functional derivative of the free energy with respect to the order parameter as a generalized thermodynamic force acting on the order parameter. Thus, according to the classical non-equilibrium thermodynamics theory (de Groot and Mazur (1962)), the second law of thermodynamics (3.6.19) can be satisfied to a linear approximation by the following phenomenological relations, which, for isotropic superconducting materials, may be expressed according to Curie's principle (1908) as

$$\frac{\partial F'_{sn}}{\partial \psi^*} + \frac{1}{2m^*}(-i\hbar\nabla - e^*\mathbf{A})^2\psi = - \gamma_R\, (\hbar\frac{\partial}{\partial t} + ie^*\phi)\psi \qquad \text{in } V \qquad (3.6.20)$$

with γ_R (>0) being a phenomenological transport coefficient. The gauge invariance of the equation is satisfied by introducing the term $-\gamma_R ie^*\phi\psi$ on the r.h.s. of eqn.(3.6.20). In particular, near the superconducting transition temperature T_c, F'_{sn} may be expressed, in accordance with the general Landau theory of second-order phase transition (Landau et al. (1984)), by

$$F'_{sn} = \alpha(T, \varepsilon)|\psi|^2 + \frac{\beta(T, \varepsilon)}{2}|\psi|^4 . \tag{3.6.21}$$

Thus, eqn.(3.6.20) can be written, near superconducting transition temperatures, as

$$\alpha\psi + \beta|\psi|^2\psi + \frac{1}{2m^*}(-i\hbar\nabla - e^*\mathbf{A})^2\psi = -\gamma_R\,(\hbar\frac{\partial}{\partial t} + ie^*\phi)\psi . \tag{3.6.22}$$

The generalized Ohm's law and Fourier's law read

$$\mathbf{q} = -\chi_q\nabla T + \Gamma_o\,\mathbf{J}_n \tag{3.6.23}$$

$$\mathbf{J}_n = \sigma_n(\mathbf{E} - \Pi_o\nabla T) \tag{3.6.24}$$

where σ_n (>0) is the normal electric conductivity and χ_q (>0) the thermal conductivity of the material. Γ_o and Π_o are thermoelectric coupling coefficients which satisfy a relation $\Gamma_o = T\Pi_o$ according to Onsager's symmetric principle. It is noticed that, for anisotropic superconducting materials, the GTDGL equation will be, in general, coupled explicitly with the temperature gradient and the normal electric current density according to the general relation between the generalized thermodynamic forces and fluxes in the theory of irreversible thermodynamics.

Substituting (3.6.20), (3.6.23) and (3.6.24) into eqn.(3.6.19), we arrive at

$$\int_V \rho\sigma T\, dV = \int_V \{\frac{2\gamma_R}{\hbar}|(\hbar\frac{\partial}{\partial t} + ie^*\phi)\psi|^2 + \frac{\chi_q|\nabla T|^2}{T} + \frac{|\mathbf{J}_n|^2}{\sigma_n}\}\, dV \geq 0 \tag{3.6.25}$$

where we have made use of the result

$$\nabla\cdot\mathbf{J}_s = \frac{2e^*\gamma_R}{\hbar}[\frac{i\hbar}{2}(\psi\frac{\partial\psi^*}{\partial t} - \psi^*\frac{\partial\psi}{\partial t}) + e^*\phi|\psi|^2] \qquad \text{in V} \tag{3.6.26}$$

from eqns.(3.6.17) and (3.6.22).

The conservation of charge may now be expressed by

$$\nabla \cdot \mathbf{J}_n + \frac{2e^*\gamma_R}{\hbar}\left[\frac{i\hbar}{2}\left(\psi\frac{\partial \psi^*}{\partial t} - \psi^*\frac{\partial \psi}{\partial t}\right) + e^*\phi|\psi|^2\right] = 0 \qquad \text{in } V \tag{3.6.27}$$

at the magneto-quasistatic approximation.

It is shown that the first term of the integrand on the r.h.s. of eqn.(3.6.25) represents the power dissipation due to the finite relaxation time ($\gamma_R \neq 0$) of the superelectrons (Tinkham (1964)), and the second and the third terms represent the power dissipation in the irreversible processes of thermal conduction and of normal electric current flow.

The inhomogeneous equation (3.6.2) for the superconducting solid can be written in terms of the magnetic vector potential **A** as

$$\nabla^2 \mathbf{A} + \mu_o(\mathbf{J}_n + \mathbf{J}_s) = 0 \tag{3.6.28}$$

provided that Coulomb's gauge condition $\nabla \cdot \mathbf{A} = 0$ is used.

The derived equations (3.6.17), (3.6.22), (3.6.26) and (3.6.28) represent the fundamental time-dependent Ginzburg-Landau equations, which are shown to be consistent with the form of the TDGL equations derived by Schmid (1966) with, however, the difference that the effects of mechanical deformation and thermal conduction process are now being taken into account.

For an isotropic thermoelastic superconductor near its superconducting transition temperature T_c, at the first-order approximation, we may have the following generalized Duhamel-Neumann law expressed by

$$t_{ij} = C_{ijkl}e_{kl} - \beta_T(T - T_o)\delta_{ij} - [\alpha_\varepsilon(T_c - T)|\psi|^2 - \frac{\beta_\varepsilon}{2}|\psi|^4]\delta_{ij} \tag{3.6.29}$$

where β_T denotes the thermal modulus, α_ε and β_ε are two material constants, which may be called the elastic-superconductive moduli, T_o is the reference temperature and T_c is the superconducting transition temperature. Eqn.(3.6.29) shows that, besides the thermal pressure, there is an additional 'superconducting' pressure acting in the superconductive body when which is in superconducting state ($T < T_c$). The elastic moduli for the isotropic superconductor in normal state may be expressed as

$$C_{ijkl} = \lambda_e \delta_{ij}\delta_{kl} + \mu_e(\delta_{ik}\delta_{jl} + \delta_{il}\delta_{jk}) \tag{3.6.30}$$

where λ_e and λ_e are Lame's elastic moduli.

It can be seen that the set of GTDGL equations are coupled with the mechanical balance equation (3.6.5) and also with the heat conduction equation, which will be derived as follows. By eqn.(3.6.19) we have

$$\int_V \rho\sigma T dV = \int_V \frac{2\gamma_R}{\hbar} |(\hbar\frac{\partial}{\partial t} + ie^*\phi)\psi|^2 dV - \int_V \frac{1}{T} \mathbf{q}\cdot\nabla T dV + \int_V \mathbf{J}_n\cdot\mathbf{E} dV \tag{3.6.31}$$

and by eqn.(3.6.9a), we can arrive at

$$\rho T\dot{\eta} = - q_{k,k} + \rho r + \mathbf{J}_n\cdot\mathbf{E} + \frac{2\gamma_R}{\hbar}|(\hbar\frac{\partial}{\partial t} + 2ie^*\phi)\psi|^2 . \tag{3.6.32}$$

Thus, to the same order of approximation as eqn.(3.6.29), the heat conduction equation may be derived, with the use of eqns.(3.6.32) and (3.6.15) in the absence of the internal heat source (r=0), as

$$q_{k,k} = - \rho C_v \frac{\partial T}{\partial t} - T_0(\beta_T - \alpha_\varepsilon|\psi|^2)\frac{\partial \varepsilon_{kk}}{\partial t} + \mathbf{J}_n\cdot\mathbf{E} + T_0(\alpha_{\varepsilon 0} + \alpha_\varepsilon \varepsilon_{kk})\frac{\partial|\psi|^2}{\partial t}$$

$$+ \frac{2\gamma_R}{\hbar}|(\hbar\frac{\partial}{\partial t} + ie^*\phi)\psi|^2 \tag{3.6.33}$$

where C_v denotes the specific heat per unit mass measured at constant strain, defined by

$$C_v = T\left(\frac{\partial \eta}{\partial T}\right)_{\varepsilon_{ij}}, \tag{3.6.34}$$

and $\alpha_{\varepsilon 0}$ and α_ε are two material constants defined in eqn.(3.2.14).

Finally, with the use of the generalized Ohm's law and Fourier's law (3.6.23) and (3.6.24), the heat conduction equation is found to be

$$\rho C_v \frac{\partial T}{\partial t} = \nabla\cdot(\chi_q \nabla T) - T_0(\beta_T - \alpha_\varepsilon |\psi|^2)\frac{\partial \varepsilon_{kk}}{\partial t} + \mathbf{J}_n\cdot\mathbf{E} - \nabla\cdot(\Gamma_o \mathbf{J}_n)$$
$$+ T_0(\alpha_{\varepsilon o} + \alpha_\varepsilon \varepsilon_{kk})\frac{\partial |\psi|^2}{\partial t} + \frac{2\gamma_R}{\hbar}|(\hbar\frac{\partial}{\partial t} + ie^*\phi)\psi|^2 . \qquad (3.6.35)$$

It is shown that a complete set of phenomenological field equations, i.e. the mechanical motion equation (3.6.5), the time-dependent Ginzburg-Landau equation (3.6.22), the heat conduction equation (3.6.35), Maxwell's equations (3.6.27) and (3.6.28) together with the phenomenological relations (3.6.29) for the mechanical stress, (3.6.17) for the supercurrent density, (3.6.23) for the heat flux and (3.6.24) for the normal conduction current density, have been derived here to describe the electrodynamics and mechanics of thermoelastic superconductors with small strains near the superconducting transition temperature and in the magneto-quasistatic approximation. The set of ten independent field equations can, in principle, be used to determine the ten unknown quantities $|\psi(\mathbf{x},t)|$, $\arg\psi(\mathbf{x},t)$, $\mathbf{A}(\mathbf{x},t)$, $\phi(\mathbf{x},t)$, $\mathbf{u}(\mathbf{x},t)$ and $T(\mathbf{x},t)$ together with relevant boundary conditions which can be given for each specific problem.

Two special cases may be of interest. The first one is the case in which the superconductor is in normal state; thus, only normal conduction currents are flowing inside the material. In such a case, one is dealing with a non-magnetic thermoelastic normal conductor. The field equations are then Maxwell's equations (3.6.1) and (3.6.2) with $\mathbf{J} = \mathbf{J}_n$, the Eulerian equation of motion (3.6.5) with the Lorentz body force $\mathbf{f} = \mathbf{J}_n \times \mathbf{B}$, and the heat conduction equation

$$\rho C_v \frac{\partial T}{\partial t} = \nabla\cdot(\chi_q \nabla T) - T_0\beta_T\frac{\partial \varepsilon_{kk}}{\partial t} - \nabla\cdot(\Gamma_o \mathbf{J}_n) + \mathbf{J}_n\cdot\mathbf{E} \qquad (3.6.36)$$

together with the generalized Ohm's law (3.6.24).
The mechanical constitutive equation (3.6.29) now becomes

$$t_{ij} = C_{ijkl}\varepsilon_{kl} - \beta_T(T - T_0)\delta_{ij} \qquad (3.6.37)$$

which is simply the classical Duhamel-Neumann law for linear thermoelastic solids.

Since superconductive materials are, in general, poor conductors, the large Joule heat, shown by the last term on the r.h.s. of eqn.(3.6.36) can raise the temperature of the material rapidly and may eventually destroy the superconducting devices. Therefore, a properly well-designed cooling system is needed in order to prevent such destruction of the superconducting device when it returns to a normal state, and to reduce the cost of the expensive cooling system.

The second case of interest is the superconductive material in its stationary superconducting state, where the material is at rest, the heat flow is steady, and the electromagnetic fields are static. In such a case, the normal conduction current is zero (no flux motion), and the mechanical equilibrium equation becomes

$$\nabla \cdot \mathbf{t} + \mathbf{J}_s \times \mathbf{B} = 0 \tag{3.6.38}$$

in the absence of mechanical body force, and the generalized Ginzburg-Landau equation reads

$$\alpha\psi + \beta|\psi|^2\psi + \frac{1}{2m^*}(-i\hbar\nabla - e^*\mathbf{A})^2\psi = 0 \tag{3.6.39}$$

with the superconducting current density $\mathbf{J}_s$ given by eqn.(3.6.17).

The heat conduction equation (3.6.35) now becomes simply

$$\nabla\cdot(\chi_q \nabla T) = 0 \tag{3.6.40}$$

with its boundary conditions that either the temperature or the normal component of the heat flow is prescribed. It can be seen that, in the stationary case, the temperature field of the superconducting solid may be determined separately from mechanical and superconducting electromagnetic field equations, provided that the thermal conductivity coefficient χ_q may be assumed to be independent of elastic strain and of order parameter as in the case which we are considering. The mechanical equilibrium equation (3.6.38) is, however, coupled with the generalized time-independent Ginzburg-Landau equation (3.6.39) through the generalized Duhamel-Neumann law (3.6.29) and the Lorentz body force term.

3.6.3 An elastodynamic model of superconductors

In this section, we shall introduce a theoretical model for the study of elastodynamic behavior of superconductors. We shall start from the following elastodynamic equation

$$\rho \frac{\partial^2 \mathbf{u}}{\partial t^2} = \nabla \cdot \mathbf{t} + \mathbf{f} \tag{3.6.41}$$

with the mechanical constitutive equation

$$t_{ij} = C_{ijkl}\varepsilon_{kl} + \alpha'(a_{ij} + a'_{ijkl}\varepsilon_{kl})|\psi|^2 + \frac{\beta_0}{2}(b_{ij} + b'_{ijkl}\varepsilon_{kl})|\psi|^4 \tag{3.6.42}$$

for the superconductor in an isothermal state with $T=T_0$, which is near the transition temperature T_c. Here, ε_{kl} denotes the infinitesimal strain tensor and **u** the displacement field. C_{ijkl} and β_T are respectively the elastic moduli and thermal modulus of the superconductor at its normal state. For isotropic superconductors, the phenomenological coefficients a_{ij} and b_{ij} are defined respectively in eqns.(3.2.4) and (3.2.6), and the coefficients a'_{ijkl} and b'_{ijkl} are defined by eqns.(3.2.8) and (3.2.9) respectively.

For an isotropic superconductor in isothermal state, we may write

$$C_{ijkl} = \lambda_{eT}\delta_{ij}\delta_{kl} + \mu_{eT}(\delta_{ik}\delta_{jl} + \delta_{il}\delta_{jk}) \tag{3.6.43}$$

with λ_{eT} and μ_{eT} being the isothermal Lame's elastic moduli.

In some cases, we may consider a thermodynamic process in which power dissipation due to the finite relaxation time of superelectrons and Joule heating from normal current may be ignored and no internal heat sources exist in the superconductor; the order parameter ψ may then be found from the G-L equation with its quasi-static form as eqn.(3.6.39). Furthermore, we may consider a case in which the heat exchange between different parts of the material body by means of heat conduction occurs very slowly such that the heat exchange is practically absent during time intervals of the order of the period of vibrational motion; then each part of the body may be regarded as thermally insulated and the motion may be treated as adiabatic. For such an adiabatic case, we then have

$$t_{ij} = C_{ijkl}\varepsilon_{kl} - \beta_T(T-T_o)\delta_{ij} + \alpha'(a_{ij} + a'_{ijkl}\varepsilon_{kl})|\psi|^2 + \frac{\beta_o}{2}(b_{ij} + b'_{ijkl}\varepsilon_{kl})|\psi|^4 \tag{3.6.44}$$

and

$$\rho C_v \frac{\partial T}{\partial t} = - T_o(\beta_T - \alpha_\varepsilon|\psi|^2)\frac{\partial \varepsilon_{kk}}{\partial t} + T_o(\alpha_{\varepsilon o} + \alpha_\varepsilon \varepsilon_{kk})\frac{\partial |\psi|^2}{\partial t} \tag{3.6.45}$$

due to $\dot{\eta} = 0$ for the adiabatic process.

Integrating eqn.(3.6.45) with respect to time, we may get

$$\rho C_v(T - T_o) = - T_o\beta_T\,\varepsilon_{kk} + T_o\alpha_\varepsilon|\psi|^2\varepsilon_{kk} + T_o\alpha_{\varepsilon o}|\psi|^2 . \tag{3.6.46}$$

For superconductors in a normal state, substitution of eqn.(3.6.46) into eqn.(3.6.44) with $|\psi|=0$ then gives immediately the well-known classical relation connecting the adiabatic elastic moduli $\lambda_{e\eta}$ and $\mu_{e\eta}$ and the isothermal elastic moduli λ_{eT} and μ_{eT} at the normal state by

$$\lambda_{e\eta} = \lambda_{eT} + \frac{T_o\beta_T^2}{\rho C_{vn}} \qquad \text{and} \qquad \mu_{e\eta} = \mu_{eT} \tag{3.6.47}$$

where C_{vn} denotes the specific heat of the superconductor in normal state. The mechanical equation of motion (3.6.41) may then be expressed in the absence of body forces as

$$\rho\frac{\partial^2 \mathbf{u}}{\partial t^2} = \mu_{e\eta}\nabla^2\mathbf{u} + (\lambda_{e\eta} + \mu_{e\eta})\nabla(\nabla\cdot\mathbf{u}) . \tag{3.6.48}$$

For plane wave propagation in the medium, we may write

$$\mathbf{u}(\mathbf{x}, t) = \mathbf{u}_o\exp\{i(\mathbf{k}\cdot\mathbf{x} - \omega t)\} \tag{3.6.49}$$

where $|\mathbf{u}_o|$ is the amplitude of the elastic wave, $\mathbf{k}$ is a wave vector, the direction of which characterizes the direction of propagation of the wave, and ω the frequency of the wave. Substitution of eqn.(3.6.49) into eqn.(3.6.48) gives the following equation

$$\rho\omega^2\mathbf{u}_o = \mu_{e\eta}k^2\mathbf{u}_o + (\lambda_{e\eta} + \mu_{e\eta})(\mathbf{u}_o\cdot\mathbf{k})\mathbf{k}\ . \tag{3.6.50}$$

If the direction of the propagating wave is perpendicular to the wave amplitude vector $\mathbf{u}_o$, eqn.(3.6.50) is reduced to

$$\rho\omega^2\mathbf{u}_o = \mu_{e\eta}k^2\mathbf{u}_o \tag{3.6.51}$$

which, for non-zero solution of $\mathbf{u}_o$, gives

$$k^2 = \frac{\omega^2}{c_{2\eta}^2} \tag{3.6.52}$$

with $c_{2\eta}$ being the adiabatic velocity of the transverse elastic wave by

$$c_{2\eta} = \sqrt{\frac{\mu_{e\eta}}{\rho}}\ . \tag{3.6.53}$$

If the direction of the propagating wave is parallel to the wave amplitude vector $\mathbf{u}_o$, eqn.(3.6.50) can be expressed as

$$\rho\omega^2\mathbf{u}_o = (\lambda_{e\eta} + 2\mu_{e\eta})k^2\mathbf{u}_o \tag{3.6.54}$$

which, for non-zero solution of $\mathbf{u}_o$, gives

$$k^2 = \frac{\omega^2}{c_{1\eta}^2} \tag{3.6.55}$$

with $c_{1\eta}$ being the adiabatic velocity of the longitudinal elastic wave by

$$c_{1\eta} = \sqrt{\frac{\lambda_{e\eta} + 2\mu_{e\eta}}{\rho}}\ . \tag{3.6.56}$$

Furthermore, by using eqn.(3.6.47), we can get the relations between the adiabatic and isothermal velocities of elastic waves in the superconductor in normal state

$$c_{1\eta}^2 = c_{1T}^2 + \frac{T_o \beta_T^2}{\rho^2 C_{vn}}, \tag{3.6.57}$$

$$c_{2\eta}^2 = c_{2T}^2 \tag{3.6.58}$$

where c_{1T} denotes the isothermal velocity of the longitudinal wave, given by

$$c_{1T} = \sqrt{\frac{\lambda_{eT} + 2\mu_{eT}}{\rho}}, \tag{3.6.59}$$

and c_{2T} is the isothermal velocity of the transverse wave, given by

$$c_{2T} = \sqrt{\frac{\mu_{eT}}{\rho}}. \tag{3.6.60}$$

If, now, the superconductor is in superconducting state, one may naturally ask what sort of relation there may exist to connect the elastic moduli measured at a normal state and the elastic moduli measured at the superconducting state. To study this problem, let us first consider the case of the superconductor in a weak magnetic field ($B < \mu_o H_{c1}$). In such a case, we may obtain, by a perturbation procedure at the first-order approximation with respective to elastic strain (see section 3.3),

$$|\psi|^2 = -\frac{\alpha'}{\beta_o}[1 + (a_{ij} - b_{ij})\varepsilon_{ij}], \tag{3.6.61}$$

$$|\psi|^4 = \frac{\alpha'^2}{\beta_o^2}[1 + 2(a_{ij} - b_{ij})\varepsilon_{ij}] \tag{3.6.62}$$

from which we get

$$\alpha' a_{ij}\frac{\partial|\psi|^2}{\partial x_j} + \frac{1}{2}\beta_o b_{ij}\frac{\partial|\psi|^4}{\partial x_j} = -\frac{\alpha'^2}{\beta_o^2}(a_{ij} - b_{ij})(a_{kl} - b_{kl})\frac{\partial\varepsilon_{kl}}{\partial x_j}. \tag{3.6.63}$$

These relations will be used in the following discussion.

3.6.4 Velocities of isothermal elastic waves in superconductors

For superconductors in the isothermal superconducting state, we may find from eqn.(3.6.42) and (3.6.63)

$$t_{ij,j} = \{C_{ijkl} - \frac{\alpha'^2}{\beta_o}[a'_{ijkl} - \frac{1}{2}b'_{ijkl} + (a_{ij} - b_{ij})(a_{kl} - b_{kl})]\}\varepsilon_{kl,j} \, . \qquad (3.6.64)$$

Introducing the isothermal effective elastic modulus tensor C^{sT}_{ijkl} measured at the isothermal superconducting state by

$$C^{sT}_{ijkl} = C_{ijkl} - \mu_o H^2_{co}[a'_{ijkl} - \frac{1}{2}b'_{ijkl} + (a_{ij} - b_{ij})(a_{kl} - b_{kl})] \qquad (3.6.65)$$

we may find for isotropic superconductors the following relation connecting the isothermal effective Lame´s elastic moduli λ_{sT} and μ_{sT} measured at the superconducting state and the isothermal Lame´s elastic moduli λ_{eT} and μ_{eT} measured at normal state

$$\lambda_{sT} = \lambda_{eT} - \mu_o H^2_{co}[a^*_o - \frac{1}{2}b^*_o + (a_o - b_o)^2] \qquad (3.6.66)$$

$$\mu_{sT} = \mu_{eT} - \mu_o H^2_{co}[a_o + a^*_1 - \frac{1}{2}(b_o + b^*_1)] \qquad (3.6.67)$$

where H_{co} is the thermodynamic critical magnetic field of the superconductor at zero strain defined by eqn.(3.2.19).

Furthermore, by eqns.(3.6.66) and (3.6.67), we may obtain the relation connecting the propagation velocities of the isothermal elastic waves in the superconductor at superconducting state and those at normal state

$$c^2_{1sT} = c^2_{1T} - \frac{\mu_o H^2_{co} g_1}{\rho} \, , \qquad (3.6.68)$$

$$c^2_{2sT} = c^2_{2T} - \frac{\mu_o H^2_{co} g_2}{\rho} \qquad (3.6.69)$$

where g_1 and g_2 are two elastic-superconductive coupling parameters defined respectively by

$$g_1 = (a_o - b_o)^2 + 2a_o + a_0^* + 2a_1^* - b_o - \frac{1}{2}b_0^* - b_1^* \tag{3.6.70}$$

$$g_2 = a_o + a_1^* - \frac{1}{2}(b_o + b_1^*) . \tag{3.6.71}$$

3.6.5 **Velocities of adiabatic elastic waves in superconductors**

Next, we consider the case of the superconductor in an adiabatic superconducting state. By noting eqns.(3.6.44) and (3.6.46), we may find

$$t_{ij} = C_{ijkl}\varepsilon_{kl} + \alpha'(a_{ij} + a'_{ijkl}\varepsilon_{kl})|\psi|^2 + \frac{\beta_o}{2}(b_{ij} + b'_{ijkl}\varepsilon_{kl})|\psi|^4$$

$$- \frac{T_o}{\rho C_{vs}}\{[-\beta_T^2\varepsilon_{kk} + \beta_T(\alpha_{\varepsilon o} + 2\alpha_\varepsilon\varepsilon_{kk})|\psi|^2 - (\alpha_{\varepsilon o}\alpha_\varepsilon + \alpha_\varepsilon^2\varepsilon_{kk})|\psi|^4]\delta_{ij} - \alpha_{\varepsilon o}^2 a'_{ijkl}\varepsilon_{kl}|\psi|^4\} \tag{3.6.72}$$

with

$$|\psi|^2 = - q_1(1 - q_2\varepsilon_{kk}) \tag{3.6.73}$$

where the coefficients q_1 and q_2 are defined respectively by

$$q_1 = \frac{\alpha_{\varepsilon o}\rho C_{vs}(T_o - T_c^o)}{\alpha_{\varepsilon o}^2 T_o + \beta_o\rho C_{vs}} \tag{3.6.74}$$

$$q_2 = b_o - a_o + \frac{T_o\beta_T}{\rho C_{vs}(T_o - T_c^o)} + \frac{(2a_o - b_o)\alpha_{\varepsilon o}T_o}{\alpha_{\varepsilon o}^2 T_o + \beta_o\rho C_{vs}} . \tag{3.6.75}$$

Here, C_{vs} denotes the specific heat of the superconductor in superconducting state. The difference between the specific heats of a superconductor in superconducting state and in normal state may be found at the first-order approximation with respect to elastic strain by

$$C_{vs} - C_{vn} = - T\frac{\partial^2 F'_{sn}}{\partial T^2} = \frac{T\alpha_{\varepsilon o}^2}{\rho\beta_o}[1 + (\frac{2\alpha_\varepsilon}{\alpha_{\varepsilon o}} - b_o)\varepsilon_{kk}] \tag{3.6.76}$$

which at the transition temperature T_c reads

$$(C_{vs} - C_{vn})_{T_c} = (C_{vs} - C_{vn})_{T_c^o}[1 + (\Delta_o + \frac{2\alpha_\varepsilon}{\alpha_{\varepsilon o}} - b_o)\varepsilon_{kk}] \quad (3.6.77)$$

with $(C_{vs} - C_{vn})_{T_c^o}$ being defined as the value of the discontinuity in the specific heat of a superconductor at the transition temperature at zero strain by

$$(C_{vs} - C_{vn})_{T_c^o} = \frac{T_c^o \alpha_{\varepsilon o}^2}{\rho \beta_o}. \quad (3.6.78)$$

Here, α_ε and $\alpha_{\varepsilon o}$ are two material constants defined in eqn.(3.2.14).

It can be seen that eqn.(3.6.77) represents a strain-dependent value of the discontinuity in the specific heat of the superconductor at the transition temperature T_c. It is noted here that one has used the perturbation solution $|\psi|^2 = -\alpha/\beta$ when the superconductor is in superconducting state and in a uniform temperature field. Some experimental evidence of the strain-dependence of specific heat for some superconductive materials may be found in the work of Testardi et al. (1971).

For T_o very close to the superconducting transition temperature, we may find approximately

$$\varepsilon_{kk} = \frac{\alpha_{\varepsilon o}^2 T_o + \beta_o \rho C_{vs}}{\alpha_{\varepsilon o} T_o \beta_T} |\psi|^2 \quad (3.6.79)$$

which relates the volume dilatation to the superelectron density of the superconductor at the adiabatic superconducting state.

By eqn.(3.6.72), we may then obtain

$$t_{ij,j} = \{C_{ijkl} + h_1 \delta_{ij}\delta_{kl} + h_2(\delta_{ik}\delta_{jl} + \delta_{il}\delta_{jk})\}\varepsilon_{kl,j} \quad (3.6.80)$$

where

$$h_1 = \frac{T_o}{\rho C_{vs}} \{\beta_T^2 + q_1\beta_T\alpha_{\varepsilon o}(2a_o - q_2) + q_1\alpha_{\varepsilon o}^2(q_1 a_o^* + a_o q_2 - a_o^2)\}$$

$$+ q_1\alpha_{\varepsilon o}(T_o - T_c^o)(a_o q_2 - a_o^*) - q_1^2\beta_o b_o q_2 + \frac{q_1^2\beta_o b_o^*}{2} \quad (3.6.81)$$

and

$$h_2 = \frac{q_1^2\beta_0(b_1 + b_1^*)}{2} + q_1\alpha_{\varepsilon o}T_0(a_o + a_1^*)(\frac{q_1\alpha_{\varepsilon o}}{\rho C_{vs}} - \frac{T_o - T_c^o}{T_o}) . \qquad (3.6.82)$$

Introducing the adiabatic effective elastic modulus tensor $C_{ijkl}^{s\eta}$ measured at the superconducting state by

$$C_{ijkl}^{s\eta} = C_{ijkl} + h_1\delta_{ij}\delta_{kl} + h_2(\delta_{ik}\delta_{jl} + \delta_{il}\delta_{jk}) \qquad (3.6.83)$$

we may find the following relation connecting the adiabatic effective Lame's elastic moduli $\lambda_{s\eta}$ and $\mu_{s\eta}$ measured at the superconducting state and the isothermal Lame's elastic moduli λ_{eT} and μ_{eT} measured at a normal state

$$\lambda_{s\eta} = \lambda_{eT} + h_1, \qquad \mu_{s\eta} = \mu_{eT} + h_2 \qquad (3.6.84)$$

for the isotropic superconductor.

Thus, the relation connecting the adiabatic effective Lame's elastic moduli $\lambda_{s\eta}$ and $\mu_{s\eta}$ and the isothermal effective Lame's elastic moduli λ_{sT} and μ_{sT} measured at the superconducting state may be found as

$$\lambda_{s\eta} = \lambda_{sT} + h_1 + \mu_o H_{co}^2[a_o^* - \frac{1}{2}b_o^* + (a_o - b_o)^2] \qquad (3.6.85)$$

$$\mu_{s\eta} = \mu_{sT} + h_2 + \mu_o H_{co}^2[a_o + a_1^* - \frac{1}{2}(b_o + b_1^*)] . \qquad (3.6.86)$$

Furthermore, we may get the relation connecting the propagation velocities of the adiabatic elastic waves in the superconductor at the superconducting state and those at a normal state

$$c_{1s\eta}^2 = c_{1\eta}^2 + \frac{h_1 + 2h_2}{\rho} - \frac{T_o\beta_T^2}{\rho^2 C_{vn}} , \qquad (3.6.87)$$

$$c_{2s\eta}^2 = c_{2\eta}^2 + \frac{h_2}{\rho} \qquad (3.6.88)$$

and also the relation for the propagation velocities of elastic waves in the superconductor in adiabatic superconducting state and in isothermal superconducting state

$$c_{1s\eta}^2 = c_{1sT}^2 + \frac{h_1+2h_2}{\rho} + \frac{\mu_0 H_{co}^2 g_1}{\rho}, \tag{3.6.89}$$

$$c_{2s\eta}^2 = c_{2sT}^2 + \frac{h_2}{\rho} + \frac{\mu_0 H_{co}^2 g_2}{\rho}. \tag{3.6.90}$$

It is shown that a set of relations connecting the propagation velocities of either isothermal or adiabatic elastic waves in superconductors in superconducting state or in normal state has been formulated here with the aid of the elastodynamic model for superconductors.

Experimentally, the behavior of sound velocities at superconducting-normal phase transitions for some superconducting materials has been studied by, for instance, Testardi (1971). **Figure** 3.6.1 and **Figure** 3.6.2 show respectively the experimental behavior of sound velocities of longitudinal wave and of shear wave in V_3Ge vs. temperature. The normal state results below T_c = 5.9 °K were obtained by applying a magnetic field of 23 kOe transverse to the propagation direction of the waves. In the figures, the zero position of $\Delta c/c$ is chosen arbitrarily.

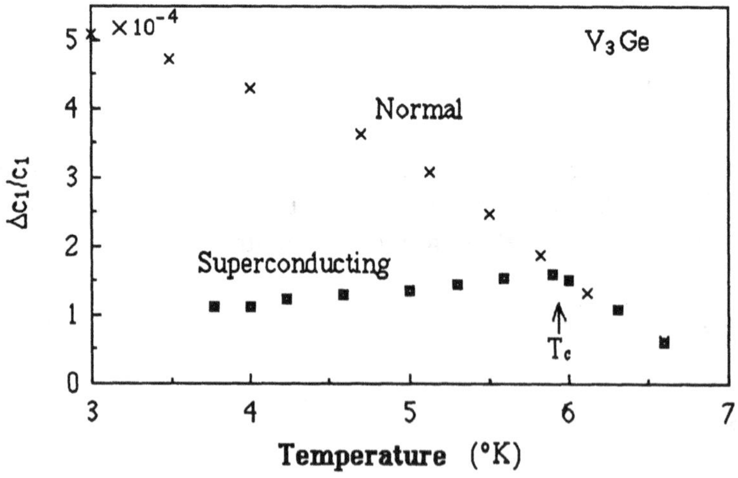

Figure 3.6.1 Sound velocity of [001] longitudinal waves in V_3Ge vs temperature.

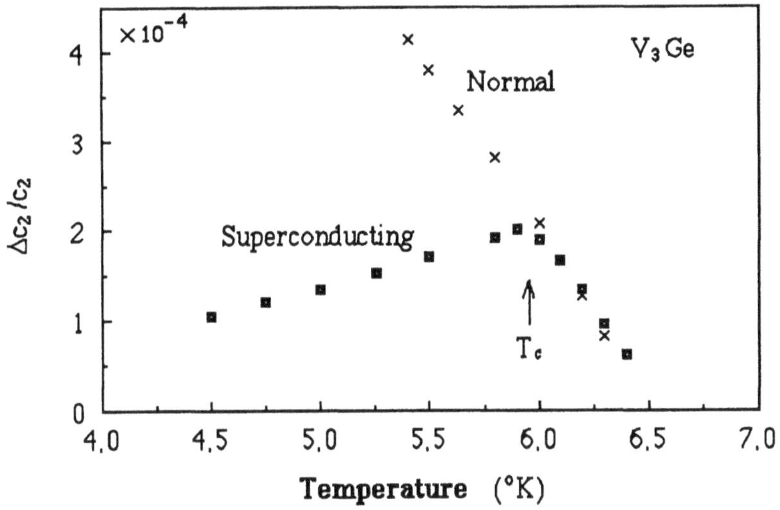

Figure 3.6.2 Sound velocity of [110] shear waves in V_3Ge vs temperature.

It is seen that there is a large discontinuity in the temperature derivative of sound velocities at the superconducting transition for the material V_3Ge. However, there are also experimental observations on some high-T_c superconductors which show no clear changes in slope of the sound velocity at T_c (see, for instance, the book edited by Ginsberg (1989)). So far, there seems to be not enough experimental data available for giving systematically a theoretical analysis about the mechanical behavior of superconductors near superconducting transition as well as the effects of electromagnetic-mechanical interaction in superconductors. Further efforts are, therefore, expected to be made through close collaboration between theoreticians and experimentalists in the field of superconductivity as well as in other relevant fields.

References

Abrikosov, A.A. (1957): Sov. Phys. JETP **5**, 1174.

Abrikosov, A.A. (1988): Fundamentals of the theory of metals, North-Holland, Amsterdam.

Adams, R. et al. (1980): Application of SQUIDs to Nuclear Gyros and Magnetometers, in: SQUID'80 (eds. by H.D. Hahlbohm and H. Lubbig), Walter de Gruyter, Berlin, pp.509-517.

Alblas, J.B. (1979): General theory of electro- and magneto-elasticity, in "Electromagnetic Interactions in Elastic Solids", CISM Course (ed. by H. Parkus), Springer-Verlag, Wien, pp.1-104.

Amaldi, E. (1968): On the Dirac Magnetic Poles, In: Old and New Problems in Elementary Particles (ed. by Puppi, G.), Academic Press, New York.

Ambegaokar, V. and Baratoff, A. (1963): Phys. Rev. Lett. **10**, 486 (1963); **11**, 104 (E).

Anderson, P.W. and Suhl, H. (1959): Phys. Rev. **116**, 898.

Anderson, P.W. (1962): Phys. Rev. Letters **9**, 309.

Anderson, P.W. and Rowell, J.M. (1963): Phys. Rev. Letters **10**, 230.

Arkadiev, V. (1945): J. Phys. U.S.S.R. **9**, 148.

Arkadiev, V. (1947): Nature **160**, 330.

Aronov, A.G., Gal'perin, Yu. M., Gurevich, V.L. and Kozub, V.I. (1986): Nonequilibrium properties of superconductors, in: Nonequilibrium superconductivity (eds. by D.N. Langenberg & A.I. Larkin), North-Holland, Amsterdam, pp.325-376.

Aslamazov, L.G. and Larkin, A.I. (1968a): Phys. Letters **A26**, 238.

Aslamazov, L.G. and Larkin, A.I. (1968b): Zh. Eksp. Teor. Fiz. **55**, 1477.

Atsumi, K., Kotani, M., Ueno, S., Katila, T. and Williamson, S.J. (eds.) (1988): Biomagnetism' 87, Tokyo Denki University Press, Tokyo.

Bardeen, J. (1950): Phys. Rev. **79**, 167.

Bardeen, J. (1951): Phys. Rev. **81**, 829, 1070.

Bardeen, J. (1962): Rev. Mod. Phys. **34**, 667.

Bardeen, J. (1977): IEEE Trans. Mag. **Mag-13**, 13.

Bardeen, J., Cooper, L.N. and Schrieffer, J.R. (1957): Phys. Rev. **108**, 1175.

Bardeen, J. and Stephen, M.J. (1965): Phys. Rev. **140**, 1197.

Barnett, S.J. (1915): Phys. Rev. **6**, 239.

Barone, A. and Paterno, G. (1982): Physics and applications of the Josephson effect, John Wiley & Son, New York.

Barone, A., Esposito, F., Likharev, K., Semenov, V.K., Todorov, B.N. and Vaglio, R. (1982): J. Appl. Phys. **53**, 5802.

Barone, A. and Paterno, G. (1982): Physics and Applications of the Josephson Effect, Wiley, New York.

Baule, G.M. and McFee, R. (1963): Am. Heart J. **66**, 95.

Bean, C.P. (1962): Phys. Rev. Letters **8**, 250.

Bednorz, J.G. and Müller, K.A. (1986): Zeitschrift fur Physik **B64**, 189.

Bhattacharya, S., Higgins, M.J., Johnston, D.C., Jacobson, A.J., Stokes, J.P., Goshorn,D.P., and Lewandowski, J.T. (1988a): Phys. Rev. Letters **60**, 1181.

Bhattacharya, S., Higgins, M.J., Johnston, D.C., Jacobson, A.J., Stokes, J.P., Lewandowski, J.T. and Goshorn, D.P. (1988b): Phys. Rev. **B37**, 5901.

Bishop, D.J., Ramirez, A.P., Gammel, P.L., Batlogg, B., Rietman, E.A., Cava, R.J. and Millis, A. J. (1987): Phys. Rev. **B36**, 2408.

Bladel, J.V. (1984): Relativity and Engineering, Springer-Verlag, Berlin.

Blatt, J.M. (1964): Theory of Superconductivity, Academic Press, New York.

Bloch, F. (1930): Z. Physik **61**, 206.

Blount, E.I. and Varma, C.M. (1979): Phys. Rev. Lett. **42**, 1079.

Bodmer, A., Essmann, U. and Träuble, H. (1972): Phys. Status Solidi (a) **13**, 471.

Bon Mardion, G., Goodman, B.B. and Lacaze, A. (1964): Phys. Letters **8**, 15.

Borges, H.A., Kwok, R., Thompson, J.D., Well, G.L., Smith, J.L, Fisk, Z. and Peterson, D.E. (1987): Phys. Rev. **B36**, 2404.

Bourne, L.C., Zettl, A., Chang, K.J., Cohen, M.L., Stacy, A.M. and Ham, W.K. (1987): Phys. Rev. **B35**, 8785.

Brailsford, F. (1966): Physical Principles of Magnetism, D. Van Norstrand LTD, London.

Brown Jr., W. F. (1963): Micromagnetism, John Wiley & Sons, New York.

Brown Jr., W.F. (1966): Magnetoelastic Interactions, Springer-Verlag, Berlin.

Buckingham, M.J. (1956): Phys. Rev. **101**, 1431.

Bulaevskii, L.N., Ginzburg, V.L. and Sobyanin, A.A. (1988): Physica **C152**, 378.

Byers, N. and Yang, C.N. (1961): Phys. Rev. Letters **7**, 46.

Böttcher, C.J.F. (1973): Theory of Electric Polarization, Elsevier, Amsterdam.

Cagniard, L.G. (1953): Geophys. **18**, 605.

Campbell, A.M. and Evetts, J.E. (1972): Flux vortices and transport currents in type II superconductors, in: Adv. in Phys. **21**, pp.199-428.

Cardona, M. and Rosenblum, M.B. (1964): Phys. Letters **8**, 308.

Carr, Jr., W.J. (1983): AC Loss and Macroscopic Theory of Superconductors, Gordon & Breach, London.

Carr, Jr. W.J. (1983): J. Appl. Phys. **54**, 5911.

Chambers, R.G. (1952): Proc. Roy. Soc. (London) **A65**, 458.

Chen, C.W. (1977): Magnetism and metallurgy of soft magnetic materials, North-Holland, Amsterdam.

Chikazumi, S. (1964): Physics of Magnetism, John Wiley & Sons, New York.

Claeson, T. (1983): Superconducting Tunnel Junctions in High Frequency Radiation Detectors, in: Advances in Superconductivity (eds. by B. Deaver and J. Ruvalds), Plenum Press, New York, pp.341-277.

Clarke, J. (1966): Phil. Mag. **13**, 115.

Clarke, J. (1977): SQUIDs for Low Frequency Measurements, in: Superconductor Applications: SQUIDs and Machines (eds. by B.B. Schwartz and S. Foner), Plenum Press, New York, pp.67-124.

Clarke, J. (1983): Fundamental limits on SQUID technology, in: Advances in Superconductivity (eds. by B. Deaver and J. Ruvalds), Plenum Press, New York, pp.13-50.

Clarke, J. (1989): SQUID concepts and systems, In: Superconducting Electronics (ed. by H. Weinstock and M. Nisenoff), Springer-Verlag, Berlin, pp.87-148.

Clarke, D.R. (1987): Advanced Ceramic Materials **2**, 273.

Cody, G.D. and Miller, R.E. (1968): Phys. Rev. **173**, 481.

Cohen, D. (1968): Science **161**, 784.

Cohen, D. (1975): IEEE Trans. Magn. **MAG-11**, 694.

Cohen, D. and Givler, E. (1972): Appl. Phys. Lett. **21**, 114.

Cooper, L.N. (1956): Phys. Rev. **104**, 1189.

Crabtree, G.W., Behroozi, F., Campbell, S.A. and Hinks,. D.G. (1982): Phys. Rev. Letters **49**, 1342.

Curie, P. (1908): Oeuvres, p.129, Paris.

Cyrot, M. (1973): Rep. Progr. Phys. **36**, 103.

Davis, J.R. and Nisenoff, M. (1977): SQUID Magnetometers for Submarine Communications at extremely low frequency, in: SQUID and Applications (eds. by H.D. Hahlbohm and H. Lubbig), Walter de Gruyter, Berlin, pp.439-484.

Dayem, A.H. and Wiegrand, J.J. (1967): Phys. Rev. **155**, 418.

Deaver, B.S. and Fairbank, W.M. (1961): Phys. Rev. Letters. **7**, 43.

De Bruyn Ouboter, R. (1977): Macroscopic quantum phenomena in superconductors, in: Superconductor Applications: SQUIDs and Machines (eds. by B.B. Schwartz and S. Foner), Plenum Press, New York, pp. 21-65.

De Gennes, P.G. (1964): Rev. Mod. Phys. **36**, 225.

De Gennes, P.G. (1966): Superconductivity of Metals and Alloys, Benjamin, New York.

De Groot, S.R. and Mazur, P. (1962): Non-Equilibrium Thermodynamics, North-Holland, Amsterdam.

De Groot, S.R. and Vlieger, J. (1965): Physica **31**, 254.

De Haas, W.J. and Voogd, J. (1931): Commun. Phys. Lab. Univ. Leiden, no.**214c**.

Deutscher, G. and de Gennes, P.G. (1969): Proximity effects in: Superconductivity (ed. by R.D. Parks), Marcel Dekker, New York, pp.1005-1034.

Dirac, P.A.M. (1931): Proc. Roy. Soc. **A133**, 60.

Dmitriev, V.M., Gubankov, V.N. and Nad, F. Ya. (1986): Experimental study of enhanced superconductivity, in: Nonequilibrium superconductivity (eds. by D.N. Langenberg and A.I. Larkin), North-Holland, Amsterdam, pp.163-209.

Dolan, G.J. (1974): J. Low Temp. Phys. **15**, 133.

Doll, R. and Nabauer, M. (1961): Phys. Rev. Letters. **7**, 51.

Douglass, D.H., Jr. (1962): IBM J. Res. Dev. **6**, 44.

Duzer, T.C. and Turner, C.W. (1981): Principles of Superconductive Devices and Circuits, Elsevier, North-Holland, New York.

Eliashberg, G.M. and Ivlev, B.I. (1986): Theory of superconductivity stimulation in: Nonequilibrium superconductivity (eds. by D.N. Langenberg and A.I. Larkin), North-Holland, Amsterdam, pp.211-251.

Eringen, A.C. (1964): Int. J. Engng. Sci. **2**, 205.

Eringen, A.C. (1967): Mechanics of Continua, John Wiley & Sons, New York.

Eringen, A.C. (1971): Tensor Analysis, in: Continuum Physics Vol.I (ed. by A.C. Eringen), Academic Press, New York, pp.1-155.

Essmann, U. and Träuble, H. (1967): Phys. Lett. **24A**, 526.

Faber, T.E. and Pippard, A.B. (1955): Proc. Roy. Soc. **A231**, 336.

Fertiq, W.A., Johnston, D.C., Delong, L.E., McCallum, R.W., Maple, M.B. and Matthias, B.T. (1977): Phys. Rev. Lett. **38**, 387.

Fertiq, W.A., Johnston, D.C., Delong, L.E., McCallum, R.W., Maple, M.B. and Matthias, B.T. (1977): Phys. Rev. Lett. **38**, 987.

Feynman, R.P. (1965): Lectures on Physics III, Addison-Wesley, Massachusetts.

Finlayson, T.R. and Milne, I. (1971): Solid State Comm. **9**, 1339.

Fischer, O. and Peter, M. (1973): in: Magnetism Vol.V (ed. by H. Suhl), Academic Press, New York, pp.327-352.

Forgacs, R.L. and Warnick, A. (1967): Rev. Sci. Instrum. **38**, 214.

Fröhlich, H. (1950): Phys. Rev. **79**, 845.

Fung, Y.C. (1965): Foundations of Solid Mechanics, Prentice-Hall, New Jersey.

Ginsberg, D.M. (1989): Physical Properties of High Temperature Superconductors I, World Scientific, Singapore.

Ginzburg, V.L. and Landau, L.D. (1950): Zh. Eksp. Teor. Fiz. **20**, 1064.

Ginzburg, V.L. (1952): Zh. Eksp. Teor. Fiz. **23**, 236.

Ginzburg, V.L. (1957): Soviet Phys. JETP **4**, 153.

Ginzburg, V.L. (1961): Soviet Phys. Solid State **2**, 1824.

Glover, III, R.E. (1971): Superconducting fluctuation effects above the transition temperature in: Superconductivity (ed. by F. Chilton), North-Holland, Amsterdam, pp.3-23.

Goodman, B.B. (1962): IBM J. Res. Dev. **6**, 63.

Gor'kov, L.P. (1958): Soviet Phys. JETP **7**, 505.

Gor'kov, L.P. (1959a): Zh. Eksp. Teor. Fiz. **36**, 1918. (Soviet Phys. JETP **9**, 1364).

Gor'kov, L.P. (1959b): Zh. Eksp. Teor. Fiz. **37**, 1407. (Soviet Phys. JETP **10**, 998).

Gor'kov, L.P. (1960): Soviet Phys. JETP **9**, 1364; **10**, 593.

Gor'kov, L.P. and Melik-Barkhudarov, T.K. (1964): Soviet Phys. JETP **18**, 1031.

Gorter, C.J. and Casimir, H.B.G. (1934): Phys. Z. **35**, 963.

Gorter, C.J. and Casimir, H.B.G. (1934): Physica **1**, 306.

Goubau, W.M. (1980): Geophysical applications of SQUIDs, in: SQUID'80 (eds. by H.D. Hahlbohm and H. Lubbig), Walter de Gruyter, Berlin, pp.603-613.

Gray, K.E. (1974): J. Low Temp. Phys. **15**, 335.

Grynszpan, F. and Geselowitz, D.B. (1973): Biophys. J. **13**, 911.

Guyon, E., Caroli, C. and Martinet, A. (1964): J. Phys. Radium **25**, 683.

Hammond, P. (1981): Energy Methods in Electromagnetism, Clarendon Press, Oxford.

Hari, R. and Ilmoniemi, R.J. (1986): Biomed. Eng. **14**, 93.

Hartwig, W.H. and Passow, C. (1975): RF superconducting Devices in: Applied Superconductivity, Vol.II (ed. by V.L. Newhouse), Academic Press, New York, pp.541-639.

Heidrich, H. and Mataew, P. (1980): in: SQUID'80 (eds. by H.D. Hahlbohm and H. Lubbig), Walter de Gruyter, Berlin.

Heisenberg, W. (1928): Z. Physik **49**, 619.

Horacek, B.M. (1973): IEEE Trans. Magn. **MAG-9**, 440.

Huebener, R.P.(1979): Magnetic flux structures in superconductors, Springer-Verlag, Berlin.

Hutter, K. and van de Ven, A.A.F. (1978): Field Matter Interactions in Thermoelastic Solids, Springer-Verlag, Berlin.

Inderhees, S.E., Salamon, M.B., Goldenfeld, N., Rice, J.P., Pazol, B.G., Ginsberg, D.M., Liu, J.Z. and Crabtree, G.W. (1988): Phys. Rev. Lett. **60**, 1178.

Ishikawa, M. and Fischer, O. (1977): Solid State Commun. **23**, 37.

Ising, E. (1925): Z. Physik **31**, 253.

Izyumov, Yu. A. and Skryabin, Yu. N. (1980): Phys. Met. Metall. **49**, 1.

Jackson, W.D. (1975): Classical Electrodynamics (2nd ed.), John Wiley & Sons, New York.

Johnk, C.T.A. (1988): Engineering Electromagnetic Fields and Waves, John Wiley, N.Y.

Josephson, B.D. (1962): Phys. Letters **1**, 251.

Josephson, B.D. (1969): Weakly Coupled Superconductors, in: Superconductivity (ed. by R.D. Parks), Marcel Dekker, New York, pp.423-447.

Kakani, S.L. and Upadhyaya, U.N. (1988): J. Low Temp. Phys. **70**, 5.

Kanamori, J. (1963): in: Magnetism Vol. I (ed. by G.T. Rado and H. Suhl), Academic Press, New York, pp.127-203.

Keesom, W.H. (1924): Rapp et Disc. 4^e Congr. Phys. Solvay, pp.288.

Khurana, A. (1987): Physics Today, July, pp.17-21..

Kim, Y. B., Hempstead, C.F. and Strnad, A.R. (1962): Phys. Rev. Letters **9**, 306.

Kim, Y. B., Hempstead, C.F. and Strnad, A.R. (1965): Phys. Rev. **139**, A1163.

Kim, Y.B. and Stephen, M.J. (1969): Flux flow and irreversible effects, In: Superconductivity (ed. by Parks, R.D.), Marcel Dekker, New York.

King, R.W.P. and Prasad, S. (1986): Fundamental Electromagnetic Theory and Applications, Prentice-Hall, N.J.

Koch, C.C. and Easton, D.S. (1977): Cryogenics **17**, 391.

Kok, J.A. (1934): Physica **1**, 1103.

Kortum, G. (1965): Treatise on electrochemistry, Elsevier, Amsterdam, London.

Kostorz, G. (1973): Phys. Status Solidi **B58**, 9.

Krey, U. (1972): Int. J. Magn. **3**, 65.

Kroeger, D.M., Easton, D.S. and Moazed, A. (1977): IEEE Trans. on Magn. Mag-**13**, 120.

Kunze, U., Lischke, B. and Rodewald, W. (1974): Phys. Status Solidi (b) **62**, 377.

Kuper, C.G. (1968): An introduction to the theory of superconductivity, Clarendon, Oxford.

Labusch, R. (1968): Phys. Rev. **170**, 470.

Landau, L.D. (1937): Zh. Eksp. Teor. Fiz. **7**, 371.

Landau, L.D. and Lifshitz, E.M. (1958): Statistical Physics, Pergamon, London.

Landau, L.D. and Lifshitz, E.M. (1977): Quantum Mechanics (3rd ed.), Pergamon, Oxford.

Landau, L.D., Lifshitz, E.M. and Pitaevskii, L.P. (1984): Electrodynamics of Continuous Media (2nd ed.), Pergamon, Oxford.

Langenburg, D.N., Scalapino, D.J. and Taylor, B.N. (1966): Proc. IEEE **54**, 560.

Langenburg, D.N. and Larkin, A.I. (eds) (1986): Nonequilibrium superconductivity, North-Holland, Amsterdam.

Ledbetter, H.M., Austin, M.W., Kim, S.A., Datta, T. and Violet, C.E. (1987): J. Mat. Res. **2**, 790.

Lifshitz, E.M. and Pitaevskii, L.P. (1980): Statistical Physics (3nd), Part 1, Pergamon, Oxford.

Livingston, J.D. (1963): Phys. Rev. **129**, 1943.

Lock, J.M. (1951): Proc. Roy. Soc. **A208**, 391.

London, F. and London, H. (1935): Proc. Roy. Soc. A, **149**, 71.

London, F. and London, H. (1935): Physica **2**, 341.

London, F. (1948): Phys. Rev. **74**, 562.

London, F. (1950): Superfluids Vol.1, John Wiley, London.

London, H. (1940): Proc. Roy. Soc. **A176**, 522.

London, H. (1962): Phys. Letters **6**, 162.

Lorentz, H.A. (1916): The Theory of Electrons and its Application to the Phenomena of Light and Radiant Heat, 2nd, Leipzig.

Lundqvist, S., Tosatti, E., Tosi, M.P. and Lu, Y. (1987): Progress in high temperature superconductivity Vol.**1**, World Scientific, Singapore.

Lynn, J.W., Shirane, G., Thomlinson, W., Shelton, R.N., and Mocton, D.E. (1981): Phys. Rev. **B24**, 3817.

Maki, K. (1964): Physics **1**, 21.

Maple, M.B. (1983): Coexistence of superconductivity and magnetism, in: Advances in Superconductivity (eds. by B. Deaver and J. Ruvalds), Plenum Press, New York, pp.279-346.

Matsubara, T. and Kotani, A. (eds.) (1984): Superconductivity in Magnetic and Exotic Materials, Springer-Verlag, Berlin.

Matveev, A.N. (1986): Electricity and Magnetism, Mir Publishers, Moscow.

Maugin, G.A., (ed.) (1984): Mechanical Behavior of Electromagnetic Solid Continua, IUTAM Symp., North-Holland, Amsterdam.

Maugin, G.A. (1985): Nonlinear Electromechanical Effects and Applications, World Scientific, Singapore.

Maugin, G.A. (1988): Continuum Mechanics of Electromagnetic Solids, North-Holland, Amsterdam.

Maxwell, E. (1950): Phys. Rev. **78**, 477.

Mazur, P. and Nijboer, B.R.A. (1953): Physica **19**, 971.

McCumber, D.E. (1968): J. Appl. Phys. **39**, 3113.

Meissner, H. (1958): Phys. Rev. **109**, 686.

Meissner, H. (1959): Phys. Rev. Letters **2**, 458.

Meissner, H. (1960): Phys. Rev. **117**, 672.

Meissner, H. and Ochsenfield, R. (1933): Naturwiss **21**, 787.

Mercereau, J.E. (1969): in Tunneling Phenomena in Solids, Chap. 31 (eds. by E. Burstein and S. Lundqvist), Plenum Press, New York.

Mercereau, J.E. (1970): Rev. Phys. Appl. **5**, 13.

Miller, R.E. and Cody, G.D. (1968): Phys. Rev. **173**, 494.

Moon, F.C. (1984): Magneto-Solid Mechanics, John Wiley, New York.

Morrish, A.H. (1965): The Physical Principles of Magnetism, John Wiley & Sons, N.Y.

Morse, P.M. and Feshbach, H. (1953): Methods of Theoretical Physics, McGraw Hill, N.Y.

Mϕller, C. (1972): The Theory of Relativity (2nd. ed.), Clarendon, Oxford.

Narlikar, A.V. and Ekbote, S.N. (1983): Superconductivity and Superconducting Materials, South Asian Publ., New Delhi.

Newhouse, V.L. (1964): Applied Superconductivity, John Wiley & Sons, New York.

Nowacki, W. (1976): Magnetoelasticity, in: Electromagnetic Interactions in Elastic Solids (ed. by H. Parkus), CISM Course, Springer-Verlag, New York, pp.158-183.

Ollendorff, F. (1974): Arch. Elektrotechnik **56**, 1.

Ollendorff, F. (1977): Arch. Elektrotechnik **59**, 133.

Olsen, J.L. and Rohrer, H. (1957): Helv. Phys. Acta **30**, 49.

Olsen, J.L. and Rohrer, H. (1960) Helv. Phys. Acta **33**, 872.

Onnes, H.K. (1911): Commun. Phys. Lab. Univ. Leiden, no.**119b**.

Onnes, H.K. (1914): Commun. Phys. Lab. Univ. Leiden, no.**139f**.

Oohira, T. (1989): Nonlinear analysis of magnetic field and stress for a thin ferromagnetic material, Master Thesis, University of Tokyo.

Ott, H.R., Fertig, W.A., Johnston, D.C., Maple, M.B. and Matthias, B.T. (1978): J. Low Temp. Phys. **33**, 159.

Pao, Y-H. (1978): Electromagnetic Forces in Deformable Continua, in: Mechanics Today **4** (ed. by Nemat Nasser, S.), Pergamon Press, Oxford, pp.209-305.

Pedersen, N.F. (1980): RF Applications of Superconducting Tunneling Devices, in: SQUID'80 (eds. by H.D. Hahlbohm and H. Lubbig), Walter de Gruyter, Berlin, pp.739-762.

Penfield, P. and Haus, H.A. (1967): Electrodynamics of Moving Media, MIT Press, Cambrige, Massachusetts.

Pillips, J.C. (1989): Physics of High-T_c Superconductors, Academic Press, Boston.

Pippard, A.B. (1950a): Proc. Roy. Soc. **A203**, 210.

Pippard, A.B. (1950b): Proc. Roy. Soc. **A203**, 98.

Pippard, A.B. (1953): Proc. Roy. Soc. **A216**, 547.

Podney, W. (1975): J. Geophys. Res. **80**, 2977.

Potter, R.I. and Schmulian, R.J. (1971): IEEE Trans. Magn. **MAG-7**, 873.

Reuter, G.E.H. and Sondheimer, E.H. (1948): Proc. Roy. Soc. **A195**, 336.

Reynolds, C.A., Serin, B., Wright, W.H. and Nesbitt, L.B. (1950): Phys. Rev. **78**, 487.

Richards, P.L. (1977): The Josephson junction as a detector of microwave and far-infrared radiation in: Semiconductors and Semimetals (eds. by R.K. Willardson and A.C. Beer), Vol. 12, Chap. 6, Academic, New York.

Robinson, F.N.H. (1973): Macroscopic Electromagnetism, Pergamon Press, Oxford.

Robinson, S.E. (1981): IEEE Trans. Nuc. Sci. **28**, 272.

Rose-Innes, A.C. and Rhoderick, E.H. (1969): Introduction to superconductivity, Pergamon Press, oxford.

Rosenfeld, L. (1951): Theory of Electrons, Amsterdam.

Rosser, W.G.V. (1964): An Introduction to The Theory of Relativity, Butterworths, London.

Rowell, J.M. (1963): Phys. Rev. Letts. **11**, 200.

Rutgers, A.J. (1934): Physica **1**, 1055.

Russakoff, G. (1970): Am. J. of Physics **38**, 1188.

Saint-James, D. and de Gennes, P.G. (1963): Phys. Letters **7**, 306.

Saint-James, D., Thomas, E.J. and Sarma, G. (1969): Type II superconductivity, Pergamon, Oxford.

Sakurai, A. (1978): Solid State Commun. **25**, 867.

Schieber, D. (1986): Electromagnetic Induction Phenomena, Springer-Verlag, Berlin.

Schiff, L.I. (1968): Quantum Mechanics (3rd. Ed.), McGraw-Hill Book Co., New York.

Schmid, A. (1966): Phys. Kondens. Materie **5**, 302.

Seeger, A. and Kronmuller, H. (1968): Phys. Stat. Sol. **23**, 371.

Seraphim, D.P. and Marcus, P.M. (1962): IBM J. Res. Develop. **6**, 94.

Shadowitz, A. (1975): The Electromagnetic Field, McGraw-Hill, New York.

Shapiro, S. (1963): Phys. Rev. Letters **11**, 80.

Shenoy, G.K., Dunlap, B.D., Fradin, F.Y., Sinha, S.K., Kimball, C. W., Potzel, W., Probst, F. and Kalvius, G.M. (1980): Phys. Rev. **B21**, 3886.

Shoenberg, D. (1952): Superconductivity (2nd ed.), Cambridge Uni. Press, Cambridge.

Simon, I. (1953): J. Appl. Phys. **24**, 19.

Sizoo, G.J. and Kamerlingh Onnes, H. (1926): Commu. Phys. Lab. Leiden Univ. No.180.

Smith, T.F. and Chu, C.W. (1967): Phys. Rev. **159**, 353.

Smith, T.I. (1965): Phys. Rev. Letters **15**, 460.

Solymar, L. (1972): Superconductive Tunnelling and Applications, Chapman and Hall, London.

Stacey, F.D. (1964): Pure and Appl. Geophys. **58**, 5.

Stephen, M.J. (1965): Phys. Rev. **139**, A197.

Stewart, W.C. (1968): Appl. Phys. Lett. **12**, 277.

Stogryn, D.E. and Stogryn, A.P. (1966): Mol. Phys. **11**, 371.

Stoletov, A.G. (1873a): Pogg. Ann. **144**, 439.

Stoletov, A.G. (1873b): Phil. Mag. (4) **45**, 40.

Suhl, H. (1978): J. Less Common Metals **62**, 225.

Swihart, J.C. (1961): J. Appl. Phys. **32**, 461.

Tedrow, P.M., Farazi, G. and Meservey, R. (1971): Phys. Rev. **B4**, 74.

Testardi, L.R. (1971): Phys. Rev. **B3**, 95.

Testardi, L.R., Kunzler, J.E., Levinstein, H.J., Maita, J.P. and Wernick, J.H. (1971): Phys. Rev. **B3**, 107.

Tiersten, H.F. (1964): J. Math. and Phys. **5**, 1.

Tiersten, H.F. (1990): A Development of the Equations of Electromagnetism in Material Continua, Springer-Verlag, New York.

Tilley, D.R. (1965): Proc. Phys. Soc. **85**, 1177.

Tinkham, M. (1963): Phys. Rev. **129**, 2413.

Tinkham, M. (1964): Phys. Rev. Letters **13**, 804.

Tomasch, W.J. and Joseph, A.S. (1964): Phys. Rev. Letters **12**, 148.

Truesdell, C. (1961): in: Principles of Continuum Mechanics, Socony Mobil Oil Co.

Tyler, F. (1931): Phil. Mag. **11**, 596.

Ullmaier, H. (1975): Irreversible Properties of Type II Superconductors, Springer-Verlag, Berlin.

Ulrich, B.T. and Tutter, M. (1980): Josephson Junction Applications in Plasma Physics, in: SQUID'80 (eds. by H.D. Hahlbohm and H. Lubbig), Walter de Gruyter, Berlin, pp.831-839.

Van Vlack, J.H. (1932): The Theory of Electric and Magnetic Susceptibilities, Oxford.

Vonsovskii, S.V. (1974): Magnetism Vol.2, John Wiley & Sons, New York.

Werthamer, N.R. (1963): Phys. Rev. **132**, 2440.

Werthamer, N.R. (1969): The Ginzburg-Landau Equations and Their Extensions, in: Superconductivity (ed. by R.D. Parks) Vol.**1**, Marcel Dekker, N.Y., pp.321-370.

Weiss, P. (1907): J. Phys. **6**, 667.

Welch, D.O. (1980): Alteration of the superconducting properties of A15 compounds and elementary composite superconductors by nonhydrostatic elastic strain, in: Advances in Cryogenic Engineering Materials **26**, 48.

Weller, W. (1968): Phys. Stat. Sol. **30**, 373.

Wilson, M.N. (1983): Superconducting magnets, Clarendon Press, Oxford.

Wolf, P. (1977): SQUIDs as Computer Elements, in: SQUID and Applications (eds. by H.D. Hahlbohm and H. Lubbig), pp.519-540, Walter de Gruyter, Berlin.

Woolf, L.D., Torar, M., Hamaker, H.C. and Maple, M.B. (1979): Phys. Letters **71A**, 137.

Wyatt, A.F.G., Dmitriev, V.M., Moore, W.S. and Sheard, F.W. (1966): Phys. Rev. Letters **16**, 1166.

Wynn, W.M. Frahm, C.P., Carroll, P.J., Clarke, R.H., Wellhoner, J. and Wynn, M.J. (1975): IEEE Trans. Magn. **Mag-11**, 701.

Xiang, X.D., Brill, J.W., Delong, C.J., Bourne, L.C., Zettl, A., Jones, J.C. and Rice, L.A. (1988): Solid State Com. **65**, 1073.

Yamomoto, Y. and Miya, K. (eds.) (1987): Electromagnetomechanical Interactions in Deformable Solids and Structures, IUTAM Symp., North-Holland, Amsterdam.

Zhou, S.A. and Hsieh, R.K.T., Dislocations in nonlinear elastic dielectrics, in: Nonlinear Mechanics (ed. by W. Chien), pp.1367-1375, Science Press, Beijing (1985).

Zhou, S.A. (1987): Material Multipole Mechanics of Electromagnetoelastic Solids with Defects, Ph.D. thesis, Royal Institute of Technology, Stockholm.

Zhou, S.A. and Hsieh, R.K.T. (1988): Int. J. Engng. Sci. **26**, 13.

Zhou, S.A. and Miya, K. (1990): Int. J. Appl. Electromag. Mater. **1**, 1.

Zhou, S.A. and Miya, K. (1991): A Nonequilibrium Theory of Thermoelastic Superconductors, Int. J. Appl. Electromag. Mater. in press.

Zaitsev, R.O. (1965): Soviet Phys. JETP **21**, 1178.

Zaitsev, R.O. (1966): Soviet Phys. JETP **23**, 702.

Zimmerman, J.E. and Mercereau, J.E. (1965): Phys. Rev. Letters **14**, 887.

Zimmerman, J.E. and Silver, A.H. (1966): Phys. Rev. **141**, 367.

Zimmerman, J.E., Thiene, P. and Harding, J.T. (1970): J. Appl. Physs. **41**, 1572.

Index

www.ingramcontent.com/pod-product-compliance
Lightning Source LLC
Chambersburg PA
CBHW050522190326
41458CB00005B/1628
* 9 7 8 0 8 6 3 4 1 2 5 7 8 *